A history of mathematics education in England

A history of mathematics education in England

GEOFFREY HOWSON

Reader in Mathematical Curriculum Studies,
University of Southampton

CAMBRIDGE UNIVERSITY PRESS

Cambridge

London New York New Rochelle
Melbourne Sydney

Published by the Press Syndicate of the University of Cambridge
The Pitt Building, Trumpington Street, Cambridge CB2 1RP
32 East 57th Street, New York, NY 10022, USA
296 Beaconsfield Parade, Middle Park, Melbourne 3206, Australia

First published 1982

Printed in Great Britain at The Pitman Press, Bath

Library of Congress catalogue card number: 82–4175

British Library cataloguing in publication data

Howson, Albert Geoffrey
A history of mathematics education in England.
1. Mathematics—Study and teaching—England
—History
I. Title
510'.7 QA14.E/

ISBN 0 521 24206 1

P.P.

Contents

Preface *page* ix

Prelude 1

1 Robert Recorde 6
- 1 The man and his times 6
- 2 Provision for mathematics education in Recorde's day 8
- 3 Robert Recorde and mathematics 13
- 4 Robert Recorde and mathematics education 19
- 5 Robert Recorde's editors 24

2 Samuel Pepys 29
- 1 Early Schooling 30
- 2 University 32
- 3 The Admiralty and Christ's Hospital Mathematical School 35
- 4 Pepys, the amateur of science 39

3 Philip Doddridge 45
- 1 The Puritan Revolution and its aftermath 45
- 2 The early dissenting academies 47
- 3 Philip Doddridge and Kibworth Academy 48
- 4 Religion and mathematics 49
- 5 Northampton Academy 52
- 6 Joseph Priestley and the later academies 54

4 Charles Hutton 59
- 1 Boy and schoolmaster 59
- 2 The Royal Military Academy, Woolwich 64
- 3 *The Ladies' Diary* 69
- 4 Hutton as patron 72

5 Augustus De Morgan 75
- 1 Early years and school 75
- 2 Cambridge 76
- 3 William Frend 79
- 4 London University 82
- 5 The Society for the Diffusion of Useful Knowledge 85

6 A challenge and a response 87
7 The Spitalfields Mathematical Society 92
8 University College and the London Mathematical Society 94

6 Thomas Tate 97
1 Thomas Tate and the popular tradition 97
2 Elementary education at the commencement of Tate's career 101
3 Tate and teacher education 105
4 Tate's *Philosophy of Education* 113
5 Retirement and the Newcastle Commission 118

7 James Wilson 123
1 Schooling 123
2 University 125
3 Rugby and the Clarendon Commission 128
4 Elementary geometry 130
5 The Association for the Improvement of Geometrical Teaching 134
6 Science, Clifton College and after 137

8 Charles Godfrey 141
1 School and Rawdon Levett 141
2 University 143
3 Alternatives 144
4 Reform 148
5 Chaos and reaction 152
6 A golden era 155
7 The examination system 160
8 The international scene 163
9 A period of consolidation and stagnation 166

9 Elizabeth Williams 169
1 Primary school education 169
2 Secondary education for girls 172
3 University 176
4 Professional training 179
5 Teaching: the inter-war years 182
6 Research in mathematics education 187
7 Teacher-training 190
8 Three reports 193
9 'Retirement' 202

Postlude 205

Appendix: a selection of examination papers, syllabuses, etc. 212
Notes 239
Name index 281
Subject index 289

Preface

To my knowledge this is the first book to be published which attempts to tell the story of the development of mathematics education in England. That this should be so is rather surprising; for one can turn to histories of the teaching of science and to a history of mathematics teaching in Scotland. Any attempt to fill such a gap is, therefore, fraught with difficulties, for the 'only book' is likely to be invested with an authority it may not deserve. Extra problems may also arise as a result of my having chosen to present the material through the medium of biographies. Emerson's claim that 'there is properly no history; only biography' could be used to justify this decision. The truth is, however, more mundane; for I abandoned a 'chronological' account thinking it would have little appeal for the general reader as opposed to the serious student. I believe also that a biographical account, even though it requires frequent scene-settings and 'flashbacks', better demonstrates the great part which individuals have always played in the advancement of mathematics education in England. The book, however, is not to be compared with Macfarlane's *Ten British Mathematicians of the Nineteenth Century*, for that author set out to identify the ten 'greatest' mathematicians of that age. My basis for selection has been different, for I have chosen subjects from various periods and traditions whose own mathematical education and whose contributions to mathematics teaching provided a framework around which I could construct a representative story. Often there were competing candidates: why, for example, Godfrey, rather than Perry, Nunn or Siddons? The answer lies not in the weight of their contributions to mathematics education, but in their attraction to me as persons and in a number of diverse associations. The reader is warned, therefore, as he should be when reading all histories, that the work has a subjective air. Yet, although the choice of descriptions and quotations is bound to reflect my own interests and beliefs, I have attempted to refrain from explicitly drawing parallels and linking the past to the present. Nevertheless, I hope the reader will think throughout about implications for today, for I have always aimed to show that history can not only help us to understand the past, but, more importantly, that it can lead us better to comprehend the present.

In writing this book I have received help from a variety of sources; indeed, too many to list in detail. I should, however, like to thank all those friends and colleagues, librarians, archivists, officers of examination boards, students (both at Southampton and in institutions at which I have acted as external examiner) and members of Cambridge University Press who have supplied me with information and counsel. Their assistance has been much appreciated. Particular help has come from Elizabeth Williams (how fascinating it would have been to have had such criticism and aid from all my 'subjects'!), Michael Price, who caused me to rethink and revise my views on that most important period, 1895–1914, and especially Peter Wallis, without whose constant urging, encouragement and advice the book would probably never have been written. As on other occasions I am also greatly indebted to Jennifer and Katharine for their patience and their help in the preparation of the final typescript, in proof-reading and in the compilation of the indexes.

GEOFFREY HOWSON

Southampton, October 1981

As a guide to the reader, the character of the notes is indicated thus: round brackets ( ) indicate sources of further reading and are primarily intended for the enthusiast or serious student, square brackets [ ] indicate biographical notes, and notes of a more general, expository nature are indicated by curly brackets { }.

Prelude

The early English schools were the offspring of the Church and their curricula were designed with one principal aim in view, the more widespread dissemination of Christianity. If Christianity and civilisation were to be brought to the Anglo-Saxons, it was essential to have priests who were trained to read the Bible – a literate priesthood was a prerequisite for evangelisation.

After the coming of St Augustine in 597 schools were established by the great monastic churches, for example at Canterbury, Westminster and York, in an attempt to supply this basic, vocational training. These schools provided instruction in the Latin language, the Latin scriptures and in the Church's liturgy and music. Gradually, at the more important centres, the curriculum widened to include astronomy and arithmetic {1}. There was a utilitarian need for the latter – if only to calculate the dates of movable feasts and to keep rudimentary accounts.

We cannot be certain exactly when mathematics first was taught in England, but Bede (674–735) in his *Ecclesiastical History* {2} tells how Theodore, Archbishop of Canterbury in the late seventh century, and his deacon, Hadrian, 'gathered a crowd of disciples . . . and . . . taught them the art of ecclesiastical poetry, astronomy and arithmetic'. Bede, himself, in his *De Temporum Ratione* (3) demonstrated the arithmetic needed to make the necessary calculations concerning Easter. Of course, at that time the West had no knowledge of the Hindu–Arabic notation and so Bede made use of 'finger-reckoning', a form of arithmetic taught in monastic schools for many centuries (cf. p. 15 below).

The body of knowledge available to Bede was that of the Romans, who are scarcely remembered for their contributions to, or interest in, mathematical thought. One Roman, Boethius (*c.* 480–524), did, however, have a considerable influence on mathematics education. He was responsible not only for producing textbooks, which were to be used in Britain until the sixteenth century {4}, but for spelling out the place that mathematics should hold in a liberal education. For it was Boethius who, to the trivium of grammar, logic and rhetoric, adjoined the mathematical quadrivium of arithmetic, geometry, music and astronomy {5}. Grammar was the basis of all, and was later to become the chief concern of the schools. Logic, rhetoric and the subjects of the quadrivium were seen as more advanced and, accordingly, it was on these that the curriculum of the medieval universities was based.

Such arithmetic as is to be found in Bede's work cannot, however, have been common knowledge. Clearly, some 'Church arithmetic' was quite advanced, but the vast majority of priests would have obtained their information about the Church calendar from tables: indeed many would scarcely have been literate. Yet those priests who were lettered were expected as part of their duties to assist in the propagation of knowledge and to teach the basic skill of reading to young boys in their villages. The Church was clearly committed to education {6}.

The education which the Church could provide was, however, extremely limited. Away from the major centres little more than Latin was taught, although there had been attempts in the late ninth and tenth centuries to develop teaching and learning in the vernacular (7). Nevertheless, the basis of an educational system existed, which in the thirteenth century was to provide the foundation for expansion.

The thirteenth century was indeed an auspicious one for European mathematics and education: it saw the definite arrival in Europe of Arabic ideas and notation{8} and also witnessed the establishment of the great medieval universities. Gradually the Arabic arithmetic, algorism, began to displace the earlier arithmetic of Boethius. The general acceptance of the new notation was slow, but its presence and that of algebra prompted the emergence of creative, Western mathematicians.

In England, the beginning of the thirteenth century was notable for the creation of the universities at Oxford and Cambridge. These were to develop slowly over a period of many years and initially bore little relation to the universities as they exist today. At first they were collections of teachers and pupils rather than organised institutions. Only gradually did they evolve in more formal ways; thus legal privileges were conferred on Oxford in 1214 (Cambridge, 1231) and its degrees were recognised by the Pope in 1296 (Paris, 1296, Cambridge, 1318). The 'universities' themselves owned no property: lectures were given in houses, and assemblies and disputations were held in the churches. Not until the fourteenth century did the universities begin to erect buildings for their own use, by which time various 'colleges' had been established to provide lodging and financial security for graduate scholars or fellows.

By 1300 the two universities had between them some 2,000 students, the majority being at Oxford. Most were teenagers; some were friars and monks, for the religious orders were represented in the two towns.

With the coming of the universities, there also came a clear division of teaching duties. Grammar was now seen as a school subject; logic and, to a lesser extent, rhetoric were studied during the four-year course for a bachelor's degree; and the quadrivium was studied by those who remained a further three years to become 'masters'. The degree of master was a licence to teach and was initially sought only by those who intended to use it for that purpose. Eventually the need arose to refuse to grant degrees to those students who were not thought to be worthy {9}, but, in general, tests of proficiency were not imposed: students were only required to have read certain books or attended specified lecture courses {10}.

Soon, also, came the first 'university mathematicians', among the most notable of whom was John of Holywood (Sacrobosco) [11]. His two texts *Tractabus De Sphaera* and *De Arte Numerandi (De Algorismo)* were much used, the latter incorporating the new Arabic symbolism. Other outstanding scholars at Oxford were Robert Grosseteste [12], later Bishop of Lincoln, and Roger Bacon [13]. Both of these advanced the cause of mathematics and of experimentation as a corrective against prevailing dogmatism, and so helped Oxford, and in particular Merton College, to become a leading centre of thirteenth-century science. Mathematical standards prevailing at Oxford are indicated, however, by Bacon's (choleric?) observation that there were few, if any, residents there who had read more geometry than Euclid's definitions and the first five propositions of Book 1 {14}.

The fourteenth century brought little that was new for mathematics. Although the work of Oresme and Buridan [15] in Paris and of Bradwardine [16] and others at Oxford prepared the way for future advances, it was not until the advent of printing in the mid fifteenth century that the means existed for the rapid dissemination of new ideas and theories. Thus, texts by Bradwardine, probably written during his time at Oxford, were printed some 150 years later in Paris. Not until then could his work be freely and cheaply circulated.

The century did see important changes in the social status of Oxford and Cambridge for their graduates now began to hold the highest offices in Church and State. The universities became not only places in which learning was cultivated for its own sake, but vital rungs in the social ladder.

Some indication of the mathematics studied in the universities is contained in the statutes of the new foundations of that time (17). Candidates for the bachelor's degree at Prague (1350) were required to have read Sacrobosco's work on the sphere, and candidates for the master's degree to be acquainted with the first six books of Euclid, optics, hydrostatics, the theory of the lever and astronomy. Lectures were given on arithmetic, finger-reckoning, algorism, elementary astrology and Ptolemaic astronomy. Other 'university' topics included perspective, proportion and the measurement of superficies (surfaces). Often, though, the teaching of mathematics and science was tolerated rather than encouraged. At Vienna (1364) the study of mathematics was seen as a more seemly occupation for feast days than drinking or fighting (18) – an attitude to be encountered again in eighteenth-century public schools.

A significant event in 1382 was the establishment of a college 'of poor scholar clerks' in Winchester, although this was hardly to have an immediate positive effect on mathematics education since over 450 years were to elapse before Winchester College acquired its first mathematics master (see p. 124 below). This new school, independent of any ecclesiastical institution, was, however, to exert considerable influence on the future development of English education.

Although an independent foundation, Winchester's aims were far from revolutionary: it aimed to provide the basic education required by those who would 'study theology, canon and civil law and arts' in the recently established sister college – New College – at Oxford. This basic education was simply stated: 'studying and becoming proficient in grammaticals or the art faculty or science of grammar'. The principal aim of the school was clear – *to prepare students for university.*

When judging the curricula and aims of the educational institutions of this period it must be remembered that a boy could enter university at the age of fourteen. The grammar school, then, was concerned with the education of boys between the ages of seven and fourteen. Older boys and, indeed, young men were still to be found in such schools, for society did not place great importance on age: 'adolescence' had no place in its vocabulary; from the age of seven onwards all were treated alike, at work or before the law. Again, it will be noted that all the references have been to male students. Certainly, educational facilities for girls did exist, but these concerned but few (19): mathematics education for girls was still some centuries away.

The foundation of Winchester College was part of a more general expansion of the school system. Some schools remained closely attached to the Church, others were independent and endowed by rich clerics, merchants or guilds. Their curriculum followed that of the older schools, although there was a move towards rhetoric and away from the old-fashioned grammar. Mathematics, though, had still to find a place in the grammar schools.

Other types of school were also coming into existence, the forerunners of elementary schools. In these, emphasis was laid on English rather than Latin, and children were taught the elements of spelling, reading and writing. From the fourteenth century some private teachers, known as scrivenors, taught not only writing but also the rudiments of arithmetic in the form of account-keeping.

Outside the formal, institutionalised system much learning, some of it arithmetical and geometrical, took place through the apprenticeship schemes practised by the trade guilds (20); an important reminder that much mathematics education takes place outside schools and other educational establishments. By the fifteenth century, then, society was becoming sufficiently complex to ensure that education was required by others, apart from would-be priests, and the founders of schools came to recognise this.

The expansion of education was accompanied, not surprisingly in the light of later experience, by a shortage of teachers. Complaints were heard of a decline in the standards of grammar teaching and it was claimed that many schools were moribund for want of teachers (21). To ameliorate the situation a teacher-training college, God's House (later enlarged in 1505 as Christ's College), was established at Cambridge in 1437, the students of which were obliged, after taking their degrees, to accept appointment as grammar-school masters.

God's House was not the only new college to be established at the two universities. England was now entering a period of prosperity which was marked by the endowment of many new colleges: King's (1441), Queens' (1448), St Catharine's (1473), Jesus (1496) and St John's (1511) at Cambridge; All Souls' (1438), Magdalen (1458), Brasenose (1509), Corpus Christi (1517) and Cardinal (1525, later to be renamed Christ Church) at Oxford.

This period of expansion and consolidation in the schools and universities was not marked by any comparable developments in mathematics education. However, the seeds were being sown for a period of great growth. Printing had been invented, the great sea voyages were giving rise to a demand for navigational mathematics, and eloquent pleas were being made for the place of mathematics within a liberal education. A young protégé of Wolsey and More, the Spaniard Vives, who had been engaged by Henry VIII as a tutor for his daughter, Mary, set out the case for mathematics in the university (22). This was a subject to 'display the sharpness of the mind' and to provide discipline for 'flighty and restless intellects which are inclined to slackness and shrink from or will not support the toil of a continued effort'. Yet, Vives was aware of the dangers of the subject and of the way in which it 'leads away from the things of life, and estranges men from perception of what conduces to the common weal'.

The dual aspects of mathematics, the practical and the contemplative, were clearly distinguished then by Vives, and he recognised the need healthily to reconcile the two. The subsequent history of mathematics education in England is largely a chronicle, on the one hand, of how this problem was ignored – with the result that a bipartite system of mathematics education was effectively created – and, on the other hand, of how individual educators have constantly sought to effect a reconciliation.

1 · *Robert Recorde*

As the writer of the first series of mathematics texts in English, Robert Recorde clearly holds a special place in the history of mathematics education in England. Yet it can be argued that Recorde should be recognised for yet another reason: that he was the first mathematics educator. Not only did Recorde teach mathematics, but his writings show clearly – both implicitly and explicitly – that he had also given serious consideration to the problems of learning and teaching mathematics.

1. THE MAN AND HIS TIMES

Robert Recorde [1] was born in Tenby, Wales, about 1512. That his exact date of birth is not known is regrettably typical, for there are many details of Recorde's life and career that are lacking. It is also significant that the sole piece of information we have concerning Recorde's age should be the result of a political trial [2]. The times in which Recorde lived were indeed turbulent ones, and the various political and religious struggles {3} affected both his personal well-being and his plans for publishing mathematical texts.

The first stages of the Reformation took place whilst Recorde was a student at Oxford. He graduated BA there in 1531 and was the same year elected a Fellow of All Souls' College, a graduate foundation for the study of theology, law and medicine. Clearly, Recorde must have devoted considerable time to the study of the last named, for, according to the Cambridge records, he had been licensed in medicine at Oxford about 1533. Recorde later moved from Oxford to Cambridge and it was there in 1545 (and after the publication of his first mathematical work, *The Ground of Artes*) that he was awarded the degree of Doctor of Medicine. It is clear that the author of the *Ground* had practical experience of mathematics teaching and there are hints in the book that he had taught at the universities, but there is no firm evidence to support this. Alternatively, that experience may have been gained acting as a tutor to one of the twenty-five children of Richard Whalley to whom he dedicated the first edition of the *Ground*.

After leaving Cambridge, Recorde moved first to Oxford and then to London where he practised as a physician, published his medical text *The Urinal of Physick* (1547), and became known at Court [4]. In January 1549

{5} he was appointed comptroller of the Bristol Mint. Whilst in that position, in October 1549, he decided to support the Protector Somerset, whose influence was waning, and refused to divert money intended for King Edward to the armies raised by John Russell (later Earl of Bedford) and Sir William Herbert (later Earl of Pembroke). This decision had important repercussions, for it signalled the beginning of a state of animosity between Recorde and the influential Pembroke.

The first round in the battle was an accusation of treason made by Pembroke against Recorde, which led to the latter's being confined for sixty days. The Mint was closed in the spring of 1550 and presumably Recorde returned to London [6]. Recorde's patron, Whalley, was another to fall from grace with Somerset. He was imprisoned in 1551 and not released until 1553 after Mary had come to the throne. For some time then it was not politic to publish a book dedicated to Whalley. The Preface of the *Ground* was accordingly skilfully adjusted: 'Good Maister Whalley' became 'the gentle reader', his 'young children' became the 'simple ignoraunt' and the private dedication to Whalley was now made 'unto the Kyng his Majestie' (7).

Edward VI was also the dedicatee of Recorde's next book, *The Pathway to Knowledge*. The actual year of its publication is in doubt, but it probably appeared (8) after Recorde had, in May 1551, taken up a new post as General Surveyor of the Mines and Monies in Ireland. It was a position which kept him fully occupied for two years, for he had oversight of the silver mines, the miners and the Irish Mint. Soon accusations of incompetence and malpractice came pouring into London (9). The French complained that he had taken a cargo of whale oil and refused to pay; the mines were said to be decaying because of 'the ill conduct of Mr Recorde'; he was dilatory in providing essential buildings and in opening new mines; he appropriated meat intended for the miners, kept the best and sold the rest for a profit; he also profiteered on the sale of corn, herring and hake, and dissuaded his men from buying shoes in England for 12d or 13d a pair so that his own shoemaker could monopolise the market at 3s 4d (40d) a pair; he borrowed money from his German Captain of the Mines, Joachim Gundelfinger {10}, and failed to repay it, . . . Recorde, for his part, had little to say, apart from complaining that the 'Almains' working for him lacked the skill of the English and Irish. In 1553 he was recalled to London and once again took up mathematical authorship. Repercussions of the Irish venture were, however, still to come.

The year 1553 (see note 3) proved a hectic one for Londoners; it witnessed Edward's death, the brief reign of Lady Jane Grey and Mary's accession as queen. It was to Mary, and in a most obsequious fashion, that Recorde dedicated his next book, *The Castle of Knowledge* (1556): this was a time when great care had to be exercised by those who wished to live to fight another day!

Unfortunately, Recorde's past now began to catch up with him. Enquiries began into his conduct of affairs in Ireland and the quarrel with Pembroke flared up again. The latter by adroit shifts of ground had retained his influence under Henry, the two Protectors Somerset and Northumberland, Lady Jane Grey and Mary. In 1557 he sued Recorde for libel and was awarded damages of £1,000 (11). What happened next is not clear: certainly, *The Whetstone of Witte* published in that year ends with a passage in which the Master (Recorde) expresses an ominous sense of foreboding. Whether for failure to pay the debt or for other reasons, Recorde was imprisoned. His will was written in King's Bench Prison and he must have died some time before 18 June 1558 when the will was admitted to probate [12].

Recorde's life, then, was not that which we now associate with the typical scholar: academic life was soon left behind and much of his mathematical work was done at times when he was carrying out duties as a civil servant or when embroiled in politics. The reports from Ireland do not cast him in a particularly favourable light, yet his works reveal him as an outstanding scholar [13], conversant with Greek and medieval texts, a knowledgeable antiquary and a gifted expositor. In particular, his contributions to mathematics education in England were enormous and seminal.

2. PROVISION FOR MATHEMATICS EDUCATION IN RECORDE'S DAY

As we have seen, at the time of Recorde's birth an embryonic 'educational system' existed in England. The universities at Oxford and Cambridge were well established and there was provision, of varying degrees of efficiency, for the education of the children of the different classes. The middle classes, farmers, lesser land-holders and prosperous tradesmen tended to supply the students of the grammar schools; the sons of the nobility and gentry might have had a tutor or been boarded in a monastery, whilst their daughters, if the need was felt, were educated by a private tutor, or occasionally were sent to nunneries for their education; for the lower classes, the song and charity schools provided a form of elementary education. Yet, despite such provisions, it has been estimated that of the population of about 5 million at the time of the dissolution of the monasteries, only 26,000 were at school. Schooling was still a rare commodity (14).

In the grammar schools, the curriculum was very restricted. Timetables indicate that the students' time was devoted exclusively to Latin grammar, Latin authors and Latin compositions and verse writing. Elementary education offered in the main the rudiments of literacy and some religious knowledge.

Soon, however, new pressures on the grammar-school and university curricula arose as a result of the cultural movement known as the Renaissance – that rebirth of interest in, and knowledge of, classical learning. There was a reaction against medieval scholasticism, and education gained a new emphasis – no longer did it seek to create priests and clerics, but rather the excellent man, the possessor of gentlemanly accomplishments.

The humanists, scholars like Erasmus (*c.* 1466–1536), believed in the powers of the human intellect and in the value of the study of the works of great men; they advocated a life devoted to happiness in this world rather than the one to come. 'It was the substitution of humanism for divinity, of this world for the next, as the object of living, and therefore of education, that differentiated the humanists from their predecessors . . . The humanist's progress consisted in the adoption of the dogma, "The noblest study of mankind is man"' (15).

This shift of emphasis, coming when an expanding educational system and a loosening of Church ties made changes in the school curriculum possible, had important results. Latin was now taught not as an end in itself, but as a means whereby one might study the literature of ancient Rome. The desire to study the works of Greek authors led to Greek being taught in many schools, and occasionally Hebrew, too, entered the curriculum. These changes were also reflected in the two universities where, following the Reformation, various regius professorships were founded – in divinity, Greek, Hebrew, medicine and civil law. Significantly, there were no chairs in mathematics.

Indeed, mathematics held little appeal for the humanists. Vives (p. 5) was something of an exception, although even he had qualifications to make concerning its aptness as a component of general education. Erasmus considered it a subject which those sufficiently far advanced might 'taste', but, in general, the humanists were not attracted by a purely abstract, non-human subject such as geometry. The effect of the humanist movement on the school curriculum can be seen in the timetable of a typical Elizabethan grammar school (Figure 1).

Yet this timetable is that of a grammar school. What was happening at the lesser esteemed and more utilitarian levels of education? As we have noted, there was by 1500 a network of 'petty schools' which provided a basic education in English. This network was to be virtually reorganised as a result of the Chantries Act of 1547 (16). In the following years elementary

Figure I. *Timetable of a typical Elizabethan grammar school, 1598 (from Schools Inquiry Commission, 1868 (17))*

Classes III, IV, and V were taught by the master; Classes I and II by the usher. In winter the school closed at 4 p.m.

	Monday	Tuesday	Wednesday	Thursday	Friday	Saturday
Class V 7–11 a.m.	Prose theme Lecture in Cicero or Sallust or Caesar's *Commentaries*	Verse theme Lecture same as Monday	Prose theme Lecture in Vergil or Ovid's *Metamorphosis*, or Lucan	Lecture in Vergil, etc., same as Wednesday	Verse theme Repetition of the week's lectures	Examination in lecture of previous afternoon
1–5 p.m.	Latin syntax or Greek grammar or Figures of Sysenbrote Home lessons and exercises given out and prepared	Latin syntax, etc., same as Monday	Latin syntax, etc., same as Monday	Half-holiday	Repetition continued. Lecture on Horace, or Lucan, or Seneca's *Tragedies*	Declamation on a given subject by several senior scholars Catechism and New Testament
Class IV 7–11 a.m.	Lecture on Cicero's *de Senectute* or *de Amicitia*, or on Justin	Lecture on Cicero, etc., as on Monday	Lecture on Ovid's *Tristia*, or *de Ponto*, or on Seneca's *Tragedies*	Lecture, etc., as on Wednesday	Verse theme, and repetition of the week's lectures	Examination in lecture of previous afternoon
1–5 p.m.	Prose theme Latin syntax or Greek grammar or Figures of Sysenbrote Home lessons and exercises given out and prepared	Verse theme Latin syntax, etc., as on Monday	Prose theme Latin syntax, etc., as on Monday	Half-holiday	Repetition continued Lecture on Ovid's *Fasti*	Catechism and New Testament

Class III 7–11 a.m.	Lecture on the letters of Ascham, or Sturm's Cicero's *Letters*, or Terence Paraphrase of a sentence	Lecture on Ascham, etc., as on Monday *Vulgaria* in prose	Lectures on Palengenius, or the Psalms of Hess Paraphrase of a sentence	Lecture on Palengenius or the Psalms of Hess	*Vulgaria* in Prose, and repetition of the week's lectures	Examination in lecture of previous afternoon
1–5 p.m.	Latin syntax or Greek grammar or Figures of Sysenbrote Home lessons and exercises given out and prepared	Latin syntax, etc., as on Monday	Latin syntax, etc., as on Monday	Half-holiday	Repetition continued Lecture on Erasmus' *Apophthegms*	Catechism and New Testament
Class II 7–11 a.m.	Lecture on *Colloquies* of Erasmus or on *Dialogues* of Corderius	Lecture, etc., same as on Monday	Lecture on the Cato senior, or Cato junior	Lecture, etc., same as on Wednesday	Repetition of the week's lectures	Examination in lecture of previous afternoon
1–5 p.m.	Translations from English into Latin Home lessons and exercises given out and prepared	Translations as on Monday	Translations as on Monday	Half-holiday	Repetition continued Lecture on *Æsop's Fables*	Writing out the Catechism in English Arithmetic
Class I 7–11 a.m.	The Royal Grammar	The Royal Grammar	The Royal Grammar	The Royal Grammar	Repetition of the work of the week	Examination in lecture of previous afternoon
1–5 p.m.	The English Testament, or the Psalms of David, in English	As on Monday	As on Monday	Half-holiday	Repetition continued Lecture on *Æsop's Fables*	Writing out the Catechism in English Arithmetic

education came to be provided in a variety of forms; but usually by private individuals. Frequently grammar schools provided for a 'class of petties' taught by an usher or an older pupil; curates and parish schoolmasters gave basic lessons, and private teachers opened schools. The exact nature that elementary education took cannot readily be described, for much would depend upon the capabilities of individual teachers. We do know that in some cases the schools, in addition to supplying the basics of reading and writing, also taught children elementary arithmetic in the form of 'casting accounts' (18); note also that in the Elizabethan grammar school timetable (Figure 1) the junior classes were taught some arithmetic – not, however, at a time particularly conducive to learning! Arithmetic, therefore, had some place in the vocational education of those who were to be 'prentices'.

The universities, like the grammar schools, put little stress upon mathematics. As Hill writes of a somewhat later period (1558–1642): 'One or two dons, and some of the abler undergraduates, might have a spare time interest in mathematics or science. But such studies were not the reason for coming to Oxford or Cambridge' (19). Certainly, there is no shortage of contemporary critics to support such a statement. Yet, there were individuals such as Sir John Cheke, the first Regius Professor of Greek at Cambridge, who made notable attempts to encourage the study of mathematics and, in particular, Euclid's geometry. Some of Cheke's protégés were also to teach at Cambridge, but their influence was to be felt more at Court where they were appointed as tutors. In general, 'the universities at large appeared indifferent or even hostile to any mathematics that went beyond the meagre medieval curriculum upon which a single lecture course was provided at Cambridge' (20).

For a brief period there were signs of change. An attempt to reorganise the studies at Cambridge resulted in the Edwardian Statutes (21) of 1549 which laid down that all freshmen were to be taught mathematics as the foundation of a liberal education, the textbooks recommended being the arithmetics of Tonstall and Cardan, Euclid's geometry, and astronomy after Ptolemy. Even this minimal amount of mathematics was removed, however, when new statutes were given by Elizabeth in 1570. Mathematics was then excluded because the commissioners 'thought that its study appertained to practical life, and had its place in a course of technical education rather than in the curriculum of a university' (22).

The need for teaching mathematics for technical, vocational reasons was, therefore, becoming accepted. It was a need answered, as we shall see, by a growing number of private teachers, mathematical practitioners or 'philomaths'. Meanwhile, in the universities the subject was 'smally regarded' (23) and students seeking a liberal education were allowed to ignore it.

3. ROBERT RECORDE AND MATHEMATICS

An analysis of Recorde's published texts demonstrates the range of mathematics which he comprehended and imagined would attract and prove of value to his readers. However, it also serves to raise other issues of interest. Recorde lived at a time when old theories, in particular those of Ptolemaic astronomy, were being challenged. It was a period in which the idea of progress within science and mathematics was gaining ground: no longer could these disciplines be looked upon as static bodies of knowledge to be handed on from generation to generation. Such a change had enormous implications for mathematics education; for if the body which constituted mathematical knowledge was continuously to be expanded and some details replaced, then this would have implications for curricula. They too would demand constant revision and the need would also arise to prepare students to cope with change and with the expansion of knowledge and understanding. The former need was, often belatedly, recognised by educators; even in our own time the implications of the latter have to be fully appreciated and acted upon.

'*The Ground of Artes*'

This, Recorde's first book and his most popular and important one, first appeared in 1543 {24}. There were to be numerous other issues and editions – at least forty-five – and under the various editors (see section 5 below) the book was gradually expanded as new topics, for example, decimal fractions, were added. The final edition appeared in 1699, a century and a half after the first.

The *Ground*, however, was neither the first important arithmetic to be written in England, nor the first arithmetical text to be written in English. Tonstall's [25] *De Arte Supputandi* had already appeared in 1522 and, although it was never reprinted in England, seven editions were published in Paris and Strasbourg. The book is based on Italian models and 'although it was written for the purpose of supplying a practical handbook, is very prolix and was not suited to the needs of the mercantile class . . . It is . . . the work of a scholar and a classicist rather than a business man' (26). It was recommended as a textbook at Cambridge in 1549 (see p. 12 above), but would seem to have had little effect upon succeeding English texts. Tonstall had, of course, written his work in Latin. The first known arithmetical text in English was an anonymous text *An Introduction for to Lerne to Reckon with the Pen* which is reputed to have appeared for the first time in 1537 (27). At least seven other editions were published before 1629. Recorde, however, in his Preface to the *Ground*, expresses his belief that 'some [readers] wyll lyke this my booke above any other Englyshe Arithmetike hytherto written' which suggests that other rivals existed.

The *Ground* is in three parts which are far from equal in length. The first part constitutes the bulk of the book and consists of the fundamental arithmetical operations on the positive integers with their usual applications. After a description of the Hindu–Arabic form of notation – in which Roman numerals still make fleeting appearances – Recorde goes on to deal with the four basic operations. Subtraction is explained by means of 'equal additions'. Although a 9×9 triangular multiplication table is given, the reader is shown how to use complementary multiplication to obtain the product of single-digit numbers greater than 5. Thus to multiply 8 and 7 one finds the respective differences from 10 and multiplies these to obtain the units digit of the product. The tens digit is found by subtracting one of the differences from the other original (i.e. 3 from 8 or 2 from 7). Again, this was a common method, and indeed Recorde offers no new techniques in this work. After this foundation work, Recorde has sections on reduction (converting weights and coins) and progression (arithmetic and geometric progressions in which the correct formulae are used but not proved: 'prove' for Recorde has the old meaning of 'test', thus one 'proves' the formula 'by adding together of all the parcels, for so will the same summe amount'). Next follow the golden rule (proportion) including the backer rule (inverse proportion), the double rule (compound proportion) and the rule of fellowship (merchants contributing unequal sums for different periods: how are the profits to be divided?). This ends the section on arithmetic 'with the penne'. We note that Recorde does not introduce operations with fractions. He avoids their use by taking integer multiples of suitably small units, for example farthings or quarters of an ell. A section on fractions was, however, introduced in the 1552 edition.

The second part covers computations with counters intended 'for theym that can not write and reade' or those that do not have 'their penne or tables ready with them'. Abacal arithmetic had, of course, been a mainstay of merchants for many centuries (indeed, in some countries it remains so today), and some idea of its universality can be gained from the manner in which Shakespeare on several occasions associated arithmetic with counters, for examples,

> *Iago* Forsooth, a great arithmetician.
> One Michael Cassio . . .;
> this counter-caster
> He, in good time, must his lieutenant be. (*Othello*, I, i)

Schools often taught the two types of arithmetic side by side or according to their pupils' abilities. Thus the schoolmaster of Bungay School, Suffolk, was in its 1592 statutes enjoined to keep school every Saturday and every half-holiday until 3 p.m. 'for writing and casting accounts with the pen and counters according to their [the students'] capacities'.

The final, and very brief, section in the *Ground* explains the 'arte of Numbrynge by the hande'. Again, the practice of showing numbers and/or reckoning on the fingers had by Recorde's time a long history (cf. pp. 1,3 above); yet it too still survives today.

'The Pathway to Knowledge'

The *Pathway* was Recorde's second mathematical work to appear. It was first published in 1551 (also in 1574 and 1602) and its purpose is conveyed in its full title:

The Pathway to Knowledg, containing the first principles of Geometrie, as they may moste aptly be applied unto practise, bothe for the use of instrumentes Geometricall, and astronomicall and also for projection of plattes [*plans*] *in everye kinde, and therfore much necessary for all sortes of men.*

This then was to be a geometry book – the first to be published in English – with an emphasis on the subject's practicality. It was the second in a series intended to cover all the elementary mathematics of value to navigators, surveyors, merchants and craftsmen. Despite its title the book is largely derived from the first four books of Euclid's *Elements* (28), although there are some significant deviations, some of a mathematical, and others of a pedagogical nature. Recorde planned the work in four parts, but 'misfortune . . . hindered it' and only two parts were printed. The first part contains definitions and various constructions (including those which comprise Book 4 of Euclid); the second includes the postulates and axioms, and also those theorems from the first three books of Euclid which do not refer to constructions. The missing third part would have included additional constructions and applications of Pythagoras; the fourth would have dealt with 'measurynge'. Recorde's definitions sometimes draw on authors who wrote after Euclid. Thus, for example, a straight line is defined not as 'breadthless length' but as the 'shortest that maye be drawenne betweene two prickes [points]. And all other lines . . . are called Croked lynes.' Again, Recorde differs from Euclid in using the equidistant property to define parallel lines.

So as to justify his claims to applicability, Recorde also includes in the first part of the *Pathway* a number of approximate constructions which would have been used by craftsmen {29}. Sometimes, Recorde takes the 'easy' way out; to construct an isosceles triangle with base angle 72° in a circle (as the first step towards inscribing a regular pentagon) he simply commands: 'Fyrste make a circle, and devide the circumference of it into fyve equall parts.' In Easton's words: 'Whether Recorde himself did not know how to inscribe a pentagon, or whether, in view of the complexity of Euclid's construction, he considered approximate methods sufficient, is not clear' (30).

The second part of the *Pathway* includes postulates ('certain grauntable requestes'), axioms ('certayn common sentences manifest to sence and acknowledged of all men') and seventy-seven theorems ('whiche may be called Approved truthes') all taken from Euclid. Each theorem is accompanied by a figure and an explanation, but not by the usual proofs.

The *Pathway* would not seem to have had any immediate marked effect on the teaching of geometry in the universities, and its reappearance in 1574 may not have been due to its own merits but rather because it offered an inexpensive, abridged alternative to Billingsley's *Euclid* (1570) (see p. 25 below). Nevertheless, it was a scholarly work, and not merely a handbook of mathematical recipes as previous English texts had tended to be, and so presumably helped to establish the respectability of texts written in the vernacular.

Yet another feature of the *Pathway* is claimed as a significant innovation by Lilley in his paper 'Robert Recorde and the idea of progress' (31). Lilley in this work sets out to explore the thesis propounded in Zilsel (32) that the most progressive elements in scientific thought and practice in the fifteenth and sixteenth centuries were developed not among scholars but rather among those who might be labelled superior craftsmen or artisans, for example, Brunelleschi, Tartaglia and Stevin. In particular, Zilsel argues that it was these artisans who first achieved a properly balanced assessment of the relations between past inheritance and present endeavour. The concept of 'progress' is one which can be traced back to the twelfth century: Newton's 'If I have seen further it is by standing on the shoulders of giants' was at that time to be found in the form 'we are like *dwarfs* standing on the shoulders of giants'. As Lilley points out, though, the use of the term 'dwarf' has important implications concerning the relative potency and value of the contributions of the two ages – doubts which Newton's formulation removed. Yet such notions were alien to humanist thought for 'they were not conscious of a process of building on the immediate past (only of restructuring the distant past) and so they were unable to regard their scholarship from the "contribution" viewpoint' (p. 9).

It was this concept of 'contributing' to a continuing progress that Lilley searched for in Recorde's work and which he claims to have found in the address to the reader which opens the *Pathway*.

Excuse me, gentle reader if oughte be amisse, straung [strange] paths ar not troden al truly at the first: the way muste needes be comberous, wher none hathe gone before . . . If my light may so light some other, to espie and marke my faultes, I wish it may so lighten them, that they may voide [avoid] offence . . . yet maie I thinke thus: This candle did I light: this lighte have I kindeled: that learned men maie se, to practise their pennes, their eloquence to advaunce, to register their names in the booke of memorie. I drew the platte [plan] rudelie, whereon thei maie builde . . . I will not cease from travaile the pathe so to trade [tread], that finer wittes maie fashion them selves with such glimsinge dull light, a more complete woorke at laiser to finisshe, with invencion agreable and aptness of eloquence.

In Lilley's opinion this address indicates that Recorde sees potential future developments as the ultimate goal of his work. Moreover:

I dare to suggest that perhaps Recorde was the first learned man to state the idea, not as a mere passing reference, but as the central idea which explains the relation of his work to the world around him. If he *was* the first to do this, his achievement was very great – far greater than his minor contributions to mathematics as such. For surely this view . . . is actually the central idea in what we call 'the scientific attitude'. If Recorde was indeed the first scholar to give expression to it, then he may have done more for science than many a man who made striking discoveries or even improved the methods of research (pp. 30–1).

'The Gate of Knowledge'

The preliminary pages of Recorde's *Castle* (1556) contain verses which suggest that Recorde published the third of his planned series of texts, *The Gate of Knowledge*, dealing with measuring and the use of the quadrant. However, no copy of it has been found; its disappearance, if it ever existed, is a publishing mystery (33).

'The Castle of Knowledge'

This introduction to astronomy 'containing the explication of the sphere bothe celestiall and materiall, and divers other thinges incident thereto. With sundry pleasant proofes and certaine newe demonstrations not written before in any vulgare woorkes' first appeared in 1556. A second edition was printed in 1596.

This book displays more than the preceding ones the range of Recorde's reading and scholarship. In it he makes frequent references to his sources, Cleomedes, Ptolemy, Proclus, Richard of Wallingford, Sacrobosco, Oronce Fine, etc., and engages in a critical examination of their works. Yet this does not result in an over-theoretical book. Various astronomical instruments are described and a number of astronomical and navigational tables given. As a result, the *Castle* probably excels its predecessors in its immediate usefulness to the mathematical practitioner and navigator.

Martin Frobisher, we are told (34), took it with him on his first search for the Northwest Passage.

In the first part of the *Castle* Recorde unequivocally supports traditional geocentricism:

5. The earthe is in the middle of the worlde, as the centre of it . . .
6. The earthe hathe no motion of itselfe, no more than a stone, but resteth quietly (p. 15).

Yet, did Recorde really subscribe to this Ptolemaic view of astronomy? There are suggestions in the fourth part that he was a Copernican, for here he cleverly uses the dialogue form in which the book is written to cast doubt upon the ancient theories and to ask that contemporary views should be given a more open hearing. That Recorde should choose to conceal his personal belief is understandable, for openly to have adopted the heliocentric hypothesis would have subjected him to ridicule and possibly religious persecution. Perhaps, then, it was as a result of such pressures that Recorde cannot be described as the author of the first English Copernican text {35}.

'The Whetstone of Witte'

This was the last of Recorde's works to appear (1557) and the only one never to have been republished. It was a work dedicated not to a person, but to a body which sought to promote overseas trade, the Company of Moscow Adventurers, to which Recorde acted as adviser. Although generally thought of as a book on algebra, the *Whetstone* is, in fact, also much concerned with arithmetic. Described as being 'the seconde parte of Arithmeticke: containyng the [e]xtraction of Rootes: The *Cossike* practice, with the rule of *Equation*: and the woorkes of *Surde Nombers*', the work begins with the kind of arithmetic to be found in Euclid and Boethius in which the emphasis lies not on commercial applicability but on various kinds of numbers and their properties.

The algebra contained in the book is based on German sources and, as Hutton points out in his *Dictionary* and his *Tracts* (vol. 2) in the course of extensive and still valuable comparative surveys of books on algebra, Recorde often quotes and takes examples from Scheubel. Recorde deals with elementary algebra as far as quadratic equations and offers some new mathematics, such as techniques for finding the square roots of algebraic expressions. However, the *Whetstone* is above all remembered for the fact that it was the first English book to use the + and − notation (following Stifel), and the first book ever to use the '=' sign to denote equality (36). As Recorde put it 'to avoide the tediouse repetition of these woordes: is equalle to: I will sette as I doe often in woorke use, a paire of paralleles, or Gemowe [twin, as in gemini] lines of one lengthe, thus: =, bicause noe 2 thynges, can be moare equalle'. With the help of this innovation Recorde was able to express equations in a purely symbolic form {37}.

Although only four of Recorde's mathematical works were definitely published, others, in addition to the *Gate*, were planned, namely works on advanced astronomy, navigation and a translation of Euclid's *Elements*. To have succeeded in producing such an encyclopaedic coverage of mathematics would have been an outstanding achievement indeed; to have come so near to it was remarkable.

4. ROBERT RECORDE AND MATHEMATICS EDUCATION

Robert Recorde not only taught mathematics, he had also effectively to construct a curriculum. In the course of doing this he frequently had to make conscious choices; often he supplied reasons for these. It is to such pedagogical considerations that we now turn.

(a) *The choice of form*

Recorde was writing for the individual reader: 'suche as shall lacke instructers' (Preface, *Ground*). Yet, he also realised that the reader might be prompted by different motives; there were those 'which studi principalli for lerning . . . but also those whiche have no tyme to travaile for exacter knowledge' (*Pathway*, Sig. z, ii, r–v). There was, then, a need to reconcile the needs of these two types of reader.

In order to facilitate the reader's learning, Recorde made several important decisions. The first was, of course, to write in the vernacular (38) (although it is likely that as a university-trained man he, himself, would have found it easier to write in Latin). A second, to which he specifically refers, was to provide and 'plainely set foorth' examples in a way that 'no booke (that I have [seen]) hath doone hytherto': this practice was intended to bring 'greate ease to the rude reader'. (It must be stressed here that Recorde did not include exercises for the reader: this was an eighteenth-century development.) A third was to construct three of his books as dialogues between the Master and the Scholar. He explains his reasons for doing this in the Preface to the *Ground*: 'I judge that to be the easyest waie of instruction, whan the scholar maie aske every doubt orderly, and the mayster maie aunswere to his question plainely.' 'There wyll be some', he assures us, 'that will fynde faute [fault], because I write in a dialogue'. Yet, notwithstanding such fault-finders, Recorde adopted the dialogue form in all of his books with the exception of the *Pathway*. That he should make an exception of this work shows a willingness to adapt form to content. Writing in English and in dialogue form were skills which Recorde was to polish. Initially, his language was formal and lacked obvious popular appeal. Later, however, his writing became livelier and more lucid. The dialogue form was, in fact, to survive for many years {39}. As a device for introducing short breaks into long and tedious narrations it has, of course, been widely employed by dramatists and librettists such as Shakespeare and Wagner. In a similar fashion, Recorde uses interjections by the Scholar or Master to emphasise the separate steps in a particular technique. The method, too, enables the author to demonstrate the Scholar's enthusiasm and growing mastery. Indeed, the varied use which Recorde makes of the dialogue is greatly to be admired.

Not everyone appreciated Recorde's books, however. Edward Worsop, writing in 1582, thought that they could not 'be understood of the common sort'. He, though, was not an unbiased observer, but the author of an alternative popular text (40). Notwithstanding such complaints, it is significant that the changes in the later editions consisted almost entirely of additions and not amendments to Recorde's original {41}.

(b) *Order and emphasis*

When developing his series of texts, Recorde had to take decisions about the order in which he presented his material and also the relative emphasis which he was to place on the understanding of content and on the acquisition of mastery.

As mentioned earlier, he did not include sets of exercises in his work {42}. Yet he did expect the reader to practise on his own:

> Howbeit I will yet exhorte you now to remembre . . . to exercise yourselfe in the practice of it: for rules without practice; is but a light knowledge: and practice it is, that maketh men perfect and prompte in all thynges (*Ground*, Sig.C, ii, v).

Later, Recorde was to return to the theme:

> *Master*. So may you if you have marked what I have taught you. But because thys thynge (as all other) must be learned [surely] by often practice, I wil propounde here ii examples to you, whiche if you often doo practice, you shall be rype and perfect to subtract any other summe lightly . . .
>
> *Scholar*. Sir, I thanke you, but I thynke I might the better doo it, if you did showe me the woorkinge of it.
>
> *M*. Yea but you muste prove yourselfe to do som thynges that you were never taught, or els you shall not be able to doo any more then you were taught, and were rather to learne by rote (as they cal it) than by reason (*Ground*, Sig.F, i, v).

Recorde then was the first British mathematics educator to emphasise the limitations of learning by rote!

But how was understanding to be gained? Was one to be given reasons and then techniques, or vice versa? Recorde was in no doubt:

M. . . . whyche is equal to the number noted at the end of the lyne, and thereby I have done well.

S. But I doe not se the reason of this.

M. No, no more do you of many thinges els, but hereafter wyll I shewe you the reasons of all arithmeticall operations for this I indg [judge] to be the best trade of teachying, firste by somme briefe preceptes to instructe a learner somewhat in the use of the arte, before he learne the reasons of the árte, and then may you afterwarde more soner [sooner] make him to perceyve the reasons: for harde it is for to occupye a yonge learned witte wyth bothe the arte and the reasons of it all at ones [once]: how be it he shall never be connyng in deede in an arte, that knoweth not the reason of every thyng touchyng it (*Ground*, Sig.D, viii, r–v).

Understanding is vital, therefore, but mastery of a technique may well precede understanding. This is a viewpoint which also helps to determine Recorde's ordering in the *Pathway*. First, we recall, comes the practical work with constructions {43}. Then follow Euclid's theorems, but with explanations rather than proofs, for Recorde's experience of teaching had convinced him that 'it is not easie for a man that shall travaile in a straunge arte, to understand at the beginninge bothe the thing that is taught and also the juste reason whie it is so' (*Pathway*, Sig.a, ii, v). Recorde mentions here that there existed precedents for Euclid without proofs, namely Rheticus and Boethius.

Asking the learner to 'practice' without supplying him with carefully graded exercises would today strike us as odd. Yet Recorde did something to ameliorate the difficulties arising from this omission by stressing the need to check answers. In the case of the *Ground* he demonstrated two methods: first, checking by inverse operations – a subtraction by an addition – and secondly by 'casting out nines' (or, in modern nomenclature, repeating the problem in arithmetic modulo nine). In his *Whetstone* he argued the use of arbitrary numbers to check algebraic operations rather than inverse operations.

The scholar, too, had to ask whether or not the answer was reasonable! To this end Recorde included the occasional whimsical example. Thus:

M. If I solde unto you a horse, havynge 4 showes [shoes], and every shoe 6 nayles, with this condition, that you shall paie for the fyrste nayle, i ob [halfpence], for the second 2 ob, for the third 4, and so forth . . . Now I aske you howe muche woulde the holle price of the hrose [*sic*] come unto . . .

S. . . . £34952, 10s, 7d, ob.

M. That ys well done, but I thinke you wylle bye no horse of the pryce.

S. No Syr, if I bee wyse (*Ground*, Sig.L, ii, r–v).

Choosing examples so that they might by their ready applicability, topicality, humour, etc. have added extra-mathematical interest for the reader is, of course, part of the textbook writer's art. Evidence that Recorde recognised this can readily be found. Suitably chosen examples can also be used to transmit political views and this too was a technique known to Recorde (44). In the first edition of the *Ground*, Recorde, an active champion of Protestantism, was able gently to mock the Established Church in the question: 'There is in a cathedrall churche 20 cannones, and 30 vicars, those maie spend by yeare, £2600 but every cannon muste have to his part 5 tymes so much as every vicar hath: how much is every mans portion saie you?' (*Ground*, Sig.M, v, v). More contentiously, in the second edition he was to add a question concerning the vexed contemporary problems of sheep and enclosures. Somerset, whom Recorde supported, had passed an Act in 1548 taxing sheep in an attempt to balance the country's economy. The Act was repealed the next year following Somerset's fall and prior to the publication of the second edition of the *Ground* which was to contain further indications of Recorde's political leanings (45).

(c) *Giving meaning to mathematics*

The provision of examples demonstrating the applicability of mathematics does in some sense provide 'meaning' for the reader. Yet 'meaning' goes beyond this. It can extend to the provision of names for mathematical objects which carry 'meaning', i.e. which actually help the learner to appreciate and recall the properties of those objects; for this can assist the learner to relate a new concept or definition to his existing knowledge. Recorde was greatly aware of this. Indeed, for him, the first writer of a series of texts in English, the choice of mathematical terms posed a major problem. The *Ground* was comparatively straightforward, for there were precedents and, indeed, its arithmetical terms were in common usage. The *Pathway*, however, set novel problems. There was no commonly accepted English terminology. In some cases, English equivalents to specialist Latin terms were lacking. What should the author do – borrow the classical terms (used by scholars both in England and elsewhere), or create new English terms which would, to the reader happier in the vernacular, convey more 'meaning' of the signified {46}? In general, Recorde opted for 'meaning' and attempted to create new 'English' terms rather than to adopt the classical ones based on Latin or Greek. Thus, for example, he speaks of 'pryckes' rather than 'points' {47}, of 'sharpe' (acute) and 'blunt' or 'brode' (obtuse) angles; tangents are 'touche lynes' and the common 'dye' [die] is used rather than 'cube'; as alternatives to equilateral, isosceles and scalene triangles, he has 'threelike', 'tweylike' [twolike] and 'nouelike' [nonelike]; and the square and rhombus are 'likesides'. The pentagon, hexagon and heptagon become 'cinkeangle', 'siseangle' and 'septangle', thus indicating how often Recorde looked to Old French rather than to Saxon when choosing English terms. Recorde's efforts did not succeed. Indeed, in the *Whetstone*, when asked by the Scholar why he did not give English names to the various types of proportion, the Master replied 'Bicause there are no soche names in the Englishe tongue. And if I should give them newe names, many would make a quarrelle against me, for obscuring the olde Arte with newe names: as some all redy have done' (Sig.B, iv, v). Recorde's failure to convince his countrymen resulted in the English taking over the classical terms almost in their entirety. The results are still with us today, as inspection of any primary-school wordlist will show. Contemporary with Recorde there was an attempt by German authors to create new mathematical terms in the German language. These, however, took root and have survived. It is not obvious why one attempt succeeded and the other failed; neither has there been any research into the effects of these differences {48}.

The pedagogical problem of definition is perhaps at its acutest in geometry, and so it is the *Pathway* that best indicates Recorde's methods. There we can find examples of the way in which he provides not only a 'rigorous' definition, in the style of Euclid, but also less formal notes intended to help the reader more fully to comprehend the concept. Thus, for example:

A *Poynt* or a *Prycke* is named of Geometricians that small and unsensible shape, whiche hath in it no partes, that is to say: nother [neither] length, breadth nor depth. [This, so far, is a slightly amplified form of Euclid's definition.] But as this exactness of definition is more meeter for onlye Theorike speculacion, then for practise and outwarde worke (consideringe that myne intente is to applye all these whole principles to woorke) I thynke meeter for this purpose, to call a poynt or prycke, that small printe of penne, pencyle, or other instrumente, which is not moved, nor drawen from his fyrste touche, and therefore hath no notable length nor bredthe (*Pathway*, Sig.A, i, r).

Recorde goes on to define a 'lyne':

a great numbre of these prickes . . . so if you with your pen will set in more other prickes betweene everye two of these, then wil it be a lyne, as here you may see — and this lyne, is called of Geometricians, Lengthe withoute breadth.

But as they in theyr theorikes (which ar only mind workes) do precisely understand these definitions . . . (*Pathway*, Sig.A, i, r–v).

Whether Recorde shared our appreciation of the distinction which he drew between the abstract nature of the geometer's system and the model of it which the artisan employed is, of course, doubtful. Nevertheless, it once again demonstrates the care with which he attempted to satisfy his two kinds of reader: scholar and user.

Recorde did not write specifically on pedagogy yet, as these extracts show, it is clear that he identified some key problems of mathematical education and sought his own solutions to them.

5. ROBERT RECORDE'S EDITORS

As mentioned earlier, Recorde's *The Ground of Artes* appeared in many editions between its first (1543) and that edited by Edward Hatton in 1699. After Recorde's death the book also appeared in editions by John Dee, John Mellis, John Wade, Robert Norton, Robert Hartwell and Thomas Willsford. Some idea of the problems and progress of mathematics education in that period can be gained from a brief survey of these mathematicians.

Without doubt, the most interesting and influential of them was John Dee, mathematician, scientist and magician. He, unlike the others, was an internationally recognised scholar, a university man and a friend of royalty and the nobility.

Dee was born in London in 1527, the son of a mercer. (The mercers who dealt in textiles were members of the new 'middle class' and it was their company that governed St Paul's School after its foundation in 1509.) He was a student at St John's College, Cambridge, where in 1545 (at the age of eighteen) he graduated BA, and three years later MA [49].

His success at university was such that he was elected a Fellow of St John's, a foundation Fellow of Trinity College (1546) and was also encouraged to travel abroad. In 1547 he visited Louvain, and the following year Brussels and Paris. In this way he both greatly extended his mathematical and scientific knowledge, and also became acquainted with such leading scholars as Gemma Frisius, Peter Ramus and Gerard Mercator. Whilst in Paris, Dee lectured on Euclid [50] and among the large audiences he attracted were some of the young Dudleys (sons of the Duke of Northumberland). Further connections with the Court were established on Dee's return to England when he was introduced by Cheke (see p. 12 above) to Sir William Cecil, Secretary of State. Thus, by the time he was twenty-five, Dee had been accepted both by the Continental mathematicians and the nobility. He was to retain and develop both of these links, and, in particular, made many more journeys abroad. The interest which the young Dudleys and others showed in mathematics was evidence of the way in which mathematics was now beginning to interest the wealthy amateur (51). The work of Cheke, Dee and others was clearly beginning to have some effect. Yet this interest did not pass without critical comment. When, in 1564, Robert Dudley, the Earl of Leicester and favourite of Queen Elizabeth, developed an enthusiasm for geometry, Ascham, Elizabeth's tutor, author of the influential *The Schoolmaster* and one who demonstrated a continual distrust of mathematics and mathematicians (see p. 31 below), was not slow to express his displeasure:

Your lordship doth very well remember my poor advice (which proceeded both of good will and also of right judgement) to have had your lordship increase your knowledge of the Latin tongue . . . it is a gift . . . more praiseworthy indeed, more profitable for use, than if you had upon your finger ends all the geometry that is in all the book of Euclid . . . I say this, my lord, because I think you did yourself injury in changing Tully's wisdom with Euclid's pricks and lines (*Letters*, 5 Aug. 1564).

In the same letter Ascham also pointed out that, anyway, spiders and bees with their subtle use of proportion and dimension far outdid humans at geometry. This then was the opposition to mathematics education shown by a noted humanist educator.

Dudley did not altogether heed Ascham's advice and was one of the first to purchase Henry Billingsley's *Euclid* (1570), the first to be published in English {52}. Exactly how much Billingsley contributed to the work is not clear; probably he was indeed responsible for the translation. However, it is now generally accepted that the book was edited by Dee who also supplied the preface. In this 'Mathematical Preface' Dee advanced arguments for mathematics' place in education, not only for its utility, but also for its power to 'lift the heart above the heavens'. Moreover, he stressed the need for mathematical books in the vernacular such as were available in Italy, Germany and elsewhere. Only with this provision would the common man come to know the subject and to make use of, or derive pleasure from, it. The 'Preface' was to have an enormous effect (53). It was reprinted (in the Euclids by Rudd (1651) and Leeke and Serle (1661)), quoted by later reformers such as Samuel Hartlib (see p. 48 below), and was sufficiently well known to be ridiculed by Samuel Butler in his *Hudibras*, a Restoration reaction to puritanism. Perhaps it was the success of his 'Preface' which led to the large number of reissues of Dee's editions of Recorde's *Ground*. Dee's was, indeed, a name to conjure with, for now he was not only a renowned author, but was established as the Queen's favourite astrologer – the man whose mathematics had helped to determine the most auspicious date for her coronation, Sunday 15 January 1559.

Dee, in fact, made remarkably few alterations to Recorde's work. The only significant changes in the 1561 edition were in the section on progression. Dee included six propositions 'of which foure . . . were invented by a friende of myne and never before this published' (54). Rather more changes appeared in the 1570 edition. Then Dee replaced some of Recorde's definitions {55}, and also included sections on the valuation of English, Flemish and French money and on the different weights and measures in use in various Continental towns. These last additions, of course, served to increase the book's commercial appeal.

Throughout the 1570s Dee continued his studies of navigation and helped to prepare the way for the eventual adoption of the Gregorian calendar. However, in the words of Hutton, great changes took place in 1581: 'Hitherto the extravagancies of our eccentrical philosopher seem to have been tempered with a tolerable proportion of reason and science; but henceforward he is to be considered as a mere necromancer and credulous alchymist' (56). The story of the decline of Dee as a serious scientist has often been charted (57); it suffices here to remark that he survived Elizabeth to die aged eighty-one in 1608.

Recorde's other editors had different social and mathematical backgrounds from Dee. They were representatives of the new breed of mathematical practitioners, usually non-university men, often self-taught, who made a living by teaching mathematics at private schools to those who needed it for their vocations. Thus their additions to Recorde's text biased it more and more in the direction of commercial arithmetic, while their other publications concerned applicable mathematics rather than Euclid. Theirs were not the type of books to be read at Cambridge (58).

The first of these editors was John Mellis, a schoolteacher and one-time Cambridge student, who produced several editions, the first in 1582. Mellis kept the Dee version substantially intact but added a new part which included 'Losse and gaine', 'Rules of payment', barter, exchange, interest and other business applications, together with a chapter on 'Sportes and pastimes done by number'. Such chapters as the last were to appear in many subsequent textbooks {59}. Mellis' other work of outstanding interest was his *A Briefe Instruction and Manner how to Keepe Bookes of Accompts* (1588), the earliest English work on book-keeping by double entry. Here again we have evidence of the growing need for arithmetic among merchants and of the manner in which that need was met by private practitioners.

After John Wade (1610), of whom little appears to be known, the next editor (1615) was Robert Norton. He too was a teacher of practical mathematics – in this case with a bent for navigation and engineering (he held an official position in the Tower of London). He translated Stevin's work on decimal arithmetic from the Dutch and also included work on decimals, together with Stevin's tables of interest and further tables of 'Board and timber measure' in his edition of Recorde which had the new title *Records Arithmeticke: Contayning the Ground of Arts . . . Also the Art and Application of Decimall Arithmeticke*. Norton, then, made a serious attempt to modernise Recorde's work. His book also casts further light on the work of the mathematical practitioner, for, in an advertisement, it quotes the syllabus offered by N. Physhe (Fiske) [60]:

Arithmeticke	Vulgar, with fraction and the rule of prop^{n}.
	Extr of the square roote cubicke, and square cubicke root, at one worke, according to Vieta.
	Equation, with Cosse or Algebar.
	The qualitie resulting of medicines compound, and the finding of the Dose or quantity of Compositions purative by their ingredience.
Geometrie	Principles thereof with practicall demonstration.
	Surveying with the use of Topographical instruments.
	Reduction of Plots, or Maps to any proportion desired.
	Measuring of solid contents as Timber, stone and changing of vessels, with the use of instruments thereto belonging.
	The art of dialling in generall.
	The Doctrine of triangles with the manifold use of the Sinus, Tagents [*sic*] and Secants.
Astronomy	The calculation of the true place of the Planet, and fixed stars, according to the most worthy Tycho Brahe, for any time, or place.
	The use of the Sphere and Globes.
Astrologie	The erection of the figure Astralogicall, for any time, or place. The use of the *primum Mobile* with the effects of the Planets and fixed stars, in every particular figure, and Mansion. Also their direction to any point of the Heavens assigned, with their profection, Revolutions, and Transitions.
Navigation	Principles thereof with instruments for that purpose.
	Direction to finde the distance of places, by Longitude, and Latitude.
	And to finde the Longitude and Latitude by instruments, or planetical observations.

We note that specific attention is now paid to problems of measuring or gauging. This application of mathematics was to grow in importance during the next three centuries and led to the publication of numerous books on mensuration and gauging, a key work being that by Hutton (p. 63 below).

The next editor, Robert Hartwell (1618 and frequently thereafter), added more material on interest and annuities, but also made a slight correction to the commercial bias by including an appendix on figurate numbers (triangle numbers, etc.). This was, of course, a reversion in part to the arithmetic to be found in Boethius and was matter which Recorde had included in his *Whetstone*. Hartwell, too, advertised the mathematics he was prepared to teach (61). Some idea of how quickly the practitioner was called upon to expand his syllabus is given by the fact that in 1623 Hartwell advertised that he taught logarithms even though Napier's *Descriptio* was only nine years old and Wright's English translation seven. Hartwell also offered to teach book-keeping by double entry methods.

After Hartwell came Thomas Willsford, another mathematical teacher and a writer on navigation and cosmography (62). The final editor (1699), Edward Hatton, supplies us with evidence of the growing country-wide need for mathematics, for Hatton was not, like his predecessors, London based, but kept a mathematical school at Stourbridge in Worcestershire. Again, his interests were primarily practical, namely commercial arithmetic and gauging. By now, however, Recorde's attractions were waning. Although in the sixteenth century Humphrey Baker's *The Well Spryng of Sciences* had provided some opposition, and in De Morgan's words (63) had helped to 'break the fall from the "Grounde of Artes" to the commercial arithmetics of the next century', there were now, as the seventeenth century came to a close, far more substantial and modern rivals to Recorde (64). In particular, Cocker's *Arithmetic* (see p. 39 below) had embarked on its remarkably successful career. No longer then did Recorde's name attract the purchaser. Hatton did not prepare any further editions of Recorde's *Ground* but instead wrote his own *Arithmetic* {65}.

Recorde's *Ground* was 'in print', therefore, for over a century and a half. It provided a model of careful exposition which was both supplemented and copied. As with his other books it was written for both scholar and practical user. Gradually, however, the interests of the two classes diverged until it was necessary to cater for their needs separately. Despite Recorde's efforts, in particular his provision of books in the vernacular which could be read by scholar and artisan, a binary system of mathematics education had arisen.

2 · *Samuel Pepys*

4th July, 1662

By and by comes Mr Cooper, Mate of the *Royall Charles,* of whom I entend to learn Mathematiques. After an hour's being with him at Arithmetique, my first attempt being to learn the multiplication table, then we parted till tomorrow.

5th July

At my office all afternoon and then my maths . . . at night with Mr Cooper; and so to supper and bed.

8th July

Cooper being there, ready to attend me; so he and I to work till it was dark.

9th July

Up by 4-aclock and at my multiplication table hard, which is all the trouble I meet withal in my arithmetique.

10 July

Up by 4-aclock and before I went to the office, I practised my arithmetique.

11th July

Up by 4-aclock and hard at my multiplication table which I am now almost master of.

12th July

At night with Cooper at Arithmetique . . .

13th July

Having by some mischance hurt my cods . . . [I] keep my bed all this morning.

14th July

Up by 4-aclock and to my Arithmetique . . .

18th July

. . . and then came Cooper for my Mathematiques; but in good earnest my head is so full of business that I cannot understand it (1).

That a highly-paid civil servant in the Admiralty (earning £350 p.a.), who was educated at St Paul's School and Cambridge University, should have to employ a tutor to teach him the multiplication table is an indication of the status of mathematics education in the seventeeth century. Perhaps even more astonishing to us is the fact that, within three years of having learned his tables, Pepys should have been elected a Fellow of the Royal Society.

Yet, without having any pretensions to being a mathematician, Pepys was to have considerable influence on the development of mathematics education in England, and consideration of his activities and of his own personal education will reveal much about the state of mathematics teaching in his age.

1. EARLY SCHOOLING

Samuel Pepys was born in London on 23 February 1633, the fifth child (although only the third to survive infancy) of the tailor, John Pepys. His father had been settled in London for some thirty years, but came from a family which had been established in the Fens for several centuries, and which at the time of Samuel's birth contained several successful lawyers and landowners. Thus, although Samuel's home was a humble one, he was soon introduced to more influential and affluent households. It was while Samuel was staying with well-to-do relatives that his formal education began – at the Free School in Huntingdon, one of the small grammar schools then to be found all over England.

Like many such schools, that at Huntingdon had a long and obscure history. It is probable that it sprang from a religious foundation and it certainly existed in the twelfth century (2). From medieval times it had an endowment which paid the salary of a master and so allowed towns-children to be educated free. Its curriculum, which would have been determined by the current master's interests, was in Pepys' time almost certainly confined to Latin grammar, the Catechism and Bible study (cf. p. 9 above) {3}. In this it resembled the majority of contemporary grammar schools; indeed at that time, pupils arriving at Oxford and Cambridge frequently did not have any knowledge of Arabic numerals, to say nothing of the elementary arithmetical operations (4).

Yet, mathematics was to be found in some pioneering schools; thus Pepys' future friend, the economist Sir William Petty, was introduced at his school in Romsey to 'all the Rules of Common Arithmetick, of practical Geometry, Dialling and the astronomical parts of Navigation' {5}.

A former pupil at Huntingdon was Oliver Cromwell, and, during Pepys' stay there, Cromwell returned to the town to recruit the nucleus of his Ironsides. It was, of course, a period of great national unrest, and, with the Civil War at its height, Pepys returned to London, where on 30 January 1649 he stood at Charing Cross and watched King Charles' dripping head displayed to the crowd.

After returning to London (at the age of eleven) Samuel entered the nearby school of St Paul's. (His father as a 'foreigner' was not a member of the Guild of Merchant Taylors to whose school Samuel might otherwise have gone.) St Paul's, which had been established about 1509, was a monument to the influence of humanism. Colet, its founder, was a friend of Erasmus, and the curriculum he devised reflected classical humanist thought:

I would they were taught all way in good literature both Latin and Greek and good authors such as have the very Roman eloquence joined with wisdom; especially

Christian authors that wrote their wisdom with clean and chaste Latin either in verse or in prose . . . And for that intent I will the children learn first above all the Catechism in English, and after, the accidence that I may, or some other if any be better to the purpose, to induce children more speedily to Latin speech (6).

Children were admitted to the school – and the age of entrance varied widely – provided they could read and write competently, and could remain there unless found 'unapt or unable to learn'.

From its establishment St Paul's was a large school: it had over 150 pupils, about the same as Eton and Winchester combined. It was also well endowed. The High Master received a salary of £400 p.a. plus free lodgings.

Apart from Lily, the first High Master, whose *Latin Grammar* dominated the teaching of that subject for three centuries, only one of the early headmasters, Richard Mulcaster, warrants particular attention. Mulcaster came to St Paul's late in life, at the age of sixty-four! In 1561, when only around thirty, he had been appointed the first headmaster of Merchant Taylors' School (where his pupils included the poet Edmund Spenser) but friction with the Governors caused him to resign, and for ten years before his appointment to St Paul's he was a country parson.

During his years at Merchant Taylors', Mulcaster wrote the book *Positions* (1581), in which he set out his plans for education and, in particular, proposed the establishment of colleges (or faculties) for mathematics and for teacher-training.

He realised that in supporting mathematical education he would be 'opposed by some of good intelligence who not knowing the force of [Mathematical Sciences] . . . are accustomed to mock at mathematical heads' (7) – a clear reference to Ascham. In his view:

such studies require concentration, and demand a type of mind that does not seek to make public display until after mature contemplation in solitude . . . The Mathematical Sciences show themselves in many professions and trades . . . whereby it is well seen that they are really profitable; they do not make outward show, but our daily life benefits greatly by them. . . . Mathematics are the first rudiments for young children, and the sure means of direction for all skilled workmen, who without such knowledge can only go by rote, but with it might reach genuine skill. . . . In the manner of their teaching [the sciences] also plant in the mind of the learner a habit of resisting the influence of bare probabilities, of refusing to believe in light conjectures, of being moved only with infallible demonstrations. Mathematics had its place before the tongues were taught, which though they are now necessary helps, because we use foreign languages for the conveyance of knowledge, yet push us one degree further off from knowledge.

Alas, after such fine words, Mulcaster never wrote his projected textbook on mathematics exemplifying his proposals. Indeed, far from introducing mathematics to St Paul's, he was responsible for placing yet another 'tongue', Hebrew, in the school's curriculum. Perhaps, though, in his late sixties his reformatory zeal was flagging – although from contemporary accounts it is clear that his zest for flogging continued unabated.

The High Master in Pepys' day was John Langley, 'a great antiquary, a most judicious divine' who had 'a very awful presence and speech that struck a mighty respect and fear into his scholars' (8). Certainly, Langley was no curriculum reformer and Pepys followed the well-trodden classical path of accidence and grammar; Livy, Virgil and Cicero; Aristotle and Thucydides; Aeschylus, Sophocles and Aristophanes. He composed classical orations and was examined in Latin, Greek and Hebrew before the assembled Governors of the Mercers' Company. Yet no room could be found for the multiplication table! At best he would probably have learned the names of the numbers, about numerals and about numeration in the decimal system. Many scholars entering university at that time did not, however, possess even this rudimentary knowledge and were unable to 'help themselves by the indices or tables of . . . books as they should use for turning to anything of a sudden' (9).

2. UNIVERSITY

The Civil War had a depressing effect on the non-military tailoring business and not only was John Pepys hard hit, but so were all the merchant tailors. As a result, an attempt was made to suppress all 'foreign' tailors and to enforce what was an early example of a closed shop. John Pepys was given the option of going out of business or paying to become a member of the Company of Merchant Taylors. He chose the latter alternative, which meant that he no longer had the money to continue Samuel's education. All now depended on the generosity of the Mercers' Company {10}; it was their award of a Robinson Exhibition tenable at Cambridge University that saved Samuel from having to join his father in the tailor's shop.

In June 1650 Pepys was admitted as a sizar {11} at Trinity Hall where his cousin, Dr John Pepys, was a Fellow. He was not, in fact, to stay at Trinity Hall long, for, as a result of the puritan purge currently being carried out in the two universities, John Sadler, a neighbour of Pepys' father, was appointed as Master of Magdalene and it was to this college that Samuel transferred.

Some idea of what life was like in puritan Cambridge can be gained from the account given by Richard Kidder, later to become Bishop of Bath and Wells, who was at Emmanuel, the most puritan college, about that time:

The discipline was very strict and the examples which the young students had in the master and fellows were very conspicuous. . . . The tutors examined their pupils very often every night, before prayers, of the study of that day. They visited their chambers twice a week to see what hours they observed and what company they kept. There was strict notice taken of those who absented themselves from prayers and great encouragement given to those who were pious and studious. The young scholars were kept to their exercises and to the speaking of the Latin tongue in the hall at their meals (12).

Mathematically, Cambridge was still in a period of transition. There was still no chair of mathematics, in contrast to Oxford where Sir Henry Savile had established a professorship in geometry in 1619 (see p. 249 below). It had been hoped that Sir Thomas Gresham, a Cambridge graduate, would have established a college in Cambridge to foster the study of mathematics and the sciences. In the event, however, he endowed a college in London to provide public lectures in such fields as law, divinity, music, geometry and astronomy. This early experiment in 'adult education' was, initially, a great success and Gresham College attracted many distinguished professors, including, in the field of mathematics, Henry Briggs (renowned for his work on logarithms, and later the first Savilian Professor of Geometry at Oxford) and Edward Gunter, famed as a table and instrument maker {13}. Even in so enterprising an institution as Gresham College the lectures were first delivered in Latin; but then as a concession to progress they were repeated the same day in English! As the names Briggs and Gunter might suggest, the College had a bias towards practical mathematics. The previously-noted, rapid spread in the use of logarithms by mathematical practitioners owed much to the lectures given there by Briggs, and the College lecturers also played a great part in the dissemination of new knowledge in the fields of trigonometry and navigation (14). As we shall see later, Gresham College was also to play a part in the foundation of the Royal Society.

The absence of a professor left Cambridge mathematics without a focus. It was possible to learn mathematics there, as the emergence of mathematicians like Seth Ward [15] and John Wallis [16] showed, yet at a time when one's personal tutor was still expected to give the student all the instruction he required, it was easy to graduate with only the most rudimentary knowledge of the subject.

The examination system, such as it was, was still based upon the medieval disputation, carried on in Latin, between the *respondent*, who put forward three propositions, and fellow students who acted as *opponents*. The discussions were presided over by the Proctors and in later years (for they were to continue into the nineteenth century) by Moderators. To obtain a degree, an undergraduate was obliged to be a member of a college in which he had to reside for three years. At the end of that time he had (in theory) to read two theses and keep at least two responsions and opponencies. There then followed a formal questioning by Proctors before the degree of bachelor was awarded. In practice, the performance of all the exercises was rarely enforced. Since mathematics was excluded from the general curriculum, it was very easy for even the keen scholar such as Pepys to leave the university in ignorance of it (17).

The Cambridge attitude to mathematics in the 1630s and 1640s is described in the much-quoted passage written by Wallis in his old age:

> [they were] scarce looked upon as Academical studies, but rather Mechanical; as the business of Traders, Merchants, Seamen, Carpenters, Surveyors of Lands, or the like, and perhaps some Almanack Makers in London. And among more than two hundred students (at that time) in our college, I do not know of any two (perhaps not any) who had more of Mathematicks than I (if so much) which was then but little; and but very few in the university. For the study of Mathematicks was at that time more cultivated in London than in the universities {18}.

Yet Wallis did learn some mathematics while at Emmanuel College, where he graduated in 1637. There were a few active mathematicians at Cambridge, such as Pell [19], from whom he could learn, and there were lectures which those interested in the subject might attend.

The Cambridge mathematician of that time was, however, beset by difficulties. Some of these were political and/or religious. Ward, for example, was expelled from his Fellowship at Sidney Sussex College for refusing to subscribe to the league and covenant {20}. Equally serious was the lack of opportunity at Cambridge, and it was for lack of a chair in mathematics that the university lost Pell to Holland and Wallis to Oxford.

Cambridge did, of course, soon acquire a chair, the Lucasian Chair, founded in 1662 under the will of Sir Henry Lucas. Its first holder, Isaac Barrow, had just been elected to a Fellowship at Trinity College at the time that Pepys entered Cambridge. It was there that he produced his translation into Latin of Euclid's *Elements*, for Latin rather than English was still the language of communication at the universities. However, Barrow, too, was a victim of political intolerance, and he was driven out of the country in 1655. He returned in 1659 and the following year was appointed Professor of Greek at Cambridge. In 1662 he was appointed Professor of Geometry at Gresham College and later that year was selected to become the first Lucasian Professor. As professor he lectured on various topics in mathematics including geometry and optics – two parts of 'pure' and 'applied' mathematics which were then greatly emphasised – in which he foreshadowed the differential calculus. He is, of course, remembered mainly because of his pupil, Newton, who succeeded him when, in 1669, he moved on to the older and more influential chair in divinity.

In Pepys' Cambridge, then, mathematics was something of interest only to the connoisseur. Yet to those who practised it, the time was an exciting and critical one. The work of Descartes, Cavalieri and others became known and encouraged the flowering of Cambridge mathematics if only for a brief fifty or so years.

3. THE ADMIRALTY AND CHRIST'S HOSPITAL MATHEMATICAL SCHOOL

There was no glamorous work awaiting Samuel when he left university – indeed it appears possible that he spent some time serving in his father's shop. In 1656 however, he was able to benefit from the appointment of his cousin, Edward Montagu, to the Admiralty Commission. Montagu was despatched to Spain and required someone to watch over his London affairs. The twenty-two-year-old Samuel was given this post which brought him into contact with many important people and gave him administrative experience. In 1658 he became a clerk in the Exchequer and remained there for two years until he returned to the Navy Office as Montagu's secretary. His life's work at the Admiralty now commenced.

It was, indeed, in connection with an Admiralty matter, concerning the supply of timber from the Forest of Dean, that Pepys realised that he was being hampered by his lack of arithmetical knowledge {21}. The solution has already been mentioned; he engaged a private tutor to remedy the deficiency.

More importantly, Pepys gradually came to realise that not only he, but the Navy as a whole, was handicapped by a lack of mathematical expertise. The restoration of the monarchy in 1660 had been quickly followed by a naval war with Holland which seriously depleted the complement of naval officers. It was necessary to make good these losses and it was increasingly clear that among the replacements should be many officers who had a good grasp of navigation and the underlying mathematics. The idea of establishing a school where future officers might receive special mathematical education appealed to Pepys and in 1671 he suggested this to Montagu (by then Lord Sandwich). It is probable that the idea first occurred in the mind of Sir Robert Clayton, former Lord Mayor of London and a benefactor of Christ's Hospital – a school which had been established in London over a century earlier for foundlings, but which had come to resemble the typical grammar school of that day (22).

With the help of Sir Jonas Moore, Surveyor General of the Ordnance [23], Pepys and others, Clayton contrived to attach a special mathematical school to Christ's Hospital. The Royal Warrant establishing the school for:

> forty poore boys . . . who having attained to competence in the Grammer and Comon Arithmatique to the Rule of Three . . . may be fitt to bee further educated . . . [and] taught and instructed in the Art of Navigation and the whole Science of Arithmatique

> until their age and competent proficiency . . . shall have fitted and qualified them . . . to be initiated into the practices of Navigation and to bee bound out as Apprentices

was issued in August 1673.

The aim of establishing a mathematical school had been attained. Yet great disappointments lay in store.

It was one thing to take the bold decision to teach mathematics, but another to think of dropping Latin from the curriculum. One might lay down that mathematics should be taught in a practical way, but where could one find teachers who would do that and not merely teach geometry and trigonometry in an academic manner?

John Pell was suggested as Master, but he, after initially accepting, refused the post. Eventually, John Leake, a man having no practical experience of navigation, was appointed. He proved incapable of keeping discipline and, in order to eke out his salary of £50 p.a., devoted most of his energies to teaching private pupils. Pepys, by then a Governor of Christ's Hospital, found 'that our mathematical boys spend their time idly without control, without correction, without any awe to their present master . . . corrupting by their examples the rest of the children and in short rendering themselves every day less and less fit for the calling they are designed for' (24). Leake was offered an increase in salary of £20 p.a. provided he stopped taking private pupils, but this was insufficient inducement and he resigned the post.

Finding a successor proved no easy matter, which is not surprising when one considers what the Governors sought:

1. That he be a sober discreet and diligent person of good life governement and conversation.
2. That he be a good schollar very wel understanding the Latine and Greek Languages to the end the boyes may be kept so, and furthered in the Latine tongue, and the Master able to answer straingers if need be in that Language.
3. That he doe write a very good Scrivenor like hand, that dureing the time the Boyes shall stay with him he may be to them as good as a Writing Master . . .
4. That he be an able and very good Mathematician well knowing in the theory and practice of all its parts and soe ready that noe strainger from abroad or Practitioners at home shalbe able to baffle him, but on the contrary shall finde his abillityes to satisfaction (p. 108).

Leake's successor, Peter Perkins, proved more successful. He was elected a Fellow of the Royal Society on the basis of his knowledge of magnetics and he prepared a textbook for the use of his pupils. Unfortunately, he died only two years after appointment and once more the search for a successor began.

The choice fell on Robert Wood, a scholar, who many years before had translated Oughtred's *Clavis Mathematica* into English {25}. He had been the mathematics tutor at the short-lived university at Durham (see p. 45 below), but thought himself above teaching schoolboys and employed an incompetent, ill-tempered usher to deputise for him.

When Trinity House, the public corporation responsible for the provision of navigational aids to mariners, complained of the boys' bad performance in their examinations, Wood could only argue (with some justification) that the standard set and the curriculum he had to follow (26) were unreasonably difficult. Following Leake's example, Wood resigned after only two years in the post.

Yet once more the search for a suitable master began. The problem, however, was basically intractable, for Society had a clear and specific need that Academia had not foreseen and could not meet. Eventually the applicants were reduced to five: four mathematical practitioners and a Fellow of Trinity College, Cambridge, Edward Paget. As Pepys wrote, 'Not one . . . was master of all that was required of him, viz. Latin, with the practice as well as the theory of navigation'. Pepys insisted that the appointee should have practical experience at sea, but he was overruled by the Court who appointed Paget on the grounds that he would 'by his great learning lead the boys to converse with each other (in Latin)', while his ignorance of the sea could be rectified 'by spending now and then a few days at sea . . . in the Channel' (27).

Paget 'devoted the next thirteen years to the neglect of his duties', while Pepys, despairing of the whole business, temporarily abandoned the school and its affairs. Nonetheless, when on a visit to Spain in 1684, he took the opportunity to visit the Spanish equivalent of Christ's Hospital (28). He was impressed by the 'sandwich' course provided of sea duty, academic studies and further experience at sea.

Despite its shortcomings, the Mathematical School at Christ's Hospital provided an example that could be emulated elsewhere. By 1685 plans were being laid for a mathematical school at Greenwich, and although this came to nought, schools specialising in mathematics were founded early in the eighteenth century at Rochester, London, Petersfield and elsewhere {29}.

Perhaps even more significantly, the Mathematical School at Christ's Hospital attracted interest overseas. Louis XIV of France followed Charles' example and instituted mathematical schools at Toulon, Rochefort and Brest, while Peter the Great of Russia sent an inspector to Christ's Hospital before he in turn established a School of Navigation and Mathematics in Moscow. Moreover, two youths who had just completed their course at Christ's Hospital were appointed tutors at the Moscow school. One was later to be transferred to Leningrad to found a school of navigation there. It has also been argued that schools established in Germany and Austria received some inspiration from the work carried on at Christ's Hospital (30).

By 1693, Pepys, now in retirement, returned to the fray at Christ's Hospital. He was disturbed about many aspects of the school's working and, in particular, about the system of examining. With Pepys' encouragement, the incompetent Paget began in 1694 to draw up a revised syllabus for the Mathematical School. The advice of Wallis and Gregory at Oxford, and Newton at Cambridge (all three of whom were by that time well known to Pepys) was sought. Newton's long reply (31) shows him in an unfamiliar rôle, that of mathematics educator, participating in curriculum evaluation and construction. Not surprisingly, we find him pressing the claims of the 'science of Mechanicks':

> he that can't reason about force and motion is as far from being a true Mechanick, as he that can't reason about magnitude and figure from being a Geometer. A vulgar Mechanick can practice what he has been taught or seen done, but if he is in error he knows not how to find it out and correct it. . . . Whereas he that is able to reason . . . about figure, force and motion, is never at rest till he gets over every rub.

Newton was also keen that the boys should become fluent in Latin, for 'the best Mathematicall books are in that language', and he warned against the hazards of 'mixed-ability' teaching:

> Care should be taken that the Kings boyes be not retarded in their learning, by joyning wth them too great a Number of other boyes of inferior parts, soe as to hinder them from getting through their scheme of learning within the time limited.

In the event, Newton's advice on mechanics was to go unheeded, but many of his other proposals were accepted and are reflected in the revised scheme of 1696 (32).

By that time Paget had been replaced by Samuel Newton which was, if anything, a change for the worse. Samuel Newton had experience of running a mathematical school in London (presumably for mariners) and his appointment could be defended on those grounds. He turned out, however, to be brutish and ill-tempered and, after Trinity House had, in December 1708, failed all five candidates from the school as inefficient and 'ignorant in their business', he was forced to resign.

Pepys, however, did not live to see this further enforced departure. He had died in May 1703 with his hopes for the Mathematical School still largely unrealised. His hopes had foundered on a succession of ill-judged appointments, on an unwillingness to subjugate academic respectability to practical needs, and, indeed, on the scarcity of teachers capable of realising his objectives.

4. PEPYS, THE AMATEUR OF SCIENCE

In addition to his work for Christ's Hospital, Pepys had another, somewhat stranger, contribution to make to mathematics education. We now consider how it came about that the greatest book on mathematics ever written by an Englishman should bear Pepys' name on its title page.

We began this chapter by recalling how Pepys, when almost thirty, had to engage a private tutor to teach him arithmetic. From that time his interest in mathematics and the sciences gradually grew. Encouraged by his own success at arithmetic, he attempted to teach it to his wife (33), and to learn the use of the various aids to arithmetical calculation then being produced. In 1667 his friend, Jonas Moore, introduced him to Napier's bones and three years before that he was shown what was probably an early slide-rule by the famous writing master and mathematics teacher, Edward Cocker (34). Pepys was much impressed by Cocker's skill as an instrument maker and as a conversationalist, but it is as the author of *Arithmetic* that Cocker is best remembered {35}. This book took over from Recorde's *Ground* as the most popular text of its time. First published in 1678, it appeared in well over sixty different editions, before, a century later, it was supplanted in popularity by Walkingame's *Tutor's Assistant* {36}. Cocker's text appeared posthumously, edited by another writing master, John Hawkins, who had probably taken over Cocker's school and some of his papers. It was an elementary work – dealing mainly with integers and vulgar fractions with some commercial applications – but it clearly answered a great need. It is often remembered as the book which Samuel Johnson was to present to the landlady's daughter on his tour of the Hebrides. In Boswell's words, 'What was this book? My readers prepare your features for merriment. It was Cocker's *Arithmetic!*' Yet, as Johnson pointed out, once you had read a book for entertainment, you had finished with it, but Cocker, 'a book of science, is inexhaustible' (37).

Even more significantly, Pepys was introduced to the society of scientists which met at Gresham College (38). For some fifteen years a group of men, a 'college of virtuosoes' (39), had met in London and in Oxford to discuss the 'new learning'. An early account of the 'college' is given by John Wallis in his *Account of some passages of his own Life*.

About the year 1645, while I lived in London (at a time when, by our civil wars, academical studies were much interrupted in both our Universities) . . . I had the opportunity of being acquainted with divers worthy persons inquisitive into natural philosophy, and other parts of human learning. . . . We did by agreements, divers of us, meet weekly in London on a certain day and hour, under a certain penalty, and a weekly contribution for the charge of experiments, with certain rules agreed upon amongst us to treat and discourse of such affairs; of which number were Dr John Wilkins (afterwards Bishop of Chester), then chaplain to the Prince Elector Palatine in London,

Dr Jonathan Goddard, Dr George Ent, Dr Glisson, Dr Merrett (Drs in Physick), Mr Samuel Foster, then Professor of Astronomy at Gresham College, or some place near adjoyning, Mr Theodore Haak (a German of the Palatinate, and then resident in London, who, I think, gave the first occasion and first suggested those meetings) and many others.

These meetings we held sometimes at Dr Goddard's lodgings in Wood Street (or some convenient place near), on occasion of his keeping an operator in his house for grinding glasses for telescopes and microscopes; sometimes at a convenient place (The Bull Head) in Cheapside, and (in term time) at Gresham College at Mr Foster's lecture (then Astronomer Professor there), and, after the lecture ended, repaired, sometimes to Mr Foster's lodgings, sometimes to some other place not far distant.

Our business was (precluding matters of theology and state affairs) to discourse and consider of *Philosophical Enquiries,* and such as related thereunto; as Physick, Anatomy, Geometry, Astronomy, Navigation, Staticks, Magnetics, Chymicks, Mechanicks, and Natural Experiments; with the state of these studies, as then cultivated at home and abroad . . .

Those meetings in London continued, and (after the King's return in 1660) were increased with the accession of divers worthy and Honourable Persons; and were afterwards incorporated by the name of the Royal Society, etc., and so continue to this day (40).

The process of evolution to 'The Royal Society of London for Improving Natural Knowledge' was, then, a gradual and protracted one.

A key event occurred at Gresham College on 28 November 1660 when, after Christopher Wren (then Professor of Astronomy there) had lectured, the twelve present proposed 'founding a College for the promoting of Physico–Mathematicall Experimental Learning'. Forty persons were invited to become members of the new Society which, it was agreed, was to have no more than fifty-five members. The limit was raised the following year, by which time the Society (whose members had strong Royalist leanings) had obtained the support of the reinstated Charles II who was 'pleased to ofer of him selfe to bee entered one of the Society'. In 1662 the Society received its Charter of Incorporation and formally became 'The Royal Society of London'.

The first President to be appointed was Viscount Brouncker, a man with whom Pepys had many connections (not all of which were amicable) and who had considerable talents as a mathematician [41]. Among the first members of Council were Robert Boyle, the chemist, William Petty, John Wallis, Christopher Wren and John Evelyn, the diarist.

To us nowadays, the most striking fact about the Royal Society in its earliest times was the preponderance of non-scientific Fellows. Until the nineteenth century these constituted about two-thirds of the membership {42}. Initially, therefore, the Royal Society provided a meeting place for the scientist and the interested layman. It served not only to disseminate knowledge within the scientific profession, but also to inform society at large. Thus, although nowadays we can hardly conceive of anyone being elected a Fellow of the Royal Society within three years of his first learning his multiplication table, yet Pepys was so honoured in 1665.

The advent of the Royal Society was not welcomed by all, for scientific work was still seen by many as tending to the subversion of religion. Nevertheless, its scientific work began to prosper; it was able to establish valuable connections with the more newly founded Académie des Sciences in France, and its *Philosophical Transactions* was soon recognised as the most important scientific periodical of the day.

The early successes were, however, followed by more difficult years. The King's interest waned and with it the membership. The meetings were poorly attended and in June 1680 Evelyn wrote to Pepys to persuade him to attend again and to ask for 'one half-hour of your presence and assistance toward the most material concern of a Society which ought not to be dissolved for want of address' (43). Sixty Fellows had to be removed for being many years in arrears with subscriptions and the Society was virtually penniless.

Pepys responded to Evelyn's appeal and began to take more interest in the Society and to place his administrative skill at its disposal. In 1685 he achieved the highest office in the Society, being elected its President. During his two years of office he was able to give great assistance to the Society as an administrator and as a patron (for example, Pepys bore the cost of sixty of the plates of Willoughby's *De Historia Piscium*).

Pepys' Presidency is, however, remembered above all for the fact that it was during this time that the Society resolved that 'Mr Newton's *Philosophiae Naturalis Principia Mathematica* be printed forthwith in quarto, in a fair letter; and that a letter be written to him to signify the Society's resolution and to desire his opinion as to the volume, cuts' etc. (44).

Unfortunately, the Society's funds were now exhausted; Willoughby's book had used up what remained and even the staff salaries were in arrears. Help came in the form of an offer by Edmund Halley [45] to finance the printing and publication of the *Principia*. Thus it was that Newton's book came to be published and to bear on its title page:

'Imprimatur, S. Pepys, Reg. Soc. Praeses. Julii 5, 1686.'

On his retirement from the Presidency of the Royal Society, Pepys continued to interest himself in mathematics and to correspond with his old mathematical acquaintances. His mathematical interests were not very deep, but he had a keen and possibly not disinterested, concern for matters of probability as shown by a substantial correspondence with Newton in 1693. The problem posed by Pepys was:

A has 6 dice in a box with which he is to fling a 6.
B has in another box 12 dice, with which he is to fling 2 sixes.
C has in another box 18 dice, with which he is to fling 3 sixes.
Q. Whether B and C have not as easy a task as A at even luck (46)?

Even more illuminating as far as mathematics education is concerned is the correspondence between Pepys and Dr Charlett, Master of University College, Oxford, who, in October 1700, sent to Pepys for his comments and suggestions the following scheme of mathematical work drawn up by David Gregory, at that time Professor of Astronomy at Oxford [47].

Dr Gregory's Scheme

Without discouraging any other person in the University, that teaches or intends to teach Mathematics, at the desire of some persons of note he undertakes to teach the different parts and sciences of Mathematics by way of Colleges or Courses, after the manner following:

If any number of scholars desire him to explain to them the Elements, or any other of the Mathematical Sciences, if they are already acquainted with the Elements, he will allow that company such a time as they among themselves shall agree upon; not less than an hour a day for three days in the week: in which time he will go through the said Science, explaining every proposition, and illustrating it with such examples, operations, experiments, and observations as the matter shall require, until all the company fully apprehend and understand it.

And because some may be desirous to give an account of their proficiency for their own satisfaction and that of their friends, he will once a week examine such as shall signify that they are willing to be examined.

These Schemes are to be in English, mixing Latin words or terms of Art when they occur and are necessary: and there shall be full liberty to every person of the company to propose such doubts and scruples as he pleases.

The Courses or Colleges that he thinks of most ordinary use, are these:

1. The first Six, with the Eleventh and Twelfth Books of Euclid's Elements.
2. The plain Trigonometry; where is to be showed the Construction of natural Lines, Tangents, and Secants, and of the Tables of Logarithms, as well of natural Numbers as of Sines, &c. The practical Geometry, comprehending the descriptions and use of Instruments, and the manner of measuring heights, distances, surfaces, and solids.
3. Algebra. Wherein is taught the method of resolving and constructing plain and solid problems, as well Arithmetical as Geometrical: to which will be subjoined the resolution of the indetermined Arithmetical (or Diophantæan) Problems.
4. Mechanics. Wherein are laid down the principles of all the Sciences concerning Motion; the five Powers, commonly so called, explained; and the Engines in common use, reducible to these powers, described.
5. Catoptrics and Dioptrics. Where the effects of Mirrors and Glasses are showed; the manner of Vision explained; and the Machines for helping and enlarging the Sight, as Telescopes, Microscopes, &c. described.
6. The principles of Astronomy, containing the Explication of all the most obvious Phænomena of the Heavens from the true System of the World, and the generation of the circles of the sphere thence arising. How also is to be taught the Doctrine of the

Globes, and their use, with the problems of the first motion by them resolved. After this is to be demonstrated the Sphærical Trigonometry, and the application thereof to Astronomy, showed in resolving the Problems of the Sphere by calculation, and the construction of the tables of the first motion depending on this.

7. The Theory of the Planets. Where the more recondite Astronomy is handled: that is, the Orbits of the Planets determined by observation. The tables for their motions described, and the method of constructing taught, and the use of these tables showed in finding the Planets' places, the Eclipses of the Luminaries, &c. Many of these Courses may be farther carried, as the particular inclination of the Class lead them: for example, subjoined to the practical Geometry, may be a Lecture of Fortification, so far as it is necessary for understanding it without actual serving in an army, or fortifying a town or camp. Under the head of Mechanics, there may be (if desired) Colleges of Hydrostatics, with all the experiments thereunto belonging; of the Laws of the Communication of Motion, whether the bodies be hard or elastic. Of the gravity of bodies lying on inclined planes; of Ballistics, or the Doctrine of Projectiles or Bombs, &c. Of the Doctrine of Pendulums, and their application to the measuring of time. After the principles of Astronomy, or 6th College, may be prosecuted the doctrine of the Sphere projected *in plano*, or of the Analemma and Astrolabes, and Dyalling; as also Navigation: and so of others.

But though he shall always be ready to gratify the request of those who desire his instruction, in these or any other parts of Mathematical learning, or in reading on and explaining any Mathematical book, he thinks that after all or most of the above set down Colleges, one may by his own study proceed as his occasions require; and he shall very readily give his advice concerning their studies, and the choice of books for that purpose.

For the Text to be explained, and to give occasion for the necessary digressions in the aforesaid Colleges, he will take a printed book, if any there be that is proper. In other cases, he will take care timeously to give those of the Class proper notes to be written by them.

He intends not, by the preceding order, to tie up his Colleges to that order: for after the Elements, at least after the two first Colleges, or being acquainted with them before, they may choose what other they please; but that the seventh necessarily presupposeth the sixth. In all these he supposes one is pretty well acquainted with the Numerical Arithmetic; and if they desire regular demonstrations of the operations of Integers or Fractions, vulgar or decimal, any Class shall have it when they please.

He reckons that any one of these Colleges will require about three months, a little more or less; and that the number of scholars proper for such a class is more than 10, and not more than 15.

Pepys' reply holds less of mathematical interest. Clearly, he felt unable to question Gregory's mathematical judgement. However, with a backward glance to the old quadrivium, he argued at length for the inclusion of music and drawing. Like many others of his time, Pepys still viewed these subjects as forming part of mathematics {48}. So far as the proposed move to the vernacular was concerned, Pepys was in two minds. He himself was entirely convinced of the desirability of such a change, but was concerned 'how far it may elsewhere be thought to affect the honour of the university' {49}.

It was Wallis, however, rather than Gregory or Newton with whom

Pepys was on the closest terms. Pepys held this great and aged mathematician in considerable esteem and in 1702 presented to Oxford University a portrait of Wallis by Sir Godfrey Kneller. The inscription it bears, including the name of the donor, is yet another reminder of the mathematical connections of Pepys. Although possessing no marked mathematical talents himself, he was quick to realise the value that mathematics held for his country and its well-being. As one author put it, when dedicating his work to Pepys: he had always given 'Countenance and Encouragement . . . to Mathematical Learning especially as it hath a tendency to Promote the Publique Good' (50). Through his work at Christ's Hospital he led the way towards the establishment of mathematics within the curriculum of the endowed schools; his work for the Royal Society enabled it to survive at a time of crisis; and his association with the two greatest English mathematicians of the age will long be remembered. By his encouragement, that 'useful part of learning, I mean the Mathematics' was 'more cultivated and improved amongst us than formerly it [had] been' (51). These words of the Vice-Chancellor of the University of Oxford serve particularly well to describe the salient features of mathematics education in the seventeenth century. It was a period which saw the subject triumphantly established in the two universities and the emergence of a number of truly great English mathematicians. Yet reaction was soon to set in and vigour be replaced by torpor. In the grammar schools mathematics was still largely ignored: for, despite the arguments of Mulcaster and others, the subject's place in a liberal education had yet to be generally accepted. The growth of trade and commerce had, however, led to an increasing demand for utilitarian mathematics; a need met by the mathematical practitioners (52). The seventeenth century saw, therefore, a considerable increase in the amount and variety of mathematics being taught. This, as we have seen, served to generate an interest in curriculum construction, but as yet it was thought that the solutions to the resulting problems could be found within mathematics itself; consideration had still to be given to the specific problems of learning and teaching mathematics.

3 · *Philip Doddridge*

It is likely that I was introduced to the work of Philip Doddridge several years before I encountered that of any other of the subjects of this book, for in the village chapel in which I was christened, almost all baptismal services began with the singing of Doddridge's hymn 'See Israel's gentle shepherd stand'. Indeed, it is for his hymns that Doddridge is now best remembered. Like Pepys, Doddridge was no mathematician, yet he was one of the leaders in an educational movement that did much to secure a place for mathematics in the curriculum at secondary and tertiary level. Moreover, his own mathematical education is well documented and it is for these reasons that he has been selected to represent his period and the dissenting academies with which his name is so closely associated.

1. THE PURITAN REVOLUTION AND ITS AFTERMATH

In the time of the Commonwealth increased emphasis came to be placed on education. The forces of conservatism were, for the time being, driven underground and advancement came to depend more on merit than on social class. In such a climate, established doctrines and ways were more readily challenged. In particular, the curricula of the grammar schools and universities were attacked as lacking social utility and relevance, and demands were made for increased mathematical and scientific studies (1). Attempts were also made to break the monopoly of Oxford and Cambridge in the field of university education in England. Plans were laid for universities at Manchester and York {2}, and for a University of London which would build upon Gresham College. In 1657 a college was actually established in the close of Durham Cathedral financed out of the funds of the dissolved cathedral chapter. But this attempt to bring 'learning and piety' to 'those poor, rude and ignorant parts' was to be quashed with the return of King Charles II in 1660 (3).

As we have seen, the restoration of the monarchy was followed by the establishment of the Mathematical School at Christ's Hospital and the Royal Society – in both of which Charles had a hand. Yet the Royal Society already existed in all but name, and it is likely that the Restoration did little to expedite the formation of the Mathematical School. More significantly, the Restoration was marked by the return of the House of Lords, the Bishops and the Church of England. Steps were immediately taken to ensure that there would be no second departure.

The first of these, the 1662 Act of Uniformity, required all clergymen 'and every schoolmaster keeping any public or private school and every person instructing or teaching any youth in any house or private family' to 'conform to the liturgy of the Church of England' in particular, the Book of Common Prayer and the Thirty-nine Articles, and to refrain from seeking 'any change or alteration of government either in Church or State'. This 'declaration and acknowledgement' was also to be subscribed by all 'masters . . ., fellows, chaplains and tutors of or in any college, hall, or house of learning, and by every public professor and reader in either of the Universities' (4). Failure to subscribe meant the loss of appointment or, in the case of schoolmasters, three months' imprisonment and a fine of five pounds. Those schoolmasters who subscribed to the Act were to be licensed by their archbishop or bishop.

It is estimated that some 1,760 clergy and over 150 dons and schoolmasters were evicted from their posts as a result of these measures (5). This did not deter many of them from establishing new schools and from continuing to teach. It was in an attempt further to curb such dissenters that additional legislation, the Five-Mile Act, was passed in 1665. This restrained any dissenting clergy from coming within five miles of 'any city or town corporate, or borough that sends burgesses to the Parliament'. Moreover, it was to be illegal for any persons so restrained from coming to any such city or town 'to teach any public or private school, or take any boarders or tablers that are taught or instructed by him or herself'.

That a teacher should subscribe to the Thirty-nine Articles of the Church of England was, in fact, to remain the law until 1869, although from 1700 onwards subscription was not insisted upon in many parts of the country. Indeed, because of the accepted need for mathematicians, in the 1713 Act to Prevent the Growth of Schism, teachers of mathematics, navigation or mechanical art were specifically excused from subscribing (6).

Not only did the Act of Uniformity curb the activities of dissenting teachers, but the statutes of the universities excluded dissenting students from Oxford and prevented their taking degrees at Cambridge. Thus it came about that, nearly two centuries later, Sylvester could not graduate from Cambridge, nor De Morgan teach there.

For the universities and the endowed grammar schools, the Restoration marked the beginning of a period of stagnation and decline. For political reasons, both types of institution became too closely connected to the Established Church and to the State, two bodies which eschewed progressive thought. Educational innovation then had to come from some other sector, from schools organised by individuals, particularly those who for reasons of faith were denied a place in the institutions of the Establishment.

2. THE EARLY DISSENTING ACADEMIES

The vicissitudes that befell dissenting clergy are well illustrated by the life of Richard Frankland, the founder of the first academy in the north of England (7). Frankland, a Cambridge graduate, was in 1660 a clergyman with a living near Bishop Auckland and, according to some sources, a tutor at the college recently established in Durham (8). As a result of the Act of Uniformity he had to relinquish his living and he returned to his home at Rathmell, near Settle in Yorkshire. There he opened his academy for pupils who were 'intended for the Law, others for Physick [medicine], but most of them for the Ministry of the Gospel' (9). His academy was, like many others established at that time, soon to be of 'no settled address'. In 1674 Frankland became a dissenting minister at Natland, Westmorland and transferred his academy there. Soon, however, he became a victim of the Five-Mile Act and in 1683 he began a series of moves which eventually led him to Attercliffe near Sheffield before, in 1689, he returned to Rathmell.

Frankland's academy catered for few pupils but its standards must have been comparable with those of Cambridge for students from Natland were allowed to take degrees at Edinburgh University after only one year's extra study {10}. More to the point of this book, however, is the fact that the academy's curriculum was firmly based on the classics – instruction was given in Latin – and no place was found in the curriculum for mathematics (11). As yet, mathematics was still to be recognised as fit study for those with religious aspirations (even though Frankland himself was described as 'an acute mathematician' (12)) and it is significant that Timothy Jollie, Frankland's former pupil, who, after Frankland's departure, had founded his own academy at Attercliffe, 'forbade mathematics as tending to scepticism and infidelity' {13}. Notwithstanding this, one of Jollie's pupils, the blind Nicholas Saunderson, was to become Lucasian Professor of Mathematics at Cambridge and author of a much-used and influential text on algebra [14].

Some dissenting masters did, however, teach mathematics even in the seventeenth century. Among these were Adam Martindale who taught mathematics, including logarithms and surveying, at a private school in Manchester attended also by some boys from the grammar school there; Thomas Brancker, a Fellow of Exeter College, Oxford, who after his expulsion opened a school at Macclesfield; and Humphrey Ditton, who later became a mathematics master at Christ's Hospital (15).

As a rule, however, the seventeenth-century dissenting academies were in the hands of Oxford and Cambridge graduates, such as Frankland, who based their curriculum on that which they had followed as undergraduates. Gradually the nature of the academies began to change. Following the passing of the Toleration Act in 1689, dissenters other than Roman Catholics and Anti-Trinitarians were given a greater measure of freedom, as a result of which they became more settled both geographically and financially. Academies were supported and controlled not only by individuals but by outside societies, Presbyterian, Congregational and Baptist. Moreover the tutors of the second and later generations were less influenced by Oxford and Cambridge. They looked for inspiration to writers such as Hartlib [16], Milton [17] and the French philosopher–mathematician Ramus [18] and, as a result, began to recast their curricula.

3. PHILIP DODDRIDGE AND KIBWORTH ACADEMY

Philip Doddridge was born in London in 1702, the son of a prosperous businessman. Dissent was in his blood; his paternal grandfather was an ejected clergyman whilst his mother's father was an exiled Lutheran preacher from Prague who had established a private school at Kingston upon Thames. Philip was the youngest of a family of twenty of whom only one other, his sister Elizabeth, reached maturity.

Like many other children from such a background, Doddridge received his first education – in Latin grammar – at a private school (run by a dissenting clergyman). Later he was to attend other private schools including that established by his grandfather. Holidays were spent with his uncle who was steward to the first Duke of Bedford. In this way, Doddridge was brought into contact with the influential Russell family and, indeed, the Duchess of Bedford later offered to take the young, by then impoverished boy {19} under her patronage, to provide him with an education at either of the two universities and with a church living to follow. Nonconformity, however, had taken its hold and it was to the dissenting ministry that he aspired. Doddridge accordingly obtained a grant from the Presbyterian fund and, with its aid, the sixteen-year-old boy entered the academy of John Jennings at Kibworth, Leicestershire.

The course of education at Kibworth was later described by Doddridge in considerable detail (20) and, although still basically classical, it included some mathematics together with its applications in the first two years of the four-year course. It should be stressed that Jennings' lectures were in Latin and that his students 'were obliged to talk Latin within some certain bounds of time and place'.

The First half-year we read Geometry or Algebra thrice a week; Hebrew twice,

Geography once, French once, Latin prose authors once, Classical exercises once. For Geometry we read Barrow's *Euclid's Elements*; when we had gone through the first book, we entered upon algebra, and read over a system drawn up by Mr Jennings for our use . . . When we had ended this system, we went over most of the second and fifth books of Euclid's *Elements*, with Algebraic demonstrations, which Mr Jennings had drawn up and which were not near so difficult as Barrow's geometrical demonstrations of the same propositions. We likewise went through the third, fourth and sixth books of Euclid; but this was part of the business of the second half-year . . .

The Second half-year we ended Geometry and Algebra, which we read twice a week. We read Logic twice, Civil History once, French twice, Hebrew once, Latin poets once, Exercises once, Oratory once, Exercises of reading and delivery once . . .

The Third half-year we read Mechanics, Hydrostatics, and Physics twice, Greek Poets once, History of England once, Anatomy once, Astronomy, Globes, and Chronology once, Miscellanies once, and had one Logical disputation in a week. For mechanics, we read a short but very pretty system . . . drawn up by Mr Jennings; and for hydrostatics, an abridgment of Mr Eames's lectures. For physics, we read Leclerc's system, exclusive of his first book, of Astronomy, and of the latter part of the fourth, of Anatomy . . . We read Jones on the Use of the Globes. Our astronomy and chronology were both Mr Jennings's, and now printed amongst his Miscellanies . . .

The Fourth half-year, we read Pneumatology twice a week. The remainder of Physics and Miscellanies once, Jewish Antiquities twice. Our pneumatology was drawn up by Mr Jennings, pretty much in the same method as our logic. It contained an inquiry into the existence and nature of God, and into the nature, operations, and immortality of the human soul, on the principles of natural reason. There was a fine collection of readings in the references on almost every head. This with our divinity, which was a continuation of it, was by far the most valuable part of our course.

4. RELIGION AND MATHEMATICS

Jollie's view of mathematics as 'tending to scepticism and infidelity' was not held by all dissenters. Nor, indeed, was it the case that all mathematicians were sceptics. Boyle, Barrow and Newton all devoted a considerable part of their time to theology and Newton, although far from an orthodox Christian, saw his great work as providing proof of the existence of God:

This most beautiful system of sun, planets, and comets could only proceed from the counsel and dominion of an intelligent and powerful Being. This Being governs all things, not as the . . . Soul of the world, but as Lord over all (21).

Such arguments, by Newton, Leibniz and others, had a considerable effect on those favourably disposed to religion. Some, like Addison and Pope, were to frame Newton's argument in poetical terms; Pope's couplet

Nature and Nature's laws lay hid in night:
God said, Let Newton be! and all was light.

being illustrative of the new approach.

Others saw an understanding of mathematics as a step towards the acceptance of God. One such was Isaac Watts, a prominent dissenting minister and educator [22]. Watts had much in common with Doddridge;

they shared a similar background and outlook and were to become close friends. Like Doddridge, Watts had been educated in a dissenting academy and there he was introduced to algebra, geometry and conics {23}. Later, Watts was to take pupils in mathematics and natural philosophy and, indeed, some of his teaching notes were edited and published in 1725 under the title of *First Principles of Astronomy and Geography*.

Watts' views on the teaching of mathematics were to be spelled out in his *Improvement of the Mind* (1741) and *Discourse on the Education of Children and Youth* (1752, posthumous publication):

If we pursue mathematical Speculations, they will inure us to attend closely to any Subject, to seek and gain clear Ideas, to distinguish Truth from Falsehood, to judge justly, and to argue strongly; and these Studies do more directly furnish as with all the various Rules of those useful Arts of Life, viz., Measuring, Building, Sailing, etc.

Even our very Enquiries and Disputations about . . . Space and Atoms, about incommensurable Quantities . . . which seem to be purely speculative, will shew us some good practical Lessons, will lead us to see the Weakness of our Nature, and should teach us Humility in arguing upon divine Subjects and Matters of sacred Revelation . . .

. . . there are many and great and sacred *Advantages* to be derived from this Sort of *Enlargement of the Mind*.

It will lead us into more exalted Apprehensions of the *great God our Creator* than ever we had before.

Watts, then, saw an important place for mathematics in the education of youth – a place that could be justified on utilitarian, liberal, moral and religious grounds. In his *Discourse* he described in more detail what mathematical knowledge was desirable. He also drew attention to the view of mathematicians commonly taken in the early seventeenth century.

Several parts of mathematical learning are also necessary ornaments of the mind and not without real advantage: And many of these are so agreeable to the fancy that youth will be entertained and pleased in acquiring the knowledge of them.

Besides the common skill in accounts which is needful for a trader, there is a variety of pretty and useful rules and practices in arithmetic to which a gentleman should be no stranger: And if his genius lie that way, a little insight into algebra would be no disadvantage to him. It is fit that young people of any figure in the world should see some of the springs and clues whereby skilful men, by plain rules of reason, trace out the most deep, distant and hidden questions; and whereby they find certain answers to those enquiries, which at first view, seem to lie without the ken of mankind, and beyond the reach of human knowledge. It was for want of a little more general acquaintance with mathematical learning in the world, that a good algebraist and a geometer were counted conjurers a century ago, and people applied to them to seek for lost horses and stolen goods {24}.

They should know something of geometry, so far at least as to understand the names of the various lines and angles, surfaces and solids; to know what is meant by a right line or a curve, a right angle and an oblique, whether acute or obtuse: How the quantity of angles is measured, what is a circle, a semicircle, an arch, a quadrant, a degree and minute, a diameter and radius: What we mean by a triangle, a square, a parallelogram, a polygon, a cube, a pyramid, a prism, a cone, an ellipsis or oval, an hyperbola, a parabola, &c and to know some of the most general properties of angles, triangles, squares and circles, &c. The world is now grown so learned in mathematical science that this sort of language is often used in common writings and in conversation, far beyond what it was in the days of our fathers. And besides, without some knowledge of this kind we cannot make any farther progress toward an acquaintance with the arts of surveying, measuring, geography and astronomy, which are so entertaining and so useful an accomplishment to persons of a polite education.

Yet the learning of mathematics was not without its perils if carried to excess:

Though I have so often commended mathematical studies, and particularly the speculations of arithmetic and geometry, as a means to fix a wavering mind, to beget a habit of attention, and to improve the faculty of reason; yet I would by no means be understood to recommend to all a pursuit of these sciences, to those extensive lengths to which the moderns have advanced them. This is neither necessary nor proper for any students but those few who shall make these studies their chief profession and business of life, or those gentlemen whose capacities and turn of mind are suited to these studies, and have all manner of advantage to improve in them.

The general principles of arithmetic, algebra, geometry and trigonometry, of geography, of modern astronomy, mechanics, statics, and optics, have their valuable and excellent uses, not only for the exercise and improvement of the faculties of the mind, but the subjects themselves are very well worth our knowledge in a moderate degree, and are often made of admirable service in human life. So much of these subjects as Dr Wells has given us in his three volumes, entitled, The Young Gentleman's Mathematics [25], is richly sufficient for the greatest part of scholars or gentlemen; though perhaps there may be some single treatises, at least on some of these subjects, which may be better written and more useful to be perused than those of that learned author.

But a penetration into the abstruse difficulties and depths of modern algebra and fluxions, the various methods of quadratures, the mensuration of all manner of curves, and their mutual transformation, and twenty other things that some modern mathematicians deal in, are not worth the labour of those who design either of the three learned professions, divinity, law, or physic, as the business of life. This is the sentence of a considerable man, viz. Dr George Cheyne [26], who was a very good proficient and writer on these subjects: he affirms, that they are but barren and airy studies for a man entirely to live upon, and that for a man to indulge and riot in these exquisitely bewitching contemplations, is only proper for public professors, or for gentlemen of estates, who have a strong propensity this way, and a genius fit to cultivate them.

But, says he, to own a great but grievous truth, though they may quicken and sharpen the invention, strengthen and extend the imagination, improve and refine the reasoning faculty, and are of use both in the necessary and the luxurious refinement of mechanical arts; yet having no tendency to rectify the will, to sweeten the temper, or mend the heart, they often leave a stiffness, a positiveness and sufficiency on weak minds, which is much more pernicious to society, and to the interests of the great end of our being, than all their advantages can recompense. He adds further concerning the launching into the depth of the studies, that they are apt to beget a secret and refined pride, an over-weening and over-bearing vanity, the most opposite temper to the true spirit of the gospel. This tempts them to presume on a kind of omniscience in respect to their fellow-creatures who have not risen to their elevation: nor are they fit to be trusted in the hands of any but those who have acquired a humble heart, a lowly spirit, and a sober, and teachable temper . . .

Some of the practical parts of geometry, astronomy, dialing, optics, statics, mechanics, &c. may be agreeable entertainments and amusements to students in every profession at leisure hours, if they enjoy such circumstances of life as to furnish them with conveniencies for this sort of improvement: but let them take great care, least they entrench upon more necessary employments, and so fall under the charge and censure of wasted time.

Yet I cannot help making this observation, that where students, or indeed any young gentlemen, have in their early years made themselves masters of a variety of elegant problems in the mathematic circle of knowledge, and gained the most easy, neat, and entertaining experiments in natural philosophy, with some short and agreeable speculations or practices in any other of the arts or sciences, they have hereby laid a foundation for the esteem and love of mankind among those with whom they converse, in higher or lower ranks of life; they have been often guarded by this means from the temptation of innocent pleasures, and have secured both their own hours and the hours of their companions, from running to waste in sauntering and trifles, and from a thousand impertinences in silly dialogues. Gaming and drinking, and many criminal and foolish scenes of talk and action have been prevented by these innocent and improving elegancies of knowledge (27).

The views of Watts were to have considerable weight in the dissenting academies and nonconformist schools. It was partly due to his influence that Doddridge came to lead the Northampton Academy, probably the most influential of all the dissenting academies, where he was able to put Watts' ideas into practice {28}. Later, when various religious groups began to establish their own boarding schools, such as the Moravians at Fulneck, near Leeds, and the Methodists at Kingswood, near Bristol, they were often to model their curricula on those of the dissenting academies rather than on the classical curriculum still pursued in the great majority of the endowed grammar schools. Thus it was that John Wesley, before founding Kingswood in 1748, took Doddridge's advice on the books he should use and was recommended, among others, to make use of Watts' *First Principles of Astronomy and Geography* {29}.

5. NORTHAMPTON ACADEMY

In June 1723, Doddridge, then aged twenty-one, completed his course at Kibworth and was made a minister there. Two years later he moved to Market Harborough, Leicestershire and there, in April 1729, with the assistance and approval of Watts, he established an academy. Later that year the academy migrated to Northampton where it was to remain until Doddridge's death in 1751. Although there was by then a greater toleration of dissenters, the academy was not without its troubles. In September 1733 it was attacked by a Jacobite mob {30} at a time when Doddridge was involved in court proceedings instituted by the Chancellor of the Diocese. The proceedings were not brought to an end until 1734 when George II intervened personally on the grounds that in his reign there should be no persecutions for conscience's sake.

The number of students taken by Doddridge varied between about thirty and sixty and they followed a four-year course which included geometry and algebra in the first year and trigonometry, conic sections and mechanics ('a Collection of important Propositions taken chiefly from Sir Isaac Newton . . . They relate especially though not only to centripetal and centrifugal forces') in the second (31). Like Watts, Doddridge was against 'pursuing too long some abstruse *Mathematical* enquiries', yet he advised even those pupils training for the ministry to keep up their algebra and thereby 'strengthen your *Rational Faculties* by that strenuous Exercise'. Mathematics could 'teach . . . attention of thought, and strength, and perspicuity of reasoning' and would 'teach us to distinguish our ideas with accuracy, and to dispose our arguments in a clear, concise and convincing manner' (32). To that end Doddridge taught Euclid in a manner intended to emphasise logical principles. Moreover, Euclid, like all other subjects, was taught not in Latin but in English.

Doddridge's educational activities were not confined to his academy. He was involved in the foundation of a society in Northampton for scientific experiment and discussion – an example of a contemporary nation-wide movement to bring science and mathematics to the artisan and middle classes (33) – and he established a charity school in Northampton for teaching and clothing the children of the poor.

The provision of educational opportunities for poor children (who constituted a majority of all children) was at that period minimal. Again, Watts had argued the need for providing such children with education in his *Essay Towards the Encouragement of Charity Schools* (1728). There he refuted the objection that 'their business is to labour, not to think' and argued that there was a need to teach the poor not only to read, but to write, and sufficient arithmetic to enable them to deal with money, to add and to subtract.

It must be remarked in passing that Doddridge contributed to curriculum development not only in the field of mathematics but with even greater effect in that of science. This was taught at Northampton with emphasis on experiment and included the study of mechanics, statics, hydrostatics, optics and astronomy. Doddridge himself also contributed papers to the scientific society at Northampton on the pendulum and on elastic and inelastic collisions. Some of his pupils were to become distinguished technologists (for example, Roebuck and Lucas [34]), and indeed the academy was (somewhat extravagantly) described as 'that university for the new scientists and technologists and for their spiritual mentors' (35).

The academy did not long survive Doddridge's death from consumption in 1751 and his pupils transferred to Daventry, near Rugby, where a new academy was established under a former pupil of Doddridge, Caleb Ashworth.

6. JOSEPH PRIESTLEY AND THE LATER ACADEMIES

Although the connections between Philip Doddridge and Joseph Priestley, the famed chemist, are somewhat tenuous, Doddridge was to have a considerable influence on Priestley, his education, his outlook and his career. A brief account of Priestley's life enables us, therefore, to see how dissenting education developed in the second half of the eighteenth century. It also leads us naturally to consideration of early attempts to establish a science of education.

Priestley was born in 1733 into a middle-class family in Yorkshire. He was his mother's first child; five others were to follow in quick succession before her death in 1740. After this Priestley moved into the care of his aunt, a strict Calvinist, and it was she who was responsible for guiding his education. This, in the first instance, he received from both public and private schools. When twelve years old he went to the local grammar school at Batley where he commenced the study of Latin and Greek. During the school holidays he was sent to the local dissenting minister, Mr Kirkby, to learn Hebrew! Soon he left Batley School and until the age of sixteen went full-time to Kirkby's school before the latter's illness led to its closure.

At that time young Priestley's health was also giving cause for concern, so much so that he relinquished the idea of entering the ministry and thought instead of a career in commerce. As a first step he taught himself French, Italian and Dutch, and on two days each week visited another dissenting minister, Mr Haggerstone, for further tuition. Haggerstone, as luck would have it, had been educated in Scotland and was a pupil of Colin Maclaurin. So it happened that he was able to introduce Priestley to 'geometry, algebra, and various branches of mathematics, theoretical and practical' (36), and to induce him to read Gravesande's *Elements of Natural Philosophy* and Watts' *Logick* (37).

Thus, in spite of the inability of the grammar school to help, Priestley soon acquired a fair command of mathematics {38}. Once again, however, we have evidence of the haphazard provision of mathematics education at that time. Moreover, we note that had Priestley been sent for tuition to an Oxford-trained Anglican parson, then the chance of his having been introduced to significant mathematics would have been considerably reduced.

As it happened, Priestley was not to make use of his new knowledge in commerce, for a sudden improvement in health caused him to withdraw from a post arranged for him in Lisbon, and to revert to the idea of becoming a minister. His aunt wished him to train for the Calvinist ministry at the Mile-End Academy (London), but Priestley was now turning away from Calvinist theology and, in particular, that Church's teaching on original sin. An alternative to Mile-End had to be found. Mr Kirby strongly recommended Doddridge's academy 'on the idea that I should have a better chance of being made a scholar'. Further, somewhat fortuitous, backing for this proposal came from Priestley's stepmother who at one time had been Doddridge's housekeeper.

Doddridge's death led to the amendment rather than the abandonment of this plan. Priestley was sent to Ashworth's academy at Daventry and was the first student to be entered there. Because of his high standard of education he was excused the first year of the four-year course.

The curriculum at Daventry was modelled on that at Northampton and in this way Doddridge's influence persisted. Priestley was most content: 'the general plan of our studies . . . was exceedingly favourable to free inquiry . . . our lectures had often the air of friendly conversations . . . We were permitted to ask whatever questions and to make whatever remarks we pleased; and we did it with the greatest . . . freedom.' It was a form of education which many university students of today might envy! Priestley continued his study of mathematics, reading Maclaurin's *Algebra* and Euclid (Books 1–6, 11, 12) (39).

On leaving Daventry. Priestley became a dissenting minister, first in Suffolk, then later at Nantwich, Cheshire. There was a need to augment his income and, like so many of his ministerial colleagues, Priestley decided to open a school. He offered to teach the classics, mathematics, etc. for half a guinea (52½p) a quarter and to board pupils for twelve guineas (£12.60) a year. Not a single pupil came. In despair, Priestley advertised twelve lectures for adults on the use of globes. He attracted ten students who paid half a guinea a head 'which did something more than pay for my globes'. Even in the small country town of Needham Market there was a ready audience for further education.

Priestley was sufficiently encouraged to plan further lectures, but the move to Nantwich intervened. There, despite that 'grave aversion to the business of a schoolmaster' which he shared with 'most other young men of a liberal education', he opened a school and found that 'in this employment, contrary to my expectations, I have found the greatest satisfaction'.

The school attracted over thirty pupils, six of them girls who were taught in a 'separate room'. As we should expect, Priestley taught mathematics, but we know no other details about content or method. What was outstanding was his teaching of science. He bought various pieces of apparatus and encouraged experimental work. The senior class gave demonstrations at parents' evenings and so 'considerably extended the reputation of my school'. It was, of course, to be many decades before such practical science lessons became the norm in grammar schools.

After three years at Nantwich, Priestley moved in 1761 to the most famous of all the dissenting academies, that at Warrington; an institution which won for the town the unlikely title (to those who know Warrington today) of the 'Athens of the North'.

The academy offered a three-year course for those who were intended for 'a life of Business and Commerce' as well as a five-year course for intending ministers. As at Northampton and Daventry, mathematics was well represented in the curriculum, although Priestley had nothing to do with its teaching, for he was appointed as a tutor in languages and belles-lettres (40). After six years at Warrington, Priestley became minister at Mill Hill Chapel, Leeds. His interests from that time onwards turned more and more to chemistry. Yet even so he was still to make significant contributions to adult education – he helped, for example, to establish a circulating library in Leeds – and to the dissemination and popularisation of scientific knowledge. When he later moved to Birmingham he was an active member of the Lunar Society (see p. 258 below).

While in Warrington, Priestley wrote his *Essay on a Course of Liberal Education for Civil and Active Life*. In this he dismissed the contemporary grammar school curriculum as 'a common topic of ridicule' (41) and argued for a fundamentally new curriculum which would seek to satisfy three main needs: religious, intellectual and utilitarian. He proposed as essential subjects Latin, English, French, mathematics, physics, chemistry, history and geography. (This 'bookish' curriculum was adopted over a century later in grammar schools and was, in fact, *exactly* the same as the curriculum I followed when taking my School Certificate (O-level) examinations in the 1940s.) He continued by saying that unless such changes were made 'we must not be surprised to find our schools, academies and universities deserted, as wholly unfit to qualify men to appear with advantage in the present age'.

Yet what distinguished Priestley from numerous other curriculum designers was that not only was he interested in education, but also that he sought to establish a science of education. In this he was greatly influenced by David Hartley, to whose works he had been introduced at Daventry. Hartley in his *Observations on Man* (1749) had outlined a theory of the association of ideas, postulating that 'complex processes – imagining, remembering, reasoning – could be analyzed into clusters or sequences of elementary sense impressions formed by individual experience: all psychological acts can be explained by a single law of association' (42). He was also to argue that people were moulded by circumstances and that they could be perfected – environment rather than heredity being the dominating influence. Priestley became an ardent disciple: Hartley, he claimed, had thrown 'more useful light upon the theory of the mind than Newton did upon the theory of the natural world'. Moreover, 'the most important application of [this] doctrine . . . is to the . . . business of education' (43).

Priestley was also to adopt an early form of behaviourism: 'only to the extent that the educator's actions have a necessary effect on his pupils can his labours have any purpose' (44). Such notions were to be taken over by R. L. Edgeworth [45], a friend of Priestley, and his daughter Maria in their book *Practical Education*, in which they also stressed that education should be considered as an experimental science: 'many authors . . . had mistaken their road by following theory instead of practice'.

Mistaken optimism concerning the speed at which education could become a science led them to express the view that,

> It is not from want of capacity that so many children are deficient in arithmetical skill and it is absurd to say 'such a child has no genius for arithmetic, such a child cannot be made to comprehend anything about numbers'. These assertions prove nothing, but that the persons who make them are ignorant of the art of teaching (46).

This is probably the first appearance in an English text of the principle that there are no bad pupils, only bad teachers, a principle which is all the more dangerous for containing more than a grain of truth.

In his contributions to education, therefore, Priestley continued in the ways of Doddridge: he vigorously championed a new, wider view. Like most reformers of his time he attracted critics and enemies. Riots which broke out in Birmingham on the second anniversary of the storming of the Bastille led to the burning of his church, house and laboratory. As a supporter of the American colonists and the French Revolution, he was politically suspect. Eventually he sought refuge in Pennsylvania where he died in 1804. An anonymous author was to write of him 'What he believed to be true he thought it his duty to propagate, without any regard to his own interest or the prejudices of mankind' (47). It was an epitaph which fitted so many of the eighteenth-century dissenters.

Yet in many ways Priestley was untypical of the later dissenters, for their views on education, and particularly on the place of mathematics within it, were not so liberal. The subject ceased to be taught in many institutions training candidates for the ministry. The attempts of Newton and Leibniz to bring science and religion together had failed and the influence of

Milton, Locke {48} and Watts decreased. Celestial mechanics did not always lead the student to God, nor did a study of incommensurables necessarily induce humility. Samuel Stennett, who operated an academy for Baptists, excluded mathematics from his curriculum, as did John Newton of the Newport Pagnall Academy who believed it tended to atheism. In general, the theological colleges which grew out of the dissenting academies had no use for the subject.

In the more secular academies, however, mathematics continued to be taught and the way was opened for the establishment of mathematics in the secondary-school curriculum {49} and its revival in the universities.

4 · *Charles Hutton*

I do not consider it would be an act of kindness to the memory of Dr Hutton to suppress . . . his early history. We know, that the lower any man's origin is, the higher and the more honourable is his subsequent elevation . . . it is perhaps the first time in the annals of British biography that a person once employed in the situation of a common workman in a colliery, rendered himself so celebrated, that a Lord Chancellor of England considered it as one of the many blessings which he had enjoyed in life, to have had the benefit of his instructions (1).

In the preceding chapters we have seen how, in the absence of any State- or Church-coordinated system, the provision of mathematics education at a pre-university level often came to depend upon private enterprise. Frequently, in order to attract financial support, this led to an emphasis being placed on utilitarian ends. In this chapter we consider a mathematician who was largely self-taught and whose contribution to education was for many years a 'private' one. Later in life he was to head the mathematics department of England's leading military academy where once again emphasis had to be laid on the subject's practical aspects. Hutton, then, was an educator biased towards a utilitarian view of mathematics. Through his books and other writings he was to have a considerable influence on English mathematics, particularly outside the two ancient universities and the established grammar schools. Yet those who read his works, especially his *Dictionary* and his *Recreations* (1803, 1814), will see that Hutton was fully appreciative of other aspects of mathematics and wished to develop in his readers the pleasure of participating in mathematical activities.

1. BOY AND SCHOOLMASTER

Charles Hutton was born in Newcastle in 1737, the youngest son of an overviewer (supervisor) in a local colliery. Before Charles was seven his father had died and his mother married again, once more to a pit overman. That year the young boy became involved in a street-fight and, not wishing his parents to know, concealed from them the fact that he had damaged his right elbow. By the time they realised it was severely dislocated it was too late to have it properly corrected. The injury was permanent and Charles was unable to earn his living from hard labour. Accordingly, when his brothers were sent down the mine, Charles was sent to the local dame school to learn to read. Later when his parents moved house, Charles was transferred to a village school where he took the next step of learning to write. Yet another move of house brought the Huttons to a village near Jesmond (on the outskirts of Newcastle) where Charles went to the school run by an Anglican clergyman, Ivison. Soon, though, it was time for Charles to earn a living and the only opportunity then open was as a hewer in the local colliery. Handicapped as he was by his elbow, he was an indifferent workman. The men were paid according to the amount of coal they hewed and wage notes show that Charles earned least of all. Before he was twenty he left the colliery and established a small school. Soon he had a stroke of good fortune, for Ivison, his former teacher, left, and Charles was able to take over his school. This soon was so successful that a move to larger premises became necessary.

Hutton now combined teaching by day with learning by night, for in the evenings he attended a school in Newcastle at which he was able to study mathematics.

As an industrial and shipping centre, Newcastle was well equipped with private schools teaching mathematics. One of the leading mathematics teachers of the early eighteenth century, James Jurin, had spent some time there and his influence was probably still strong. Jurin, an old boy of Christ's Hospital and a Cambridge graduate, had been appointed Headmaster of the Newcastle Grammar School in 1709 {2}. It is probable that he included mathematics in his teaching there, but there is no direct evidence of this and, as we shall see, mathematics was certainly not included in the school's curriculum later that century. Jurin did, however, run classes for adults in mathematics and was probably the first man to offer such courses outside London.

In 1711 he advertised a course in geometry for 'young gentlemen or others' to be held for one hour a day on three days each week. The demand was so great that a second course had soon to be started and this was followed by courses on algebra and mechanics. The latter was advertised as being 'greatly useful to all sorts of Persons, but especially to Gentlemen concerned in Collieries and Lead-Mines'. It was to be given to a class of ten or twelve three times a week for twelve or eighteen months, the fee being an initial two guineas (£2.10) followed by a charge of half a guinea (52½p) per month.

The syllabus as advertised by Jurin was:

1. So much of the Principles of Geometry, Arithmetic and Algebra as shall be necessary for this undertaking.
2. The general laws of Motion, and the Principles of Mechanics deduced from them.
3. The Doctrine of Percussion, or the Effects which follow from the Stroke of Bodies upon one another.

4. The Natural Motion of all heavy Bodies.
5. The Motion of Bodies upon inclined Planes.
6. The Theory of all Kinds of Engines simple and compound, with a particular Explication of the Engines used in Collieries, and the method of Examining their Advantages or Defects.
7. Hydrostatics, under which Head will be demonstrated by Experiment; the chief Properties of Water and other fluids. . . .; The Method of Calculating the Weight or Pressure of Water against the Banks of Rivers, or Milldams . . . and other Surfaces, and consequently determining the strength requisite for those Bodies to support the Pressure . . .
8. Pneumatics, the Weight and Spring of the Air, its Rarefaction and Condensation. . . . The Air-pump, and Condenser, together with the Barometer, Thermometer, Hygrometer . . . their Nature and Uses explained . . .
9. Hydraulics, or the Doctrine of Water and other Fluids in Motion. The Method of estimating the Swiftness of Water running in any Canals open or closed, as in Rivers, or Mill-Races, . . . Conduit Pipes etc. with the Quantities of Water that they discharge. A particular Application to the draining of Collieries . . . for clearing and keeping a Colliery clear of Water. Of the Force of Fluids, as Air, or Water to carry about the Sails or Wheels of Mills and other Engines, and the best proportion of the Machines driven by them, or by Horses.
10. Lastly, The Important Theory of the Friction of Rubbing of Machines, with the Impediment caused by the stiffness of the Ropes, for want of which the greatest Engineers have been disappointed in their undertakings, and the best concerted Machines have been rendered useless . . . it will be explained in an easy manner, partly by Experiment, and the Application of it to the Calculation of Machines will be demonstrated (3).

Jurin's syllabus clearly demonstrates some of the mathematical needs of the Newcastle community and, indeed, has a most forward-looking air with its references to optimisation procedures and to teaching via experiment and practical demonstration. It is, doubtless, unrepresentative of the type of mathematical instruction to be had in a provincial city in the early eighteenth century, yet it does provide evidence that at least certain sections of the public were being brought into contact with some quite advanced mathematics {4}.

During the years that followed Jurin's departure from Newcastle, the demand in the town for mathematical tuition grew, and various schools were established to meet it. As we remarked earlier, Hutton himself learned mathematics at one such school, run by a Mr James. When James left, it was a natural step for Hutton to close his school in Jesmond and start one in the city of Newcastle itself.

An advertisement in the local newspaper informed the townspeople of the new school and its syllabus:

TO BE OPENED

On Monday April 14th 1760, at the Head of the Flesh Market, down the Entry formerly known by the name of the Salutation Entry, Newcastle, A WRITING AND MATHEMATICAL SCHOOL, where persons may be fully and expeditiously qualified for business, and where such as intend to go through a regular course of Arts and Sciences, may be completely grounded therein at large, viz., in Writing, according to its latest and best improvements; Arithmetic, in all its parts; Merchant's Accounts (or the true Italian method of book-keeping); Algebra; Geometry, elemental and practical; Mensuration; Trigonometry, plane and spherical; Projections of the Sphere; Conic Sections; Mechanics, Statics, and Hydrostatics; the Doctrine of Fluxions, etc. together with their various applications in Navigation, Surveying, Altimetry and Longimetry; Gunnery, Dialling, Gauging, Geography, Astronomy, etc., etc., etc. Also the use of the Globes, etc. Likewise Shorthand according to a new and facile character never yet published. By

C. HUTTON

¶ For the accommodation of such gentlemen and ladies as don't choose to appear at the public school, I propose (at vacant hours) to attend them in their own apartments (5).

Hutton's syllabus certainly looks impressive, but it was in fact that offered by almost every private teacher of the time (6). It is doubtful whether many of these teachers actually taught the whole of the syllabus or were capable of doing so; some indeed made a point of stressing that students would not necessarily have to cover all the topics mentioned. In Hutton's case, one might well assume that the syllabus represented what he felt might interest and attract potential students and what, given time for preparation, he felt capable of mastering.

Be that as it may, Hutton's advertisement succeeded in attracting pupils, including at least one gentleman who did not 'choose to appear at the public school'. This was a local man, Robert Shafto of Benwell Hall, who proposed that Hutton should help him to revise all the mathematics that he (Shafto) had read at college. Shafto was to have a considerable influence on Hutton. Not only were he and his children pupils, but he made his excellent library available to his tutor and so provided Hutton with further opportunities for increasing his knowledge. Hutton acknowledged his gratitude by dedicating to Shafto his first publication, *The Schoolmaster's Guide, or a Complete System of Practical Arithmetic.* This book first appeared in 1764 and was to remain a standard school textbook throughout Hutton's life. Its object as set out by Hutton was 'to introduce a regular and rational method of teaching this most necessary science into the generality of schools, and to ease masters of great part of the trouble which necessarily attends their business without such help'. The book contained 'all the useful rules of vulgar and decimal arithmetic, with their definitions, laid down in a clear and concise manner; together with a great variety of choice practical examples, with the answers, all new, under each rule, and a preface containing some useful hints towards a proper method of teaching this useful art'.

The content then was traditional, but Hutton did succeed in this and his other elementary works in introducing 'a mode of clearness, precision and simplicity . . . not to be met with in earlier or even contemporary writers on the same subjects' (7).

Not only did Hutton provide written hints on the teaching of arithmetic, but in 1766 he went even further and advertised that 'Any schoolmaster, in town or country, who are desirous of improvement in any branch of the mathematics, by applying to Mr Hutton, may be instructed during the Christmas vacation.' The course must have been a financial success, for next year a similar in-service training course was offered by a teacher in the nearby town of Chester-le-Street. Some idea of the number of private schoolmasters working in the Newcastle area is given by the fact that fifty-nine of them are listed among the subscribers who helped to launch Hutton's next book, his *Treatise on Mensuration* {8}. This book, when advertised in 1767, was announced as comprising some twelve or fourteen parts to be published at monthly intervals from January 1768 at 6d a part, a fairly common method of publication in those times. It actually ran to fifty parts. Apart from its mathematical interest, the book is remarkable for the fact that its diagrams were engraved in wood by a young apprentice engaged in his first professional work. The apprentice, Thomas Bewick, was later to become one of the greatest masters ever of the medium of the woodcut. Bewick is mainly remembered now for his woodcuts of birds and animals, but his connection with Hutton's book has guaranteed that it still fetches an enhanced price on the second-hand market!

Mensuration did not differ substantially in content from its rivals; indeed Hutton's books in general were distinguished not so much for their originality as for their utility, clarity and readability. Thus, according to Olinthus Gregory, one of Hutton's protégés {9},

> the author treats copiously and elegantly of plane trigonometry, the determination of heights and distances, the areas of right-lined and circular figures, the mensuration of prisms, pyramids, spheres, etc., polyhedrae, solid rings, conic sections, their quadrature, the cubature and complanation of solids, formed by the rotation of conic sections upon their axes and other lines, the method of equidistant ordinates and sections, the centro-baryc method of determining the measure of planes by means of their centre of gravity, etc.

While at Newcastle, Hutton was able to benefit from the absence of any mathematics master in the grammar school there. The benefits of introducing boys to the study of mathematics were being increasingly recognised although many of the grammar schools did not have the staff, or on occasion, the inclination to do this. Often boys who wanted to learn some arithmetic or geometry paid for tuition in much the same way that they would now pay in an independent school for music lessons. Schools would boast a writing master, or scrivenor, who would teach the rudiments of mathematics and, as the name implies, the art of writing {10}. In some cases, as at Newcastle, an alliance was formed between the grammar school and a private mathematical school and boys from the former attended the latter for lessons in mathematics {11}. It was to Hutton's school that the grammar schoolboys at Newcastle were sent. Thus it was that John Scott, who later became Lord Eldon, came to Hutton for tuition. As Lord Chancellor, Eldon was to be responsible for a judgement that had considerable effect on grammar schools, and which made it more difficult for them to cast off the classical curriculum and encompass the teaching of mathematics and modern languages {12}.

Hutton's pupils were not, however, all male, for the future Lady Eldon also attended his school. That a girl should learn mathematics was unusual, but not, as we shall later see, exceptionally so.

As the school attracted more and more pupils, so Hutton was forced to move from building to building in order to accommodate them all, until finally he had to have a school-room specially built to meet his needs.

Hutton's reputation in Newcastle was now assured and in 1770 he was employed by the Mayor and Corporation to survey the town. The following year there occurred a great flood during which the bridge over the Tyne collapsed. The bridge's collapse provided the impetus for Hutton's next major work *The Principles of Bridges; containing the Mathematical Demonstrations of the Properties of the Arches, the Thickness of the Piers, the Force of the Water against them, etc. together with practical Observations and Directions drawn from the whole.*

The ability to apply his mathematical knowledge to practical matters set Hutton above the other mathematics teachers and writers of his region and no doubt was responsible for Shafto's suggestion to him that he should move to London and, in particular, apply for the vacant post of Professor of Mathematics at the Royal Military Academy, Woolwich.

2. THE ROYAL MILITARY ACADEMY, WOOLWICH

Earlier we saw how the need for a steady supply of trained navigators eventually led to the establishment of mathematical schools in London and elsewhere. As the need for trained naval personnel became apparent, that for literate and numerate soldiers was also recognised.

In 1662 a request was made for a schoolmaster to be supplied to the troops quartered in Madras (13). The schoolmaster was to divide his attention between the soldiers and their families. Gradually the idea of supplying education to the troops spread and in the mid eighteenth century the first army schoolmaster was appointed in England. From then on it became increasingly the case that non-commissioned officers were expected to become 'perfectly proficient in their writing, keeping accounts, spelling, etc.', and private soldiers were encouraged 'to take the earliest opportunity of going to school, as . . . those who make the greatest proficiency there will be the first for promotion' (14).

For officers, though, and, in particular, staff officers, some more specialised knowledge of mathematics, fortification and drawing was necessary. To meet that need the Royal Military Academy (RMA) was established at Woolwich in 1741 to train future artillery and engineer officers {15}. A naval academy was similarly founded about that time in Portsmouth. In addition to the RMA there were military schools in the London area which prepared young gentlemen for the services, but these were of doubtful quality, so much so that the future Duke of Wellington was sent to a French academy to receive his pre-service education.

The RMA, however, did much to encourage the growth of mathematics education in England. True, its achievements pale into insignificance alongside those of the Ecole Polytechnique {16} (which boasted such mathematicians as Lagrange, Laplace, Monge, Poisson, Fourier, Cauchy, Chasles and Hermite on its staff), but it was to provide a favourable environment for non-university trained mathematicians such as Thomas Simpson [17], Hutton, Gregory and Peter Barlow [18], and, later, employment for the great mathematician, J. J. Sylvester, at a time when, on religious grounds, he was denied a post at Oxford and Cambridge.

When the post of Professor of Mathematics became vacant in 1773, Lord Townshend, who as Master General of Ordnance had responsibility for the RMA's affairs, took the unusual step of filling the post by public examination. The decision was indicative of the growing interest in competitive examinations. Thus, at Cambridge the Senate House Examination was supplanting the disputations as the determining factor in a man's ranking {19}, and the custom of awarding classes had begun. Adam Smith was to speak of the way in which 'rivalship and emulation render excellency . . . an object of ambition' and arguments were advanced for using examinations as a means for selection (20). Competitive examinations for public office were still, however, many years away, as was the decision to make entry to the RMA conditional upon the passing of a public examination {21}. Nevertheless, the idea of the examination as a means of certification and classification, and as a motivating force, was beginning to gain ground.

Hutton went up to London for the examination armed with letters of introduction from Shafto to Lord Sandwich, the previous Master General of Ordnance. There were eleven candidates for the post and they were examined for several days by a board including Bishop Horsley, the editor of Newton's works, and Nevil Maskelyne, the Astronomer Royal. About half the candidates were deemed to have passed the test 'in such a way as to give entire satisfaction', but Hutton was clearly superior to the others and it was he who was appointed to the post in May 1773.

By now Hutton had a national reputation. His texts were being reprinted in new editions and were in use throughout the country. In 1774 he was elected a Fellow of the Royal Society and in 1779 the University of Edinburgh bestowed on him the degree of Doctor of Laws.

The papers he published on the convergence of series and on ballistics led to his winning the Royal Society's Copley Medal, and in 1778 he communicated his calculations on the mean density of the earth on the basis of Maskelyne's observations on Mount Schiehallion. This work was to establish his name internationally.

In 1779 Hutton was elected Foreign Secretary of the Royal Society. The appointment, or at least Hutton's laying down of it, was to help provoke a crisis in the Society's affairs. Hutton was at that period, as at most others in his life, extremely busy. His tables, which were to be widely used for many years, were first published in 1781 (22), and he was engaged in many other mathematical activities. As a result, his duties as Foreign Secretary were somewhat perfunctorily performed and in 1783 after a committee had reported on the matter, he was obliged to resign his post. The motion that the Society's thanks should be given to Hutton for his past work gave rise to a stormy discussion and in the end was carried by the slender majority of five. The President, Sir Joseph Banks, was initially opposed to the motion and indeed it was he rather than Hutton who was at the centre of the storm (23). Banks was newly appointed as President and his approach to the revivification and improvement of the Society did not appeal to all. He did not hesitate to blackball those candidates for fellowship whom he thought unworthy of election: among those he personally rejected were Clarke, Professor at the Royal Military College, Marlow, and Des Barres, an army officer and military surveyor {24}. The rejection of mathematical practitioners such as these did not endear Banks to Hutton, Horsley, Maskelyne and others. Moreover, there was the suspicion that he had a 'predilection for the disciples of Linnaeus over those of Newton'. An attempt was made in February 1784 to reinstate Hutton as Foreign Secretary, and so undermine Banks' position as President, but the motion was defeated by a majority of thirty-eight. Not abashed, the dissidents then attempted to have Hutton elected as Secretary. Again they failed and as a result 'the mathematicians seceded from the Royal Society'.

In 1786 Hutton began to suffer from pulmonary disorders. The RMA was situated near the river and the dampness began to affect his chest; his predecessor Simpson had in fact died from a chest complaint. Hutton decided then to move, and bought land on the hill south of the river overlooking Woolwich. There he built himself a house and also others for letting. No sooner had he done this than it was decided to move the Academy from the damp riverside to the hilltop. A magnificient new building was erected, but, in the eyes of George III, its attractiveness was spoiled by the presence of Hutton's houses. These were therefore sold to the Crown who promptly demolished them, leaving Hutton with a hefty profit from his speculation, sufficient to guarantee his financial future. Thus a physical disability turned him to mathematics and ill-health made him rich.

Yet by then Hutton's books must have in any case provided him with a steady income, for his tables and elementary texts were frequently reprinted. In 1795 there appeared his most important publication, a two-volume *Mathematical and Philosophical Dictionary* on which he had worked for some '10 or 12 years'. Attempts had already been made to produce encyclopaedic accounts of pure and applied mathematics in English. Early in the eighteenth century an English abridgement of Ozanam's *Dictionnaire Mathématique* (1691) had been published (25), and this was followed by John Harris' *Lexicon Technicum or Universal Dictionary of Arts and Sciences* {26} – to which Newton contributed – and Edmund Stone's smaller *New Mathematical Dictionary* {27}. Hutton, in his great work, attempted single-handed to bring Harris' dictionary up to date. In doing so he provided a masterly survey of mathematics as it then existed and also many valuable biographies of mathematicians – indeed the work stands as a notable pioneering contribution to the history of mathematics.

Hutton's reputation as an author and teacher was now so high that in its 'Notices of works in hand' the *Monthly Magazine* (28) could confidently forecast of his next work *A Course in Mathematics* that 'From Dr H's talents and long experience in his profession, there is every reason to expect that this will not only be a most useful and valuable work, but will completely supersede every other of the same description.' The two-volume *Course* was indeed to prove extremely popular – it appeared in numerous editions over a period of fifty years, several in North America, and was even translated into Arabic.

That Hutton's book should be used in North America is not particularly surprising. The British influence on American education was extremely strong during the eighteenth and early nineteenth centuries. Thus, for example, although elementary texts by Americans had been published, the most popular arithmetic texts in North America during that period were both by Englishmen, Dilworth and Daboll; the latter's text having been 'adapted' to local needs by the inclusion of some examples involving dollars and cents (29). (One is reminded here of the use of textbooks by Hall and Knight, Durell, etc. in the newer countries of the Empire and

Commonwealth during the early and mid twentieth century (30).) In the field of geometry, Simson's and Playfair's Euclids [31] were both used, but soon these gave way to translations of Legendre's *Elements*, one being that by Thomas Carlyle {32} (p. 131 below). This book was to become the mainstay of American geometry teaching, and indeed from 1820 until the arrival in America of Sylvester, French rather than English mathematics was to dominate American thinking (33). The existence of an Arabic translation of Hutton's *Course* is an indication of the growing interest in mathematics outside Europe and of the political influence now being exerted overseas by Britain. From the beginning of the nineteenth century onwards, British texts began to be translated into a variety of languages, principally for use in the rapidly growing Empire.

Hutton's *A Course in Mathematics* indicates the mathematics taught at Woolwich in his time, and it is interesting to contrast its contents with those of the contemporary Tripos papers at Cambridge (34).

The *Course* was intended to be fully comprehensive and thus to remove the need for 'procuring and keeping . . . a number of odd volumes, of various modes of composition and form' which were 'wanting the benefit of uniformity and reference' (35). It was not the first attempt to provide such a course: Jonas Moore {36} and John Ward {37} had published earlier 'guides', while William Emerson [38] had written numerous volumes on different branches of mathematics which were later gathered together and published under the title *Cyclomathesis*. Unlike Moore's and Ward's courses though, Hutton's included the calculus.

Hutton determined to produce a course of applicable mathematics, 'rejecting whatever seemed to be matters of mere curiosity; and retaining only such parts and branches, as have a direct tendency and application to some useful part of life'. In particular, this meant deviating 'a little from the tedious and rigid strictness of Euclid'. Thus Hutton followed Euclid in beginning with definitions and axioms, but he used algebraic methods when dealing with ratio and proportion and selected only such Euclidean results as he thought of interest. (He departed from Euclid, for example, in his definition of a straight line as one which 'lies all in the same direction'.)

The full *Course* comprised:

Arithmetic – the four rules, rule of three, fractions, decimals, roots and powers, proportion, interest, permutations and combinations;

Algebra – the four rules, powers and roots, surds, series, arithmetic and geometric progressions, logarithms, simple, quadratic and cubic equations, interest and annuities (for the gunner's benefit there were included tables of pyramidal numbers and formulae for finding the number of shots or shells in various shapes of pile);

Plane Trigonometry, Mensuration, Surveying, Conic Sections, Dynamics, Projectiles, Hydrostatics, Hydraulics; Fluxions with some examples from dynamics including variable forces.

Although Hutton provides examples of the applications of calculus, it is still somewhat surprising to the modern reader to see how often he avoids its use on occasions when we should naturally turn to it. Indeed, for another century, a solution by calculus 'was considered rather mean if the problem could be solved by other complicated algebraical or geometrical methods, however cumbersome' (39).

The *Course* was Hutton's last major work. Like the *Dictionary* and the *Treatise on Mensuration* it was to have a great effect on mathematics education, not only through its many editions, but also for the influence it had on succeeding writers. Countless examples could be given of nineteenth-century authors who in their writings give credit to Hutton, and cite these three works as sources for their material {40}. Hutton's writing set new standards which others could emulate. He was no Wallis or Newton, but it could be argued that he was the most competent English, pre-nineteenth-century mathematician to become so closely involved in the preparation of textbooks for the general student of mathematics.

3. 'THE LADIES' DIARY'

Reference has been made to the way in which mathematics was popularised via the public lecture (for example by Jurin) and the scientific society (such as that in Northampton). Great as the influences of these two agents were, it can be claimed that neither was so successful in involving society at large in mathematical activities as were the many periodicals which flourished in the eighteenth and early nineteenth centuries (41). Outstanding among these was *The Ladies' Diary* which was to appear annually for almost a hundred and fifty years and which Hutton edited (along with other almanacs) from 1774 to 1817.

The Lady's Diary: or the Woman's Almanac first appeared in 1704 and survived with minor changes of title until 1840. The 1707 annual included certain mathematical questions and from then on the setting of such questions became a regular feature; that one had to wait a full year before the solutions appeared seems to have had little effect on their popularity. Initially the questions were always posed in verse, and solutions were expected to be in verse also! As a contemporary writer expressed it, this was a condition 'not favourable to the development of Mathematical genius'.

As with most journals, *The Ladies' Diary* reflected the aims and abilities of its editors. One, Robert Heath (1745–53), 'made the *Diary* a vehicle for low scurrility' and 'was virulent beyond all bounds against Thomas Simpson' (42). Not only that, but he published elsewhere, under his own name, material submitted for publication in the *Diary*. He was dismissed and succeeded, appropriately enough, by the vilified Simpson who thus began a long association between the *Diary* and the Royal Military Academy. This liaison appears less strange when one consults the *Diary* and sees that although ladies were represented (43), the vast majority of problem setters and solvers were men: schoolmasters, service officers, clergymen and even the occasional Cambridge don. Simpson, himself, wrote many articles under a variety of pseudonyms. Thus what was advertised as 'Designed for the Use and Diversion of the Fair Sex' was clearly read and studied by many men.

Shortly after becoming its editor, Hutton compiled *The Diarian Miscellany; consisting of all the useful and entertaining parts, both mathematical and poetical, extracted from the Ladies' Diary from the beginning of that work in the year 1704 down to the end of the year 1773. With many additional Solutions and Improvements.*

The work appeared in five volumes, the first three of which were mathematical. That it must have been a profitable venture can be inferred from the fact that there was soon a rival on the scene, *The Diarian Repository* by a 'Society of Mathematicians' (44). The 'Society' was one Samuel Clark, who according to Hutton could hardly be commended for his work: 'From the many gross errors, and the numerous omissions and absurdities in that "Repository of Errors", as it was commonly called, it soon fell under the necessity of coming to a sudden and premature end.'

Some idea of the flavour of the *Diary* questions is given by the following examples (all set by men) taken from the 1809 edition, which was edited by Hutton {45}:

What is the area of a right-angled triangle, the radius of its circumscribing circle being 20, and the area of the inscribed circle a maximum?

What two numbers are those whose sum in being increased or diminished by their difference, or the difference of their squares, the sums and differences shall be all squares?

If a be any prime number, then it is a divisor of the sum of all the consecutive squares $1^2 + 2^2 + 3^2 \ldots [\frac{1}{2}(a-1)]^2$. And if the product of all those squares be divided by a, the remainder will be always ± 1. Required the demonstration.

While one might query the claim that such questions might be considered as a 'valuable monument of the Mathematical genius of the English' (46), several well-known mathematicians were prepared to advance what might seem at first sight to be somewhat preposterous claims on the *Diary's* behalf.

Thus John Playfair wrote in 1808 (when reviewing Laplace's *Mécanique Céleste* in the *Edinburgh Review*):

A certain degree of Mathematical science and indeed no inconsiderable degree, is perhaps more widely diffused in England than in any other country in the world. The *Ladies' Diary* with several other periodical and popular publications of the same kind, are the best proofs of this assertion. In these many curious problems, not of the highest order indeed, but still having a considerable degree of difficulty, and far beyond the mere elements of science, are often to be met with: And the great number of ingenious men who take a share in proposing and answering these questions, whom one has never heard of anywhere else, is not a little surprising. Nothing of the same kind, we believe, is to be found in any other country.

Kirkman (now remembered for his 'schoolgirls'' problem {47}, but more importantly an English pioneer of group theory, knot theory and combinatorics) wrote in 1849:

I confess it to be my belief . . . that an incomparably greater share of the glory of kindling and cherishing a pure and lasting love of mathematical science in *men* as well as *boys*, must be attributed to the immortal Lady Dia, than to all our Lyceums, Athenaeums and Philosophical Societies, and to all our Imperial Boards of peace and war (48).

Perhaps Kirkman had special reason to think highly of such publications, for his schoolgirls' problem was initially set by him in the *Lady's and Gentleman's Diary . . . designed principally for the amusement and instruction of Students in Mathematics* for 1850. It arose out of Kirkman's particular solution ($n = 15$, $p = 3$, $q = 2$) to the problem set by the editor of that annual in 1844:

Determine the number of combinations that can be made out of n symbols, p symbols in each; with this limitation, that no combination of q symbols which may appear in any of them shall be repeated in another.

Kirkman's problem was later to be investigated by, among others, Cayley, Sylvester, Steiner and Benjamin Peirce.

Among contemporaries and successors to *The Ladies' Diary,* one deserves particular mention. Thomas Leybourn's *Mathematical Repository* (1795–1804 and 1804–35) listed among its contributors Hutton (to whom the first volume was dedicated), Babbage, Barlow, Gregory, Herschel, Horner, Peacock and Mrs Somerville. (Much of Horner's work on the numerical solution of equations appeared in this periodical.)

Of the other similar publications it is of significance that not all were published in London; mathematical periodicals and publications devoting space to mathematics emanated, for example, from York (see p. 98 below), Wrexham, Newark, Birmingham, Liverpool, Hull, Leeds, Blackburn, Holbeach (Lincolnshire), Bolton, Cambridge, Alnwick, Newcastle and Exeter – ample evidence of the widespread interest in the subject, despite the fact that it still had to win a secure place for itself in the endowed schools {49}.

4. HUTTON AS PATRON

Hutton as a writer and a teacher had a major influence on mathematics in his time and in the fifty years following his death. In this section we look at certain actions in the later part of his life, which, although of comparatively minor significance, are not without interest and serve to illuminate some further aspects of contemporary mathematics education.

Mary Somerville is the first woman mathematician and scientist to have been named in this book. Although their lives overlapped, it was, however, only after Hutton's death that Mrs Somerville made her name as a scientist. Born in Scotland, she was educated near Edinburgh at a fashionable boarding school for girls where she soon showed a liking for Euclid and worked hard at her Latin in order that she might read Newton's *Principia*. In 1816 she and her second husband moved to London where she was introduced to many of the leading scientists, mathematicians and educators of the time, men such as Herschel, Airy, Whewell and Brougham (see chapter 5). The last named suggested to Mary Somerville that she should produce an English version of Laplace's *Mécanique Céleste*, the latest successor to the *Principia*. Her *Mechanism of the Heavens* appeared in 1831 and established her reputation, now commemorated in Somerville College, Oxford. She was not, however, the first British woman to publish a book on mathematics, being forestalled in the closing years of the eighteenth century by a near neighbour of Hutton's, Mrs Margaret Bryan.

By the end of that century private schools for girls existed in many towns but these usually taught only reading, writing and the social arts of dancing and needlework {50}. (Some girls, however, like the future Lady Eldon, learned mathematics at private schools run primarily for boys.) Mrs Bryan owned a private school for girls at Blackheath in south-east London, but hers was a school with a difference, for she accorded great weight to the study of mathematics and its applications. In 1797 she emulated so many of her male counterparts and published part of the course she gave to her 'young ladies' as *A Compendious System of Astronomy*. To do this it was first necessary to find subscribers to the book. Here Hutton came to Mrs Bryan's aid, for not only did he agree to subscribe, but he also gave her valuable encouragement to proceed with her writing {51}. The first edition of the book was published privately by Mrs Bryan and in her preface, written in a most flowery style, she pays more than ample tribute to Hutton's help. The book is of note too for the engraved portrait it contains of the author and her children; few, if any, writers on mathematics have been so strikingly beautiful. Encouraged by the success of her first book, Mrs Bryan later produced two others, *Lectures on Natural*

Philosophy (1806) and *An Astronomical and Geographical Class Book for Schools* (1815). Apart from her books, however, little is known of this pioneer of mathematics education for girls {52}.

In pre-railway days, travel between the north of England and London was by no means easy or comfortable. Even so, it is surprising to us today to learn that Hutton, after leaving Newcastle in 1773, never returned there. Yet he retained some connections with the north-east. In particular, Hutton established a fund to support public lectures in Newcastle, and also gave money to a school there. More interestingly, he provided financial support for the Protestant Schoolmasters' Association, a welfare body that had been established in the north-east in 1774. As we have noted, there were many private teachers operating in that part of England and throughout the country. Very few of these were successful enough to earn a middle-class income and there was also, of course, a great element of competition between them. Teachers faced a real threat of poverty, and illness and old age were things to be dreaded (see also pp. 101–2 below). The Schoolmasters' Association was, therefore, established to protect members and their widows should financial hardship arise. Hutton, now a wealthy man in London, continued to support this fund and so helped to provide for those less successful and fortunate than himself.

Hutton also retained an interest in the activities of the Newcastle Literary and Philosophical Society. The growth of such societies with their libraries, meetings and journals {53} was of great importance in the field of what we now refer to as adult or further education. Their membership was usually dominated by the professions – by clergymen, lawyers, doctors and teachers – and so they were essentially middle-class bodies, with more 'liberal' and less vocational aims than the Mechanics' Institutes. They provided an important means whereby mathematical and scientific knowledge could be disseminated. It was to the Newcastle Society that a marble bust presented to Hutton in 1822 by a number of subscribers (each of whom received a medal inscribed with Hutton's head) was later donated.

The presentation of the bust was one of the last honours to be paid to Hutton, for by 1822 he was nearing the end of his life. It was now fifteen years since he had retired from Woolwich with a state pension of £500 p.a. He died the following year. During the later years of his life, his affluence had allowed him to indulge in his great hobby of collecting books. His library was, in his own words, 'esteemed the best mathematical collection in the kingdom'. It was Hutton's hope that the collection would be purchased by the British Museum, but this was not to be. The reason, it was reported, was the continued animosity of Sir Joseph Banks, with whom Hutton had quarrelled several years before. The library was sold by auction. Fortunately, some of the books did find their way to the British

Museum. Others went to the Newcastle Literary and Philosophical Society where they still serve to commemorate the achievements of the local pit-boy who became one of the leading mathematicians of his age.

5 · *Augustus De Morgan*

The first half of the nineteenth century witnessed many significant changes in mathematics education. Mathematics teaching at Cambridge was revivified; the hold of Oxford and Cambridge on university education was broken; a new type of secondary school arose in which mathematics was awarded a place of honour in the curriculum; and the particular problems of teaching and learning mathematics gradually became more explicitly and professionally scrutinised. The case for a state system of education was argued as were the objectives which such a system should seek. Closely concerned with many of these changes and developments, as well as with the general diffusion of mathematical ideas, was Augustus De Morgan.

1. EARLY YEARS AND SCHOOL

De Morgan, the son of an army colonel, was born in Madurai, southern India, in 1806. He spent only seven months in India, but the stay was to prove of great significance for he contracted an eye infection and lost the sight of his right eye. His father brought the family back to England, but he himself soon returned to India. Indeed, Augustus was to see little of his father. They had the years 1810–12 together, during which time his father taught Augustus 'reading and writing', but then Colonel De Morgan set out for India once more. There he became ill with a liver complaint and died on the voyage back to England.

By that time Augustus had begun to attend a series of private schools run by single ladies or clergymen. At the first he studied reading, writing and spelling; at the second be began to learn arithmetic. At the third, Augustus, now aged nine, was taught Greek and Latin by a Unitarian minister. Then followed periods at yet two more local schools at which he was taught Latin, Greek, Euclid, algebra and a little Hebrew, before, at the age of fourteen, he was sent away to board at a school at Redland, near Bristol, owned by a Reverend John Parsons, formerly a Fellow of Oriel College. Not surprisingly, with so much chopping and changing, it was a friend of the family rather than one of his early teachers who first noted that De Morgan possessed a germ of mathematical ability {1}. Given encouragement, De Morgan began to study the subject with enthusiasm and soon left his teacher behind. A fellow schoolboy (2) wrote:

I well remember that I was advanced in 'Bland's *Quadratic Equations*' when De Morgan took up that well-known elementary book, 'Bridge's *Algebra*' for the first time . . . He read Bridge's book like a novel. In less than a month he had gone through that treatise and dashed into Bland, and so got out of sight, as far as I was concerned.

This account, suggesting a school in which children pursued independent work unhampered by authority, is somewhat misleading. As in other schools of the time, most emphasis was placed on the memorising of Latin and Greek verses 'to strengthen the memory'. Every day forty lines had to be learned, Latin and Greek alternately, and the whole two hundred lines had to be repeated consecutively on Saturday. Moreover, although De Morgan pursued his mathematics, this was not encouraged. His master, Parsons, wished to produce classicists rather than mathematicians, while Augustus's mother wanted the boy to concentrate upon a career as an Evangelical clergyman. She had a skeleton in the family cupboard to hide – her grandfather, James Dodson, had been a mathematics teacher at Christ's Hospital: Dodson was, in fact, a mathematician of some repute, a friend of De Moivre, and a Fellow of the Royal Society [3]. His descendants, however, looked down upon a practitioner of mathematics and when, as a boy, Augustus asked one of his eight aunts (seven of whom married officers of either the Military or Civil Service in India) 'who James Dodson was', he was told, 'We never cry stinking fish' (4).

His mother saw the ministry as a suitable goal for Augustus and accordingly ensured that he never missed the chance of hearing a good sermon – twice in the week and three times on Sundays. At school, too, frequent church attendance was compulsory. One result was that De Morgan was left with a lifelong inability to listen for any time to anyone speaking; he said that listening brought back memories of dreary sermons and immediately set his mind wandering. Another result was that the school pew at St Michael's Church, Bristol, was for many years decorated with the first two propositions of Euclid and one or two simple equations pricked into the wood with the point of a shoe buckle and signed with the initials A. De M.

His mother's pressure and an over-exposure to church services, far from having the desired effect, led De Morgan seriously to question some of the Church's doctrines, and to take exception to its institutionalised nature. The results were far-reaching.

2. CAMBRIDGE

In February 1823, at the age of sixteen, De Morgan entered Trinity College, Cambridge {5}. It was a time of transition in the university's mathematical life. In purely mathematical terms the isolation of English mathematicians from their Continental contemporaries was coming to an end: gradually analytical methods and the notation of the differential calculus were finding their way into the course. Educationally, the days of the incompetent professor were numbered, although the Lucasian Chair was held in De Morgan's student days by Thomas Turton, a mathematical nonentity (later Dean of Westminster and Bishop of Ely), who is reputed never to have lectured.

The first steps in the battle to adopt the Continental notation had been taken by Robert Woodhouse, a Cambridge logician and later Lucasian Professor, who in 1803 published his *Principles of Analytical Notation*. In this he explained the differential notation and urged its employment. His writings, however, had little effect in the university as a whole and it was not until 1812 that progress began to be made as a result of the three undergraduates Peacock, Babbage (both of whom were to become Cambridge professors, the latter renowned for his 'analytical machine') and John Herschel (who was to become famous as an astronomer). These three formed the Analytical Society which aimed to persuade the university to adopt the Continental practice and which, as a first step, translated Lacroix's *Differential and Integral Calculus* {6}.

By 1817 Peacock had been appointed Moderator in the university examinations and he took this opportunity to introduce the new notation, despite the fact that his fellow Moderator continued to use the fluxional one. Gradually, however, the use of the latter notation became rarer, although for several years it continued to be exclusively employed by the older members of staff. De Morgan, himself, did not regret being at Cambridge at such a time of change: many years later he wrote 'Thank heaven that I was at Cambridge in the interval between two systems, when thought about both was the order of the day even among undergraduates. There are pairs of men alive who did each other more good by discussing x [x?] versus dx and Newton versus Laplace, than all the private tutors ever do' (7).

An insight into Cambridge mathematics in the 'fluxional days' before De Morgan went up there is given by some correspondence between Sir Frederick Pollock, Senior Wrangler in 1806, and De Morgan (8):

July 24, 1869

MY DEAR SIR FREDERICK, As we neither of us are strong on the legs, and yet can use our fingers, I employ mine to beg you will employ yours, at any leisure you like, in answering a question.

I want a tolerably distinct account of the reading of a senior wrangler of your day. How much had he read, and in what books? A very general view will do. Anything as to distinction between algebra and geometry will be valuable.

. . . There are all kinds of legends current about old reading. A trustworthy account would be of historical value.

A. DE MORGAN

July 29, 1869.

MY DEAR DE MORGAN, . . . I shall write in answer to your inquiry, *all* about my books, my studies, and my degree, and leave you to settle all about the proprieties which my letter may give rise to, as to egotism, modesty, &c. The only books I read the first year were Wood's Algebra (as far as quadratic equations), Bonnycastle's ditto, and Euclid (Simpson's). In the second year I read Wood (beyond quadratic equations), and Wood and Vince, for what they called the branches [the theory of equations]. In the third year I read the *Jesuit's* Newton and Vince's Fluxions; these were all the *books*, but there were certain MSS. floating about which I copied . . . I have no doubt that I had read less and seen fewer books than any senior wrangler of about my time, or any period since; but what I knew I knew thoroughly, and it was completely at my fingers' ends. I consider that I was the last *geometrical* and *fluxional* senior wrangler; I was not up to the *differential* calculus, and never acquired it. I went up to college [from St Paul's School] with a knowledge of Euclid and algebra to quadratic equations, nothing more; and I never read any second year's lore during my first year, nor any third year's lore during my second; my *forte* was, that what I *did* know I *could produce at any moment* with PERFECT *accuracy*. I could repeat the first book of Euclid word by word and letter by letter. During the first year I was not a '*reading*' man (so called); I had no expectation of honours or a fellowship, and I attended all the lectures on all subjects – Harwood's anatomical, Woollaston's chemical, and Farish's mechanical lectures – but the examination at the end of the first year revealed to me my powers. I was not only in the first class, but it was generally understood I was *first* in the first class . . . Now, as I had taken no pains to prepare (taking, however, marvellous pains while the examination was going on), I knew better than any one else the value of my *examination qualities* (great rapidity and perfect accuracy); and I said to myself, 'If you're not an ass, you'll be senior wrangler;' and *I took to 'reading' accordingly* . . .

My experience has led me to doubt the value of competitive examination. I believe the most valuable qualities for practical life cannot be got at by any examination – such as steadiness and perseverance. It may be well to make an examination part of the mode of judging of a man's fitness; but to put him into an office with public duties to perform merely on his passing a good examination is, I think, a bad mode of preventing mere patronage. My brother is one of the best generals that ever commanded an army, but the qualities that make him so are quite beyond the reach of any examination. Latterly the Cambridge examinations seem to turn upon very different matters from what prevailed in my time. I think a Cambridge education has for its object to make good members of society – not to extend science and make profound mathematicians. The tripos questions in the Senate-house ought not to go beyond certain limits, and geometry ought to be cultivated and encouraged much more than it is.

Euclid and conic sections studied geometrically improve, enlarge, and strengthen the mind; studied analytically, I think not. But I must have exhausted your patience – a virtue which may be *tried* but not *examined* . . .

FRED. POLLOCK

Pollock's years at Cambridge were, however, to end in great disappointment, for, as he went on to tell De Morgan:

I longed to get among the Fellows; but when I did, I was utterly disgusted at the rubbishing conversation that prevailed and I then longed to get away.

Unlike Sir Frederick, the young De Morgan made a discouraging start to his Cambridge career, for he was almost immediately thrown into the examination room with men two years his senior who had been up at Cambridge considerably longer. The result, a second class, was disappointing, but De Morgan's tutor assured him that his 'merits and exertions richly deserved a first'. By 1825 his abilities were becoming recognised and he was awarded a scholarship by Trinity College. He was now reading widely as well as playing his flute with the music society. For a while he toyed with the idea of training to become a physician, but at his mother's urging dropped the idea and turned his thoughts to law, for by now it was clear that he would not be a clergyman.

In 1827 he took his degree and was ranked Fourth Wrangler. He had been expected to do better than this by his teachers who included Airy (who succeeded Turton as Lucasian Professor and who later became Astronomer Royal), Peacock and Whewell [9], and his low ranking was attributed to his wide mathematical reading which often led him away from the narrow examination path. De Morgan, like G. H. Hardy some eighty years later (see pp. 143–4 below), realised that this arbitrary ranking was unjust and fought to have the system changed. Unlike Hardy, his efforts met with no success whatsoever.

During his time at Cambridge, De Morgan gradually turned away from the established Church and its beliefs. He could no longer subscribe to the Thirty-nine Articles and, as a result, he was debarred from an MA and a Fellowship. Although he would have liked a career teaching mathematics this was denied him, at least at the university level, and so he entered Lincoln's Inn to begin his legal studies {10}.

3. WILLIAM FREND

Shortly after coming to London, De Morgan was introduced to William Frend (1757–1841), a man who was to have a great influence on him. E. T. Bell describes De Morgan as a 'born nonconformist' (11), yet even De Morgan's strength of will and reforming zeal could scarcely match that of Frend {12}.

Frend, too, was a Cambridge mathematician and had been Second Wrangler in 1780. After graduating, Frend became a clergyman, was appointed to a living at Madingley near Cambridge and to a Fellowship at Jesus College. His future within the Establishment looked bright; indeed he could afford to turn down an offer of £2,000 p.a. (plus a pension of £800 p.a.) to act as tutor to the Archduke Alexander of Russia.

Gradually, however, his religious views changed and, like De Morgan later, he was converted to Unitarianism. His first action was to denounce the abuses of the Church and to condemn its liturgy. This led in 1787 to his being removed as a tutor in the university. Six years later he published a tract entitled *Peace and Union* which denounced the war with France. He was prosecuted by the university, tried before the Vice-Chancellor's Court and sentenced to non-residence. Frend then moved to London where he wrote and taught private pupils (including Lady Byron, before her marriage). There he also helped to establish an assurance company for which he acted as actuary. At the time De Morgan graduated, Frend had just retired from the company because of ill-health (taking with him the not-inconsiderable pension of £800 p.a.).

Frend not only provides a fine example of what happened to a man who clashed with the Establishment, but he is also of interest for other reasons.

As a mathematician he was inferior to De Morgan: he was held back by an unwillingness to use negative numbers! He denounced those algebraists 'who talk of a number less than nothing, of multiplying a negative number into a negative number and thus producing a positive number . . . This is all jargon at which common sense recoils . . . it finds the most strenuous supporters among those who love to take things upon trust, and hate the labour of serious thought' (13). Yet there was a certain modernity in Frend's outlook, for he rejected all arguments to explain the nature of negative numbers based on borrowing, owing, etc. As he explained, 'when a person cannot explain the principles of a science without reference to metaphor, the probability is that he has never thought accurately upon the subject'. It is the case, of course, that no convincing definition of negative and complex numbers could be given in purely mathematical terms at that date. As a result, in his book *The Principles of Algebra* (1796), Frend eschewed the use of negative and 'imaginary' numbers. It was on the basis of the algebraic theories exemplified in his writings that Frend made an unsuccessful bid for the Lucasian Chair when it became vacant in 1798 {14}. Frend's doubts about the legitimacy of working with negative numbers were held by others. Masères of Clare College, Cambridge in 1759 wrote a *Dissertation on the Use of the Negative Sign in Algebra* and later supplied an appendix to Frend's book. Indeed, De Morgan, himself, in his *On the Study and Difficulties of Mathematics* (1831) wrote

> The imaginary expression $\sqrt{(-a)}$ and the negative expression $-b$ have this resemblance, that either of them occurring as the solution of a problem indicates some inconsistency or absurdity. As far as real meaning is concerned, both are equally imaginary, since $0-a$ is as inconceivable as $\sqrt{(-a)}$.

It is of interest also that Frend attempted to cater for those in London who could not attend the universities, by giving courses of public lectures 'on mathematics of natural philosophy, on a plan similar to that pursued in the University, Cambridge' {15}. Indeed it was as an educator rather than as a mathematician that Frend was to have his greatest influence on De Morgan.

Frend, from bitter experience, realised only too well the great loss and deprivation suffered by those who were debarred from a place in the two universities because of their religious beliefs. He looked forward to a time when all forms of religion would be held equal at Oxford and Cambridge, and when the conditions of admission would no longer include any reference to religious doctrine. Clearly, though, no rapid changes in procedure could be expected from the two universities and thus if dissenters were to receive a university education in England it was necessary to establish a new foundation, more permanent and esteemed than the dissenting academies of the previous century.

Frend outlined his ideas for such an institution in letters published about 1819. Shortly after their publication, Frend met Brougham [16], Birkbeck [17] and others to discuss what positive action might be taken to establish an independent university.

Gradually a group of liberal politicians and wealthy merchants came together and in 1826 established the new London University which was to provide an education in 'Mathematical and Physical Science, Classics and Medicine'. The necessary finance was raised partly by subscription, partly by shares, and the management of the university was vested in a council of twenty-four drawn from the proprietors. Ill-health prevented Frend from taking an active part in the university's establishment, but he became a shareholder and, following the election of Council, an auditor. The Deed of Settlement of the new university avoided specific mention of religion, but it was generally acknowledged that the university would pay no regard to religious beliefs in either pupil or teacher.

It was in the days preceding the establishment of London University that De Morgan first met Frend and was introduced to his family, including his daughter Sophia whom, ten years later, De Morgan was to marry {18}.

Not surprisingly, De Morgan was greatly attracted to a university free from religious bias and also saw it as offering an alternative to law. He was accordingly one of the thirty-two candidates for the university's first mathematical chair. Although not yet twenty-two and much younger than the other applicants, De Morgan was appointed to the post. That he should have been the unanimous choice of the appointing committee speaks highly of his ability and of their wisdom and courage.

The appointment was not received with enthusiasm by all. Indeed, Frend, who might well have been expected to be overjoyed, wrote to De Morgan to question his decision to set aside a promising, remunerative career at the Bar for the poorly-paid post of Professor of Mathematics. (De

Morgan was never to earn more than £500 p.a. as his professorial salary and in many years towards the end of his career he earned far less.)

De Morgan's answer left no doubts as to his position:

You seem to fancy that I was going to the Bar from choice. The fact is that of all the professions which are called learned, the Bar was the most open to me; but my choice will be to keep to the sciences as long as they feed me. I am very glad that I can sleep without the chance of dreaming that I see an 'Indenture of Five Parts' or some such matter, held up between me and the *Mécanique Céleste* (19).

4. LONDON UNIVERSITY

On 5 November 1828, De Morgan gave his introductory lecture 'On the study of mathematics' at London University. This was the first of his many attempts to describe the position which mathematics should hold in a person's mental education. It was a theme which he was gradually to develop and continuously to expound throughout his career. Nearly a hundred pupils attended his first class and to these he lectured twice every day, from nine till ten in the morning and from three to four in the afternoon. To eke out his earnings, De Morgan was later forced to take private pupils in the interval between giving lectures and marking exercises. One of these pupils was Lady Lovelace, Lady Byron's daughter, who is now remembered for her friendship with Charles Babbage (20).

An idea of the mathematics taught by De Morgan can be obtained by studying the textbooks he used, many of which he himself wrote or translated. Thus, for example, he used his translation of Bourdon's *Algebra* (1828) until it was superseded by his own book. Perhaps the most notable of his texts was his *Differential and Integral Calculus* (1842), the first calculus book based on the theory of limits to be written in English. His style of teaching and his pioneering contribution to 'in-service training' are described in an obituary notice written by a former pupil, Sedley Taylor of Trinity College, Cambridge (21).

As Professor of Pure Mathematics at University College, London, De Morgan regularly delivered four courses of lectures, each of three hours a week, and lasting throughout the academical year. He thus lectured two hours every day to his College classes, besides giving a course addressed to schoolmasters in the evening during a portion of the year. His courses embraced a systematic view of the whole field of Pure Mathematics, from the first book of Euclid and Elementary Arithmetic up to the Calculus of Variations. From two to three years were ordinarily spent by Mathematical students in attendance on his lectures. De Morgan was far from thinking the duties of his chair adequately performed by lecturing only. At the close of every lecture in each course he gave out a number of problems and examples illustrative of the subject which was then engaging the attention of the class. His students were expected to bring these to him worked out. He then looked them over, and returned them revised before the next lecture. Each example, if rightly done, was carefully marked with a tick, or if a mere inaccuracy occurred in the working it was crossed out, and the proper correction inserted. If, however, a mistake of *principle* was committed, the words 'show me' appeared on the exercise. The student so summoned was expected to present himself on the platform at the close of the lecture, when De Morgan would carefully go over the point with him privately, and endeavour to clear up whatever difficulty he experienced. The amount of labour thus involved was very considerable, as the number of students in attendance frequently exceeded one hundred.

De Morgan's exposition combined excellences of the most varied kinds. It was clear, vivid, and succinct – rich too with abundance of illustration always at the command of enormously wide reading and an astonishingly retentive memory. A voice of sonorous sweetness, a grand forehead, and a profile of classic beauty, intensified the impression of commanding power which an almost equally complete mastery over Mathematical truth, and over the forms of language in which he so attractively arrayed it, could not fail to make upon his auditors. Greater, however, than even these eminent qualities were the love of scientific truth for its own sake, and the utter contempt for all counterfeit knowledge, with which he was visibly possessed, and which he had an extraordinary power of arousing and sustaining in his pupils. The fundamental conceptions of each main department of Mathematics were dwelt upon and illustrated in such detail as to show that, in the judgment of the lecturer, a thorough comprehension and mental assimilation of great principles far outweighed in importance any mere analytical dexterity in the application of half-understood principles to particular cases. Thus, for instance, in Trigonometry, the wide generality of that subject, as the science of undulating or periodic magnitude, was brought out and insisted on from the very first. In like manner the Differential Calculus was approached through a rich conglomerate of elementary illustration, by which the notion of a differential coefficient was made thoroughly intelligible before any formal definition of its meaning had been given. The amount of time spent on any one subject was regulated exclusively by the importance which De Morgan held it to possess in a systematic view of Mathematical science. The claims which University or College examinations might be supposed to have on the studies of his pupils were never allowed to influence his programme in the slightest degree. He laboured to form sound scientific Mathematicians, and, if he succeeded in this, cared little whether his pupils could reproduce more or less of their knowledge on paper in a given time. On one occasion, when I had expressed regret that a most distinguished student of his had been beaten, in the Cambridge Mathematical Tripos, by several men believed to be his inferiors, De Morgan quietly remarked that he 'never thought —— likely to do himself justice in THE GREAT WRITING RACE.' All *cram* he held in the most sovereign contempt. I remember, during the last week of his course which preceded an annual College examination, his abruptly addressing his class as follows: 'I notice that many of you have left off working my examples this week. I know perfectly well what you are doing: YOU ARE CRAMMING FOR THE EXAMINATION. But I will set you such a paper as shall make ALL YOUR CRAM of no use.'

De Morgan, however, did not seek an easy popularity and was a martinet so far as discipline was concerned. Unlike those professors who habitually began their lectures late in order to 'give the lads time to assemble', De Morgan began promptly and he expected his students to be equally punctual. To ensure they were, the lecture room door was locked promptly on the hour and admittance refused to latecomers. Once, when some of those who were locked out attempted to burst the door open, a policeman was summoned and they were brought to order by the threat of being reported to Council and subsequently expelled!

Not every professor was able to handle students with such authority and it was this that led to De Morgan's first brush with Council and, consequently, to a change of heart which has had a continuous and most important effect on teachers in English universities.

In the 1829–30 session a number of students in the anatomical class made complaints about their teacher, Professor Pattison, and questioned his competency. These complaints eventually reached the Council who appointed a committee to investigate them. The significant feature of the case was that the professors did not have tenure of office. Perhaps with an eye to the abuses that occurred at Oxbridge, the founding proprietors had decided that professors should be treated in exactly the same way as would employees in their various industrial and commercial enterprises.

In a letter to the Committee of Council, De Morgan set out the case for tenure:

> In order to induce men of character to fill the chairs of the University, these latter must be rendered highly independent and respectable. No man who feels (rightly) for himself will face a class of pupils as long as there is anything in the character in which he appears before them to excite any feelings but those of the most entire respect. The pupils all know that there is a body in the University superior to the Professors; they should also know that this body respects the Professors, and that the fundamental laws of the institution will protect the Professor as long as he discharges his duty, as certainly as they will lead to his ejectment in case of misconduct or negligence. Unless the pupils are well assured of this they will look upon the situation of Professor as of very ambiguous respectability . . .
>
> With the public the situation will be altogether as bad. Wherever the Professor goes, he will meet no one in a similar situation to his own – that is, no one who has put his character and prospects into the hands of a number of private individuals. The clergyman, the lawyer, the physician, the tutor or Professor in the ancient Universities, will all look down upon him, for they are all secured in the possession of their characters. Nothing but the public voice, or the law of the land, can touch them, and a security as good must be given to the Professor of the London University before he can pretend to mix in their society as their equal . . .
>
> The University will never be other than divided against itself as long as the principle of expediency is recognised in the dismissal of Professors. There will always be some one who, in the opinion of some of his colleagues, is doing injury to the school by his manner of teaching; and there will always be attempts in progress to remove the obnoxious individual . . . No man will feel secure in his seat; and, consequently, no man will feel it his interest to give up his time to the affairs of his class. And yet this is absolutely necessary . . . for from the moment when a class becomes numerous the preparation, arrangement, and conduct of a system of instruction is nearly the business of a life; at least, I have found it so. If a Professor is easily removable, he will endeavour to secure something else of a more certain tenure; he will turn his attention to some literary undertaking, or to private pupils, while he remains in the institution, in order

that he may not be without resource if the caprice of the governing body should remove him – and this to the manifest detriment of his class, which, when it pays him well, ought to command his best exertions. In addition to this, he will always be on the watch to establish himself in some less precarious employment . . . In this way the University will become a nursery of Professors for better conducted institutions of all descriptions, since no man, or body of men, desirous to secure a competent teacher in any branch of knowledge, will need to give themselves the trouble to examine into the pretensions of candidates as long as any one fit for their purpose is at the University of London. The consequence will be a perpetual change of system in the different classes of the University, and the eventual loss of its reputation as a place of education. These evils may be very simply avoided by making the continuance of the Professors in their chairs determinable only by death, voluntary resignation, or misconduct either in their character of Professors or as gentlemen, proved before a competent tribunal, so framed that there shall be no doubt in the public mind of the justice of their decision.

De Morgan's eloquence did not sway the Committee; in July 1831, Council agreed to dismiss Professor Pattison – although the Resolution doing so concluded with the words that 'nothing which has come to their knowledge with respect to his conduct has in any way tended to impeach either his general character or professional skill and knowledge'. This was too much for De Morgan who accordingly resigned his professorship.

De Morgan had lost the battle, but the effect on the University of his and other resignations was so great that when he later made his return it was with the tenure he desired; he had won the war.

5. THE SOCIETY FOR THE DIFFUSION OF USEFUL KNOWLEDGE

The years 1831–6 when he was no longer a professor saw no lessening in De Morgan's mathematical activities. He was now Secretary of the Astronomical Society and closely concerned with Herschel, Airy and others in raising the status of astronomy in Britain. In an attempt to earn money, however, he was forced to take many private pupils and to turn to writing a stream of books and articles; in so doing he greatly widened his influence on mathematics education in England.

The name of Brougham has already been mentioned in connection with the establishment of London University. Brougham, however, was also a champion of popular education. As a result of his efforts Parliament had been persuaded to appoint in 1815 a Select Committee on the Education of the Lower Orders. The Committee recorded the fact that there was an unabated demand for education among the poor, a demand which was by no means being met. Indeed, as Brougham pointed out in 1820, 'at that date England might be justly looked on as the worst-educated country of Europe' {22}. In an attempt to remedy this state of affairs, Brougham introduced an Education Bill which aimed at providing schools in places where they were needed. Schools would be financed by rates and local taxes and by the fees (2d or 4d a week) paid by those parents who could afford them. The masters would be members of the Church of England and the curricula and examinations in the hands of its clergy, although to appease the nonconformists it was suggested that religious teaching should be non-sectarian. Not surprisingly, the Bill was opposed by both the Established Church and the nonconformists, and consequently was withdrawn. Brougham, undeterred, sought other ways to improve popular education: his widely-circulated pamphlet *Practical Observations on the Education of the People* (1825) led to the foundation of a society which would spread scientific and other knowledge by means of cheap and clearly written treatises, the Society for the Diffusion of Useful Knowledge (SDUK). Many were, of course, to see its title as a clear challenge to the Society for Promoting Christian Knowledge (SPCK) which since 1698 had established schools in which the poor might have a sound religious (Church of England) and secular education {23}. As a result there was opposition to the Society, particularly from those who, like Thomas Arnold, thought of London University as 'Anti-christian' (24).

The Society's aims, however, appealed to De Morgan who was to become not only a regular contributor to its publications but also a committee member during the final years of its existence.

It was the Society which published some of De Morgan's best-known books, such as his *On the Study and Difficulties of Mathematics* (1831), *Arithmetic* (1835), *Algebra* (1835) and *A Differential and Integral Calculus* (1842). The importance of the diffusion of mathematical knowledge had been stressed by Brougham in his *Practical Observations*:

> a most essential service will be rendered to the cause of knowledge, by him who shall devote his time to the composition of elementary treatises on the Mathematics, sufficiently clear, and yet sufficiently compendious, to exemplify the method of reasoning employed in that science, and to impart an accurate knowledge of the most fundamental propositions, with their application to practical purposes . . . He who shall prepare a treatise . . . calculated to strike the imagination . . . may fairly claim a large share in that rich harvest of discovery and invention which must be reaped by the thousands of ingenious and active men, thus enabled to bend their faculties towards objects at once useful and sublime.

De Morgan was but one author to take up Brougham's challenge. Among those who responded was Olinthus Gregory, Hutton's successor at Woolwich, with his significantly named *Mathematics for Practical Men*. If Britain was to continue its progress, both industrial and economic, then, as Brougham and others realised, mathematical knowledge had to become more commonplace.

The method of publication of De Morgan's *Calculus* well illustrates the

Society's method of working. Although the book was not published until 1842, its contents had already appeared in serial form spread over a period of six years. Sets of chapters had appeared as thirty-two page 'Numbers' of the Library of Useful Knowledge and it is interesting to see how De Morgan designed the book so that it satisfied the constraints of serialisation!

In addition to its 'Library' the Society also published the *Quarterly Journal of Education* (1831–5) and the famous *Penny Cyclopaedia* to both of which de Morgan made many contributions. Those to the former included articles on the teaching of arithmetic and geometry (see below) and on mathematics at Oxford and the Ecole Polytechnique. His scores of articles for the *Penny Cyclopaedia* (over one hundred, for example, beginning with the letter 'C') covered an enormous range of mathematics and its history.

Soon, however, the Society was to come under attack from the Radicals. At a time of great social unrest, of riotous assemblies, rick-burning and machine-breaking, it was seen as avoiding issues of real importance and taking refuge in the publication of obscure mathematical and scientific treatises (25). When it did move into the field of political economy its advice was patronising:

> When there is too much labour in the market, and wages are too low, do not combine to raise the wages . . . but go out of the market . . . We say to you, get something else; acquire something to fall back upon. When there is a glut of labour go at once out of the market; become yourselves capitalists (26).

The new workers' organisations reacted sharply: unionists 'will read nothing which the Diffusion Society meddles with. They call the members of it Whigs, and the word whig with them, means a treacherous rascal, a bitter, implacable enemy' (27). The workers wanted bread – De Morgan, alas, could only offer superior mathematical cake.

6. A CHALLENGE AND A RESPONSE

In the late 1820s a new type of school came into being in England. Because of the 'expense . . . of the grammar schools and private schools in highest repute, and the indifferent character of many grammar and private schools where pupils are boarded at low rates, combined with the opinion that few of these schools furnish all the kinds of instruction that are now requisite' (28), a number of 'proprietary' schools came to be founded. These were schools funded, organised and managed by a committee – the proprietors. Thus, whereas in most private schools the headmaster was the sole proprietor, in the proprietary schools he was a salaried employee appointed by the managing board. Early proprietary day-schools included Liverpool Institute (1825), London University (later University College)

School (1828), King's College School (1829) and City of London School (1837). Later came the proprietary colleges such as Cheltenham (1841), Marlborough (1843) and Rossall (1844). These schools, as we have indicated, were less expensive than the large public schools. As creations of the middle class, whose income was largely derived from commerce and industry, they sought also more utilitarian ends. Their emphasis was not so much on producing gentlemen, but captains of industry, commerce and the services. The curriculum offered by the schools remained classically based, but 'modern studies' and, in particular, mathematics were given much greater emphasis. Often the schools were organised after the age of fourteen or so into 'classical' and 'modern' divisions.

Such a development was welcomed by De Morgan:

> Why are so many proprietary schools erected? The reason is, that parents, who have neither time to choose nor knowledge to guide them in the choice of a place of instruction for their children, find it easier to found a school, and make it good, than run the doubtful chance of placing their sons where they may learn nothing to any purpose (29).

Yet if mathematics was now to find a more secure place in the curriculum, it was essential that the aims of teaching it should be more clearly spelled out and methods of teaching described which would be likely to attain those objectives. De Morgan attempted to do this in a series of articles he wrote for the SDUK's *Quarterly Journal of Education* (30). The first of these 'On mathematical instruction' appeared in the journal's second number {31} and it is this article which will be used to illustrate De Morgan's views on mathematics education {32}.

First, it is essential to identify the goals of mathematics education in schools. Is it there to serve utilitarian ends?

> There are places in abundance where bookkeeping is the great end of arithmetic, land-surveying and navigation of geometry and trigonometry. In some [institutions], a higher notion is cultivated; and in mechanics, astronomy, etc., is placed the ultimate use of such studies. These are all of the highest utility; and were they the sole end of mathematical learning, this last would well deserve to stand high among the branches of knowledge which have advanced civilization; but were this all, it must descend from the rank it holds in education.

To argue for mathematics on utilitarian grounds is not sufficient. Law, medicine and architecture are also 'of the highest utility to the public, and profitable to those who adopt them as a profession'. Yet these, rightly, do not find a place in the school curriculum. Just as a lawyer commences his specialist education once his general education has been completed 'so would it be with him whose calling requires a knowledge of mathematics, were it not that an important end is gained by their cultivation, which is quite independent of their practical utility, viz. the exercise of the reasoning powers'.

De Morgan, then, like most of his contemporaries saw the principal contribution of mathematics to general education as a vehicle for the enhancement of the faculty of reasoning. Mathematics was, of course, the best agent for such training because of the 'strictly logical nature of the steps by which conclusions are deduced from . . . principles' and the fact that 'the data or assumptions . . . are few, undeniable, and known to the student from the beginning' {33}.

It is a consequence of such an ordering of priorities that

> the actual quantity of mathematics acquired . . . is . . . of little importance, when compared with the manner in which it has been studied, at least as far as the great end, the improvement of the reasoning powers, is concerned . . . We might be tempted to say, let everyone learn much and well; well in order that the habits of mind acquired may be such as to act beneficially on other pursuits; much in order to apply the results to mechanics, astronomy, optics [etc.] which can never be completely understood without them.

Time, however, is short and for that reason our priority must be 'to secure a habit of reasoning in preference to the knowledge of a host of results'. It was the latter which was in De Morgan's view preferred 'in most of our schools'. Emphasis was laid on the amount of mathematics – books of Euclid, or chapters of Bonnycastle's *Algebra* – covered. 'All this, if well learnt, would constitute a useful portion of mathematical knowledge, and would enable an intelligent pupil . . . [when necessary] to proceed in his studies without the aid of a teacher.' However, matter was not 'well learnt': mathematics teaching suffered from too many obvious defects.

The first of these was the emphasis on rote learning. Arithmetic had been broken down into a multitude of rules, many of them so unintelligible that by themselves 'they could be Hebrew'. Thus inverse proportion was described as when 'less requires more, and more requires less'. The pupil was not expected to understand the reasons for the rules, but merely to be able to apply them. If he could not he was flogged, a procedure based on the principle 'of curing a disorder in a part which cannot be got at, by producing one in another which can'. Teachers feared to teach principles, for to do so required knowledge and understanding. It was much easier to teach rules and with the help of the various books of worked solutions avoid any troublesome questions. One algebra book was advertised as saving masters from being 'pestered with questions' {34}. The result of all this was that after several years of working 'unmeaning and useless questions by the slatesfull' the student leaves school 'master of a few methods, provided he knows what rule a question falls under'. Should. however, he be faced with something new or a problem requiring a combination of rules, he is lost 'for not having any principles, the necessity of one step different from that laid down in the rule is a total extinguisher upon the success of his efforts'.

Geometry teaching suffered from an even worse preoccupation with rote learning:

the propositions are . . . said by rote, for the convenience of those who find their memory in a better state than their reason . . . Great attention is paid to the phraseology of the book {35} . . . But the prime feature of the system, though now somewhat obliterated, was the necessity for recollecting the numbers of all the propositions; for it could clearly be of no advantage to know that three angles of a triangle are equal to two right angles, unless it was also known that this is the thirty-second of the first book {36}.

Algebra teaching suffered from all the defects of that of arithmetic, but because of algebra's 'higher rank' had acquired 'some defects peculiarly of its own. If arithmetic were unintelligible, algebra is made to render that obscure which before was easy'. Symbolic work was introduced too early: 'this refined branch of the science should not be the first taught, but . . . the pupil should be led, in the track of invention, to its several parts, so as to arrive at each at the precise moment when he can understand its origin and use. A contrary course will ensure years of travelling in the dark.' All too frequently pupils could not interpret the symbols put before them: thus, for example, the sign for subtraction was used with no further explanation to denote a negative quantity.

It may be said, that pupils raise no objections. This is the most fatal argument that could be adduced. We know well that pupils always receive implicitly what their masters tell them, and why is it that they are led to the study of Mathematics? Precisely that they may learn to raise objections, and how to raise them in the proper place, when false logic and absurd definitions make objections desirable.

Again, the general level of incompetence of teachers was demonstrated by the need for the key to Bonnycastle's *Algebra* to contain detailed explanations of how one might multiply $a - x$ by itself, or solve the equation $4 - 9x = 14 - 11x$.

In answer to a defence sometimes set up, that the [teaching of algebra] is *practical*, we observe that much of what is done has no reference to any practical end whatever. The great body of the algebraical work of a school consists in questions . . . which never occur in practice – above all, in the solution of certain conundrums called problems, producing equations, of the *practical* nature of which the reader shall judge . . . [For example, from Bonnycastle's *Algebra*]

'A person has two horses and a saddle worth £50; now if the saddle be put on the back of the first horse it will make his value double that of the second, but if it be put on the back of the second, it will make his value triple that of the first; what is the value of each horse?'

Now if all this be meant for improvement in theory, no one will deny that the reasons of all the rules should be previously understood; but if they be practical questions, we need only say . . . that the Newmarket gentry have a better way of determining the value of their horses than by involving them, saddles and all, in a simple equation.

This then was the state of mathematics teaching: what might be done to remedy affairs?

First De Morgan stressed the importance of language; of being able to understand that which is taught and of being able to distinguish accurately between different phrases: 'It is useless to present reasoning in any shape until the language used is perfectly familiar. No one can learn new words and comprehend new combinations of words at the same time. Hence a perfect acquaintance with the English sentence is the first thing to be taught.' Older students must learn to distinguish between 'all equilateral triangles are equiangular' and 'all equiangular triangles are equilateral'. They will need to know 'the difference between a fact and a deduction from two or more, and also that good reasoning may be instituted upon data which are imaginary, such as the definition of a point and a line in Euclid' {37}. Without such a preparatory grounding in logic 'the boy has to contend with two difficulties at once, new things and new methods; and education is not what it should be, a search after that which is not known, by the light of that which is'.

More specifically, arithmetic teaching should commence with a clear explanation of methods of numeration 'illustrated by reference to other systems besides the decimal' and supported with the use of counters or similar apparatus:

The object of the master ought to be to make his pupil understand the process before him. The latter ought therefore to be questioned on every part of his work, and encouraged to mention all the difficulties which have occurred to him. Above all, the boy ought never to be suffered to imagine that he is stupid because he does not immediately see what is put before him . . . To discourage a beginner, by making him fancy himself beneath the rest of his species, is the surest way of losing time and trouble . . .

Numerical exactness is of the utmost importance, and will be sooner arrived at by the pupil who understands the principles than by any other {38}.

Geometry, so De Morgan argued, should not be deferred one moment longer than is absolutely necessary. His view of geometry teaching was still circumscribed by Euclid's *Elements* but he wanted to move away from much contemporary practice. The beginner should not be burdened with demonstrations but should be given time to absorb the new vocabulary and could first have visual 'demonstrations' of the leading facts of the first three books of Euclid which would serve 'to excite curiosity, and give an idea of the utility of the science'.

The beginner could not be expected to cope with the refinements of the opening pages of Euclid, 'the multitudes of definitions and axioms'. After all

the order of the propositions is not necessary to correct reasoning . . . It would not be contrary to good logic to assume the whole of the first book of Euclid, and from it to prove the second . . . We therefore recommend the following principles in teaching geometry:

Never to state a definition without giving an ocular demonstration of one or more facts connected with the term employed.

To defer every axiom, until that point is arrived at, where it becomes necessary {39}.

To impress upon the mind of the pupil that the reasoning is not affected by the assumption of a [result] to be proved afterwards, provided the proof of it is independent of the proposition which it was used in proving, and its consequences.

To accustom the beginner to retrace his steps, and going backwards from any proposition, to continue the chain, until he arrives at the point which he set out by assuming . . . {40}

It would much benefit the pupil if solid geometry were introduced . . . early . . . and by a judicious use of the *real* figures, instead of perspective drawings, the subject might be amazingly simplified.

When teaching algebra the prime need is to ensure that symbolism is fully comprehended:

The new symbols of algebra should not be all explained to the student at once. He should be led from the full to the abridged notation in the same manner as those were who first adopted the latter {41}. For example, at this period he should use *aa, aaa* etc. and not a^2, a^3, and should continue to do this until there is no fear of that confusion of $2a$ and a^2, $3a$ *and* a^3 which perpetually occurs {42}.

[In general,] It is not [sufficiently acknowledged] that [schools] cannot expect to make learned men; but they may make good learners, and at the same time produce such a desire for knowledge as shall lead the individual to devote himself to study where it is not a matter of compulsion . . . The great mistake lies in a notion that [the schools] are to teach the greatest possible number of bare facts before the pupil arrives at the age of sixteen; whether he will leave school with the desire of adding one more bit of knowledge to his stock, or with the power to do so if he has the will, does not seem to be considered of any importance.

Nowadays many of the details of De Morgan's writings are no longer relevant – attempts to teach Euclid, for example, have long been abandoned. Yet underlying the details are beliefs and principles that have not aged, and which still bear repeating. When, for example, will the full significance of the last paragraph quoted be accepted?

7. THE SPITALFIELDS MATHEMATICAL SOCIETY

On coming to London from Cambridge, William Frend had been attracted to join the Mathematical Society which had flourished in Spitalfields, London since 1717. This was only one of a number of societies of a similar nature which existed throughout the country for the studious artisan {43}. It is, however, one of the best remembered. When, in 1845, it had shrunk to only nineteen members, De Morgan was asked to superintend the removal of its collection of over 2,000 books to the library of the Royal Astronomical Society. Writing to Herschel, De Morgan told of how 'their library is a good one' and how, far from finding a collection of artisans, he found the society to consist of 'an FRS. [Fellow of the Royal Society], an F.Ant.S. [Fellow of the Antiquarian Society], an F.Linn.S. [Fellow of the Linnaen Society], a barrister, two silk manufacturers, a surgeon, a distiller, etc.; and we found that all had really paid attention to some branch of Science'. As so often has happened in education, what was originally intended for the working-class artisan had become an exclusively middle-class institution.

Originally (as De Morgan tells us in his article on the Society in his *Budget of Paradoxes* (44)), it was a plain society whose members met once a week, each with 'his pipe, his pot and his problem'. The Society had various rules and regulations of which the most important was 'it is the duty of every member, if he be asked any mathematical or philosophical question by another member, to instruct him in the plainest and easiest manner he is able'.

Certainly in its early days the Society attracted the artisan: two of its silk-weaving members are still widely remembered, John Dollond who was to become the greatest maker of optical instruments in his time, and Thomas Simpson, later Professor of Mathematics at Woolwich Academy (see p. 65 above).

In addition to its problem solving and teaching sessions (for a member was required to teach any other member to the extent of his ability or be fined one penny), in its heyday the Society also mounted a series of lectures which almost proved its undoing. In 1798 the Society promoted certain lectures for which it charged a small admission fee and which attracted audiences of two to three hundred. An informer brought an action against the Society for a sum of the order of £5,000 for giving unlicensed lectures to the public on philosophical subjects. The case was successfully defended by the Society but not before some of the members had died through anxiety over the outcome!

The Society continued to mount public lectures during the first quarter of the nineteenth century. These were widely advertised in the press and attracted good audiences, on occasions over three hundred strong (45). However, from 1825 onwards there was a marked decline in the Society's fortunes, possibly due to the establishment of rival institutions such as the London Mechanics' Institute (see p. 99 below). Membership dropped catastrophically – 81 in 1804, 54 in 1839, 30 in 1841 and 19 in 1845. These remaining members were offered life membership of the Royal Astronomical Society in return for the transfer of their library and so, after a life of over a century, the Spitalfields Mathematical Society ceased to exist.

8. UNIVERSITY COLLEGE AND THE LONDON MATHEMATICAL SOCIETY

In October 1836, De Morgan's successor at London University and his family, who had spent the summer vacation in the Channel Islands, set out to cross from Guernsey to Jersey in a small sailing boat. The sea was unusually rough and the boat capsized with the loss of all on board. London University was due to reopen immediately and was without a mathematics professor. De Morgan's offer to fill the gap until December of that year was eagerly accepted and, once there, the authorities made every effort to keep him. Reassurances were given about tenure and for the second time De Morgan was appointed professor at London University.

De Morgan's return coincided with great changes in the institution's standing. As we have seen, London University was essentially the creation of Whigs and nonconformists. In order to counter its influence, a rival institution, King's College, supported by the Tories and the Established Church, opened in the Strand in October 1831. Initially, neither of the rivals was empowered to grant degrees, an unsatisfactory situation. After attempts to persuade the colleges to unite had failed, a royal charter of November 1836 created the University of London as a body with powers to examine candidates and to award degrees. To avoid confusion with this new body, the older college adopted a new name, University College. This move, which placed the examining in different hands from the teaching, was to have a great and far from beneficial effect.

Back at what was now University College, De Morgan's industry continued unabated. He wrote many mathematical papers on probability (he was also at this time consultant actuary to the Family Endowment Assurance), on the foundations of algebra, and on logic (for which he is now best remembered) {46}. He was also able to indulge his passion for book collecting and for writing on the history of mathematics and on early mathematical texts {47}.

He was involved too (with Babbage, Airy, Herschel and others) in an attempt to establish decimal coinage in Britain (48). By 1847 the efforts of those who wished to decimalise money were rewarded by the introduction of the florin piece, or one tenth of a pound. All seemed set fair for a changeover, so much so that textbook writers began to include chapters on the decimal coinage in their books:

It may be desirable to say here a few words upon the subject of *Decimal Coinage* which has been for some time under the consideration of the Government, has been recommended for adoption by a Committee of the House of Commons, and is likely, therefore, before long to be introduced in England {49}.

The expectations proved to be over a century premature. A witty speech by Robert Lowe (of 'payment by results' fame) ensured that decimal coinage was rejected by Parliament and De Morgan's *Reply to the Facetiae of the Member for Kidderminster* (1855) could do nothing to remedy the state of affairs.

De Morgan was more fortunate, however, in his attempts to establish a mathematical analogue to the Astronomical Society. He, himself, was surprisingly not a Fellow of the Royal Society; he had rejected the possibility since he regarded it as being too much of a social club (which in those days was perhaps a just criticism).

He saw a clear need for a national association of mathematicians and for a periodical devoted to mathematical research (50). The initial move in the formation of such a society came, however, not from De Morgan but from two of his pupils, A. C. Ranyard and De Morgan's son, George. These two wrote in October 1864 to a group of mathematicians to request their attendance at the first meeting of the 'University College Mathematical Society' to be held on 7 November and to be addressed by Professor De Morgan. In the event, illness prevented De Morgan's being present and his place was taken by T. A. Hirst [51], at that time mathematics master at University College School. A decision was taken to establish a society, but with the revised title of the London Mathematical Society. The LMS (as it soon came to be known) met for the first time in January 1865 when De Morgan was elected President and delivered his inaugural address in which he set out his ideas on what the aims of the Society should be and also returned to the attack of his old enemy, the Cambridge examination system (52).

The LMS soon became a national rather than a local society and quickly attracted to itself all the leading British mathematicians of the day (with the curious exception of Stokes) {53}. It began publication of the *Proceedings* immediately, the first issue carrying De Morgan's inaugural address.

By the end of the first year the Society had sixty-nine members; a year later, one hundred. Not all of these early members were professional mathematicians – amateur mathematicians recruited from the ranks of the lawyers and the clergy were still much in evidence as in the days of the old Spitalfields Mathematical Society. Mathematical research papers were still accessible to the interested amateur {54}.

The successful establishment of the LMS (his part in which is still acknowledged by the award every three years of the Society's De Morgan Medal {55}) was the climax of De Morgan's career. From 1866 onwards his career turned into tragedy. For the second time he found himself at odds with the Council of University College. On this occasion the distinguished scholar, the Reverend James Martineau, was rejected by Council as Professor of Mental Philosophy and Logic, in spite of the fact that his appointment had been recommended to them by Senate. That the Council of University College should reject a man because of his religious beliefs was too much for De Morgan to accept. He again felt impelled to resign his chair: 'the College had left me, not I it'. Thirty-eight years after first entering the College, De Morgan finally left.

Yet another blow was to fall the following year when, in October 1867, George, his son, died whilst convalescing in Ventnor. De Morgan himself was too ill to travel from London to visit his son. The following year De Morgan's health broke down completely. He was now largely confined to his home where he still endeavoured to work on his posthumously published *Budget of Paradoxes*. In August 1870 his daughter, Christiana, died. By now De Morgan's decline in health was almost total and he himself died early in 1871. His widow was later to write that 'The state of mind in which he had lived and in which he died, is shown by a sentence in his will:

> I commend my future with hope and confidence to Almighty God; to God the Father of our Lord Jesus Christ, whom I believe in my heart to be the Son of God, but whom I have not confessed with my lips, because in my time such confession has always been the way up in the world.'

6 · *Thomas Tate*

We have seen how in the first half of the nineteenth century there were rapid and considerable developments in the teaching of mathematics at Cambridge University and within what we should nowadays term the secondary sector of education. There was also considerable mathematical activity outside schools, colleges and other such educational institutions. In a wider educational context, the period saw the State's first involvement in elementary education and teacher-training. The need for a national educational system was gradually being realised, as was the desirability of having teachers who had not only achieved mastery of academic material but had also received instruction in the art of teaching. Thomas Tate's interest in mathematics was aroused in an unconventional way and he himself was to carry on the 'popular tradition' in mathematics education before being appointed as effectively the country's first mathematics teacher-trainer, and becoming renowned as an educator and textbook author.

1. THOMAS TATE AND THE POPULAR TRADITION

Thomas Tate was born in Alnwick, Northumberland, in February 1807, the second son of a builder. Both he and his elder brother, George, attended the Borough School in Alnwick, but only George appears to have become a pupil at the Dukes (Grammar) School (1). It may well be that the elementary school catered more for Thomas' scientific tastes, since its master, James Ferguson, was to show his interest in, and ability for, mathematics by producing a new edition of Hutton's *Arithmetic* (2). Probably Thomas, too, was intended to become a builder, for we are told that he studied in Edinburgh for a short time with an architect {3}. George, although a gifted topographer and naturalist (4), remained in Alnwick throughout his life, being first a linen-draper, then postmaster. He (5) ascribed Thomas' interest in mathematics and science not to Ferguson's influence but to Thomas' becoming one of the first members of the Alnwick Scientific and Mechanical Institution which opened in 1825 (6). In later years the two brothers contributed much to this institution's work; George was its Secretary for thirty years and Thomas frequently returned to Alnwick to lecture {7}. It was by giving evening classes in Alnwick that Thomas obtained his first experience of teaching (8). Exactly when this happened is not certain, for little is known of Tate's early years.

What we do know is that he went on to lecture on mathematics and chemistry at Newcastle before moving to York {9} where he was appointed as lecturer in chemistry at the York Medical School which opened in October 1834 (10). This was the ninth medical school to be established in the English provinces within eleven years and the fourth in Yorkshire. Not surprisingly, therefore, it attracted only a dozen or so students and Tate, whose salary depended upon the number enrolled for his chemistry course, had to find means to supplement his income. So we find in the *Yorkshire Gazette* of 15 October 1836 an advertisement:

Mr Tate, lecturer on chemistry in the York School of Medicine, respectfully informs the inhabitants of York and its vicinity, that he will give private lessons on the various branches of mathematics, including the elements of arithmetic, Euclid, algebra, analytical geometry, the differential and integral calculus – together with their application to physical problems, nautical astronomy, engineering, perspective, and the use of the globes. Terms, 2s per lesson {11}.

In offering private lessons Tate was, of course, following in a long tradition of mathematics tutors; we recall how Hutton first began his career in Newcastle. Other similarities between the two self-taught men existed, for both actively encouraged the pursuit of mathematics in the home; Hutton by editing *The Ladies' Diary* and Tate the weekly 'mathematical column' in the *York Courant* (12).

The mathematical column of the *York Courant*, which first appeared in October 1828, aimed to disseminate knowledge of, and foster an interest in, that incomparable science 'which not only perfects us in all other sciences but . . . teaches us to perform almost every act of manual labour in the most expeditious and advantageous manner'. The reader was urged to participate in mathematical activity:

Go on, then, . . . give your reasoning facilities full scope; and do not be deterred by the sluggish notions of those who content themselves to skim on the surface, and who are not able to appreciate [mathematics'] beauties; dip deep; your persuit is honorable – it is necessary – it is useful [*sic*].

Like other mathematical columns of its time (cf. p. 71 above), the *York Courant* printed problems submitted by readers followed, some weeks later, by the best solutions to have been received. Although intended as 'a source of much rational entertainment' the questions were not mere 'brainteasers', but ranged over much of mathematics including, for example, conic sections and the differential and integral calculus; sometimes, for example, Tripos questions were reprinted. Further indication of the standard can be deduced from the fact that problem solvers made reference to such books as Barlow's *Theory of Numbers,* Bridge's *Conics* and *Mechanics*, Euclid's *Elements*, Hutton's *Mensuration* and *Course of Mathematics*, Newton's *Principia* and Wood's *Algebra*. Early contributors were almost entirely non-graduates and included schoolboys, schoolmasters, surveyors, a doctor, a parson and an apothecary. Originally, most came from Yorkshire, but soon the column attracted correspondents from much further afield. In October 1834 a special 'Juvenile' section was begun {13}.

The column did far more than amuse the mathematical dilettante; it offered the self-taught a means of self-appraisal, a stimulant to further study and an introduction to a circle of similar students of mathematics. Thus, for example,

One of the present professors at the Royal Military College at Sandhurst . . . when a young man . . . attended a fair . . ., and seeing a copy of Euclid's *Elements* laid open on a second-hand bookstall, he was so interested with the appearance of the diagrams that, although unable to read, he purchased the book and soon became an eminent mathematician. The *York Courant* and the *Ladies' Diary*, in both of which his name frequently occurs, afforded him a ready means of testing the progress of his studies (14).

Yet providing a test of skill for would-be professionals ran counter to the general aim of a column wishing to 'blend utility with entertainment' {15}. In 1846 it was announced that

it having been intimated to us from several quarters, that a newspaper is not the proper vehicle for MATHEMATICAL QUESTIONS, and that its columns ought to be occupied with intelligence more generally interesting, we have, in deference to the opinions of the great majority of our readers, resolved to exclude the Mathematical department at the end of the month.

It would be easy to draw many conclusions – most of them unjustified – from the life and death of this mathematical column. Certainly, its like does not exist today. One could argue that the contributors represented a mathematics-loving group divorced from the academics to be found in universities. The claim would possibly be justified. Yet significantly it did not result in the mathematics following any distinctive or truly practical bent. Indeed, as the years wore on the column fell a victim to academicalism. What began as a column for the intelligent, well-educated man, ended as one for the enthusiast who was willing to struggle with ever more involved mathematics to satisfy his competitive instincts. Nevertheless, an investigation of the column tells us much about the mathematics which interested, attracted and amused the early-nineteenth-century 'amateur' of mathematics.

One might well have expected Tate, after his Alnwick experiences, to have become a leading figure in the York Mechanics' Institute, but this does not appear to have been the case. Unfortunately, the story of the York Institute in the 1830s admirably illustrates the malaise which was so soon to strike the Mechanics' Institute movement. After the opening of the first institute in London in 1824 (see p. 257 below), institutions sprang up throughout England with the principal aim of offering 'the working class facilities of acquiring a knowledge of the principles of science involved in their respective occupations. . . . By means of classes, in which the elements of mechanics, mathematics, and chemistry are taught . . . by public lectures, and by the use of a well selected library' (16). This aim was opposed by many, including a number of Church of England parsons, who felt that a knowledgeable working class was not necessarily desirable. Often, then, it was the dissenters who led the way in forming Mechanics' Institutes and it was a Unitarian minister who in 1827 was responsible for initiating the York Mechanics' Institute (17). By 1834 the Institute was afflicted with apathy. Science for artisans was abandoned in favour of popular lectures on literary and entertaining subjects for the lower middle classes, and in 1838 the Institute was reconstituted as the York Institute of Popular Science and Literature. A similar tendency was noted elsewhere: at Liverpool 'the working mechanics, who generally prefer the sterner studies, have given place to others who attend the dancing and the essay and discussion classes, as more congenial to their tastes' (18). In some places, however, the original aims were retained and, for example, a student at the Huddersfield Mechanics' Institution could write (about 1850?):

It is now about three years since I first became a member of the Institution; I could not then sign my own name, and I had never set a figure in my life. During these three years I have gone through two Arithmetics, Nesbit's *Mensuration*, the first six books of Euclid, and as far as Logarithms in Bonnycastle's *Algebra* (19).

In general, though, there were too few working-class pupils who were willing to fight to overcome their lack of elementary knowledge and to enter a largely unknown world inhabited mainly by people who were all too clearly of another class. The hopes of the founders of the Mechanics' Institutes went largely unfulfilled.

Tate did give some evening classes in York; to boys aged between twelve and fifteen. These lessons on the science of common things admirably illustrate his belief (see section 4) that ideally elementary education begins with the environment and is directed towards utilitarian ends:

The subjects, selected for instruction, were not only useful in themselves, having a relation to the occupations of life, but also so simple as to be within the comprehension of . . . young pupils. Recondite facts of science . . . were usually avoided when they did not admit of graphic or experimental illustrations.

The following is a list of the subjects: . . .

What is the best kind of gravel for making a path? The properties of the lever. . . . How to sink a well. How to make a pump. How to economise labour. To explain the use and construction of a wheelbarrow. . . .

The boys were so interested in the lessons, that they would . . . leave their games to attend the class. Many of the parents did not, at first, quite understand what their children had to do with science; but when they found the teacher had been explaining how to make a fire, how to prevent the chimney from smoking, etc., they became as much interested in the lessons as their children – and thus the parents speedily became powerful auxiliaries in carrying on the work of education (20).

Tate believed strongly that increased mathematical and scientific knowledge would produce more educated and efficient workers who, as a result, would become better paid and enjoy better living conditions. In an attempt to achieve his goals he used a variety of methods and demonstrated once more that much 'education' takes place outside schools and colleges {21}. However, it was to the improvement of science and mathematics education in elementary schools that in 1840 he was asked to turn his attention.

2. ELEMENTARY EDUCATION AT THE COMMENCEMENT OF TATE'S CAREER

> To every class we have a school assign'd,
> Rules for all ranks and food for every mind.
>
> (G. Crabbe, *The Borough*, 1810)

Crabbe's verses aptly describe the provision of education in England in 1810 and for many decades to follow. Yet such education as existed came from a variety of independent sources; for in England, unlike Prussia and other German states, France, Switzerland, The Netherlands and Denmark, the State accepted no responsibility for public education. Indeed, there was considerable opposition to the concept of state education. Some thought it could only lead to the destruction of civil liberty {22} whilst those semi-public agencies which provided education feared that state education would lead to a weakening of their power and authority.

In fact, considering the absence of any form of compulsion, a remarkably high proportion of children attended a school of some kind, for some period or other {23}, but many schools were small {24} with teachers of doubtful quality. Crabbe's poem is known today because of the opera it inspired, Britten's *Peter Grimes*. Yet Ellen Orford, the schoolmistress, is in the latter work presented as a character at some remove from Crabbe's careworn, unmarried mother who was to become an unhappy wife and widow before

> The parish-aid withdrawn, I looked around,
> And in my school, a blessed subsistence found.

Teaching was all too often the last resort of the destitute or the lame:

[In the 1830s] anybody was considered good enough for a schoolmaster. If a tradesman failed in business, he was thought to be learned enough for a schoolmaster; a feeble sickly youth who was not considered strong enough to practise any regular trade, was thought to be sufficiently qualified . . .; if a mechanic happened to get a limb fractured he would . . . save himself from starvation by opening a school {25}.

Yet although the private schools were so frequently a disgrace, there were a number of alternatives developing.

The Sunday School movement started in 1780 and spread quickly throughout the country. By the 1830s it was estimated that over one million adults and children attended such schools every Sunday. Their aims, however, were primarily religious. Some taught writing and a few the rudiments of arithmetic, but most schools felt the latter topic ill-fitted to their aims and/or sabbath activity. One could not look to them for a mathematical education.

The National Schools, offspring of 'The National Society for promoting the Education of the Poor in the Principles of the Established Church throughout England and Wales' established in 1811, again had principally religious aims, although their curriculum included arithmetic and, as we shall see, occasionally other mathematics. These schools were, however, clearly identified as agents of the Church of England and soon other denominations were to open rival institutions. The most important of the other societies was the 'British and Foreign School Society'. In its schools 'general Christian principles' were taught, but in an undenominational manner. On its establishment in 1814 this society had taken over the affairs of the Royal Lancasterian Association named after Joseph Lancaster, who, with Andrew Bell, had been responsible for spreading the use of the monitorial system throughout Britain and even into Europe. This teaching device, the use of a small number of able children to act as intermediaries between the schoolmaster and the mass of pupils, was one which circumvented the shortage of trained teachers and thus had a ready appeal. That it condemned the majority to the rote-learning of half-digested gobbets of information was for a time ignored. Certainly, it militated against serious or successful teaching of mathematics. Indeed, as J. A. Roebuck, an educational pioneer, pointed out in 1833, elementary education 'usually [signified] merely learning to read and write' and that occasionally 'by a stretch of liberality, it is made to include arithmetic' (26). Roebuck argued against the belief 'that a rough sort of knowledge of reading and writing is sufficient for the poor man' and proposed a motion in Parliament calling for 'the universal and national education of the whole people'. The motion was rejected. Compulsory education, teacher-training colleges, certificated teachers, local education authorities and a 'Minister of Public Instruction' were all eventually to come as Roebuck had urged. However, at the time there was only one small, but nevertheless significant, victory to record; the House voted £20,000 to further the work of the 'National' and 'British' societies.

Often, though, those who argued for elementary education had narrowly conceived ends in mind. Bell's views were representative:

It is not proposed that the children of the poor be educated in an expensive manner, or all of them be taught to write and to cipher. Utopian schemes, for the universal diffusion of general knowledge, would soon . . . confound that distinction of ranks and classes of society on which the general welfare hinges . . . there is a risque of elevating by an indiscriminate education the minds of those doomed to the drudgery of daily labour, above their condition, and thereby rendering them discontented and unhappy in their lot (27).

As the greater stress which the dissenters placed on utility would lead one to expect, the Lancasterian schools tended to place more emphasis than the National schools on the acquisition of arithmetical skills. Thus Lancaster proposed ten 'grades' for arithmetic (28).

Class I	learnt to cypher and to combine figures [e.g. tens, hundreds],
Classes II–V	learnt to add, subtract, multiply and divide simple numbers respectively,
Classes VI–IX	learnt the same rules for compound numbers [fractions],
Class X	learnt reduction [pounds to farthings, tons to ounces, etc.], practice [if one pint costs . . . how much does 23 gallons cost?] and the rule of three [proportion].

Not, it must be said, a very enlightened curriculum.

Yet in some schools, depending on the interests and abilities of the schoolmaster and the local parson, more was attempted. The special needs of infants, too, were becoming recognised and in 1824 the London Infant School Society was formed. One leader in infant education was Samuel Wilderspin, whose methods, influenced by Pestalozzi, were described in a series of publications. He, for example, advocated the use of the abacus when teaching number and also the early introduction of geometry. His teaching methods are revealed by the description of a typical lesson:

A large board with geometrical figures is placed before the gallery. The master points to a straight line. 'Q. What is this? A. A straight line. Q. What does parallel mean? . . . Q. If any of you children were reading a book that gave an account of some town which had twelve streets, and it said that the streets were parallel, what would you understand?' etc. etc. Instead of diagrams a jointed strip of metal – the gonograph – was sometimes used. In these lessons the children often answered simultaneously, and though to begin with only the older ones might be able to enter into the work, yet as the same lesson recurred it was argued that the younger children would gradually pick up the necessary ideas (29).

The approach bears some similarities to Recorde's, but now the occasional more open question is included. Yet, as is clearly illustrated, the principle of teaching was constant repetition and with it, one presumes, the adoption of set responses to be declaimed by the class in unison.

Some educators, however, were soon to pursue new paths. An outstanding example was Richard Dawes, vicar of King's Somborne, Hampshire, whose 'National' school achieved widespread renown. The school in 1847 had 110 pupils, a master, a second master (aged eighteen), a mistress (who taught the infants) and four apprentice or pupil-teachers {30}. 'In all the classes . . . the children appear to have, according to their standing a good knowledge of arithmetic; they are, moreover taught English grammar, geography and history' {31}. Mathematics, too, meant more than arithmetic. Algebra was taught to twenty-one boys (including two of the pupil-teachers) and Euclid to eleven:

I have rarely, indeed, heard boys answer so well in Euclid as some of them did. . . . The great interest of Mr Dawes' experiment in mathematics teaching appears to lie in his having established the possibility of teaching Euclid with success in an elementary school; and of giving to farmers . . . and labourers . . . the advantage of that incomparable discipline of the mind which results from the habits of geometrical reasoning. . . . Having examined the boys as to their knowledge of some of [the] fundamental principles [of mensuration], I can bear witness to the fact that they are 'not taught these things in a parrot-like way, but led to understanding them as a matter of reasoning'. I have found algebra and geometry introduced in only one other National school, and the teaching of it was there attended with the like success, contributing greatly to the interest of the children {32}.

Gradually, then, the notion of what was desirable within elementary education and, indeed, what was possible came to be extended. It reveals again the influence and freedom the individual school has had throughout most of the history of English education, a freedom, however, which was soon to be challenged. The quotations, above, also serve to introduce us to a member of that newly established body, Her Majesty's Inspectorate. Although Roebuck's proposals had been defeated, they were not forgotten and in 1837 a Select Committee of Parliament was appointed to consider how 'useful education' might best be provided for 'the children of the poorer classes'. The Committee reported the following year; it favoured the increased provision of schooling, but proposed no state intervention other than an increase in the money voted to the two societies. The Government, however, felt that other measures were needed, and in April 1839 established the 'Committee of the Privy Council on Education' (or, more briefly, the 'Committee of Council') 'to superintend the Application of any sums voted . . . for the purpose of promoting public education'. For over sixty years this Committee was to function as an embryonic ministry of education.

Within a few days the Committee published its first proposals: that an inspectorate should be established, that state aid should be granted to societies other than the National and the British, and that a State Training College should be established. This last proposal was violently attacked, but the other two were accepted. National inspection, and with it a primitive form of accountability, commenced.

Attempts were made then to widen the base of state-assisted education and to improve its quality. There was, however, no move towards compulsory education. Yet not only were one-third or so of children receiving no formal education whatever, but of those who did attend school, few did so for long. In the 1840s the average length of school attendance was between one and two years {33}. Much remained to be done before England could claim to have even a barely acceptable elementary-school system.

3. TATE AND TEACHER EDUCATION

Parliament's attention had been drawn in February 1839 to 'the insufficient number of qualified schoolmasters, the imperfect method of teaching which prevails in, perhaps, the greatest number of schools' and the absence of inspection, and so it was not surprising that on its establishment the Committee of Council should immediately turn its mind to the problems of teacher education. The proposal made, for the institution of a 'normal' school (named after the French Ecole Primaire Normale) in which those who intended to teach in 'schools for the poorer classes may acquire the knowledge necessary to the exercise of their future profession' was, however, to provoke an angry reaction (34). The suggestions it contained relating to religion pleased no one, and for one transitory moment the members of the various religious sects were united. An attempt in Parliament to dissolve the Committee was defeated by only five votes in over five hundred, and the proposal for a state normal school was hastily abandoned.

In fact, some rudimentary forms of teacher-training were already in existence. Lancaster had opened one 'training' institution for teachers in connection with his school at Borough Road (London) and another in Somerset for country teachers. Other schools specifically trained teachers to use Bell's system. Soon the National and British societies designated specific institutions as training schools. As a rule, though, such schools trained only two or three teachers a year. At Borough Road:

> The students are required to rise every morning at five o'clock and spend an hour before seven in private study. . . . At seven they are assembled together in a Bible class . . .; from nine to twelve they are employed as monitors in the school . . .; from two to five they are employed in a similar way: and from five to seven they are engaged under a master who instructs them in arithmetic and the elements of geometry, geography, and the globes, or in other branches in which they may be deficient. The remainder of the evening is generally occupied in preparing exercises for the subsequent day. One object

is to keep them incessantly employed from five in the morning until nine or ten at night (35).

No wonder that 'in some instances . . . their health has suffered'!

We note that the Borough Road students were given practice in teaching and also further academic education, but that there was no real attempt to teach them 'how to teach' or what we should now term 'educational studies'. In this respect, David Stow's Glasgow Normal Seminary, which opened in 1836, was innovative; for Stow sought not only to further his students' academic education, but also to acquaint them with principles upon which methods of instruction could be based and to give them practice in putting these principles into execution. Kay-Shuttleworth, the first Secretary of the Committee of Council, was an admirer of Stow's work and its influence can be detected in the Committee's initial proposal. Once that proposal was rejected, Kay (as he was then called) decided to take matters into his own hands and with a friend, Carleton Tufnell, set up a training school at Battersea.

The Battersea course was planned to last three years and included general education in addition to vocational instruction and training. The students, who practised their craft at the adjacent 'village school', were initially drawn from those intending to teach pauper children, but within some three years the more general aim of preparing elementary school-teachers was adopted and in 1843 the management of the school was transferred to the National Society. By that time a number of other diocesan training schools had been created, including those at Chester, Exeter, Oxford and Salisbury. In Chelsea, St Mark's College was opened in 1841 and Whitelands Training College (for women) early in 1842. The Congregational Board of Education (1843) which, because of its opposition to state interference in education, did not accept the offer of state financial assistance, opened Homerton Training College in 1850, the first 'mixed' college. By then there were sixteen colleges and ten years later thirty-four (36).

It was, however, to Battersea that Thomas Tate was appointed; he joined the staff on 22 March 1840, a few weeks after the arrival of the first students.

Battersea's goals were to be set out and elaborated upon on several occasions by Kay-Shuttleworth (37). In 1843 he summarised his objectives as being:

1. To give an example of Normal Education for School-masters comprising the formation of character, the development of the intelligence, appropriate technical instruction, and the acquisition of method and practical skill in conducting an Elementary School.
2. To illustrate the truth that, without violating the rights of conscience, masters

trained in a spirit of Christian charity, and instructed in the discipline and doctrines of the Church, might be employed in the mixed schools necessarily connected with public establishments, and in which children of persons of all shades of religious opinion are assembled (p. 426).

Demonstrating that those trained by the Church of England could provide suitable undenominational education was, of course, crucial to the acceptance and establishment of a state system of education. Kay was also aware of the need not merely 'to send forth simply clever teachers'. In his opinion those German institutions he had seen concentrated too much on 'instruction as distinct from education'. It was the Swiss Normal Schools he determined to emulate; for them, 'formation of character' was 'the great aim of education'.

The way he set about this makes fascinating reading: the students were not to be pampered and were to be helped to appreciate the 'peasant's lot'. Thus, the only domestic help employed in the college was a cook; all other household duties were to be carried out by the students. They too were responsible for looking after the school's livestock, and for turning a 'wilderness' into a fruit and vegetable garden. (Having gained physical strength in the garden they graduated to gymnastics and 'walking excursions in the country'.) Dress had to be plain and 'should not degenerate into foppery'. Students were thus 'prepared for a life of laborious exertion'.

When extensions to the college were required, then a do-it-yourself exercise commenced to which Tate contributed 'his mechanical knowledge'. The tutors were expected to live in and to submerge themselves entirely in the work of the college. There was no 'high table'; their deliberately frugal meals were taken 'in the midst of their scholars'. The timetable originally devised for the students is shown in Figure 2 and illustrates their spartan schedule. It refers to the period before pedagogical work and practice teaching began, for, as at the Zurich Normal Seminary, general academic education preceded pedagogical training and practice.

Kay's report of January 1841 contains a full description of the academic curriculum. In it we find accounts of the way in which Tate tackled his various tasks, and also a hint as to why he had been appointed to the post. The debt to Pestalozzi is clearly indicated.

In . . . *arithmetic* it has been deemed desirable to put [the students] in possession of the pre-eminently synthetical method of Pestalozzi . . . The use of such a method dispels the gloom which might attend the most expert use of the common rules of arithmetic, and which commonly afford the pupil little light to guide his steps off the beaten path illuminated by the rule . . .

The opposite practice of dogmatic teaching is so ruinous . . . to the intellectual habits, and so imperfect a means of developing the intelligence, that it ought, we think,

Figure 2. *The daily routine at Battersea Training School*
(from Kay-Shuttleworth, J. P. *Four Periods of Public Education*, Longman, 1862, pp. 362–3)

Half-past 5	Rise, wash, dress, and make beds						
Quarter to 6	Household work, viz., scouring and sweeping floors, cleaning grates, shoes, knives, &c., pumping water and preparing vegetables						
Quarter to 7	March into garden and commence garden-work, feed pigs, poultry, and milk cows						
Quarter to 8	March from garden, deposit tools, and wash						
8	Reading of Scriptures and prayer. (In spring half an hour was commonly occupied in a familiar exposition of the passage of Scripture read.)						
After prayer	Superintendents present reports.						
Half-past 8	Breakfast						
		Monday	Tuesday	Wednesday	Thursday	Friday	Saturday
9 to half-past 9	Classes united	Reading in the Bible and religious instruction. Old Testament history	Reading in the Bible and religious instruction, The Gospels	Reading in the Bible and religious instruction. The Acts of the Apostles	Reading in the Bible and religious instruction. The Epistles	Committing to memory texts of Scripture	Committing to memory texts of Scripture, or examination on the Scriptural reading of the week
Half-past 9 to half-past 10	First class	Mechanics	Arithmetic	Mechanics	Arithmetic	Mechanics	Weekly examination
	Second class	Arithmetic	Mechanics	Arithmetic	Mechanics	Arithmetic	
Half-past 10 to 11	First class	Mental arithmetic	Etymology	Mental arithmetic	Etymology	Mental arithmetic	Ditto
	Second class	Etymology	Mental arithmetic	Etymology	Mental arithmetic	Etymology	
11 to 12	Classes united	Geography	Geography	Music	Geography	Geography	Music
12 to 1	Garden work, feeding the animals, &c., &c. March to the house at 1 wash, and prepare for dinner						

Quarter-past 1	Dinner						
		Monday	Tuesday	Wednesday	Thursday	Friday	Saturday
2 to 3	Classes united	Mechanical drawing	Map drawing	Mechanical drawing	Common and isometrical perspective	Map drawing	Weekly examination
3 to 4	First class	Algebra	Use of the globes	Mensuration	Use of the globes	Algebra	Ditto
	Second class	Grammar	. . .	Algebra	. . .	Grammar	
4 to 5	First class	Natural history of birds	Grammar	Object lesson	Grammar	. . .	Ditto
	Second class	Ditto	Committing to memory arithmetical tables and rules of grammar, or mechanical formulae	. . .	Committing to memory arithmetical tables and rules of grammar, or mechanical formulae	Committing to memory	

5	March to garden-work, feed pigs, poultry, &c., and milk cows
6	March from garden, wash, and prepare for supper
Quarter-past 6	Supper
7	Drill and gymnastic exercises
8	Copying music or notes on geography, or mechanical formulae, in the upper class-room. During this period the History of England is read aloud. Another class practising singing in the lower class-room
9	Reading of Scriptures and prayer
20 minutes past 9	Retire to rest

Sunday

After divine service one of the sermons of the day is written from memory. In the evening the compositions are read and commented upon, and the Catechism or some other portion of the formularies of the Church is repeated, with texts of Scripture illustrating it.

Some of the elder students teach in the village Sunday-school at all expense of time, to be avoided. With this conviction, the method of Pestalozzi has been diligently pursued.

Whilst these lessons have been in progress, the common rules of arithmetic have been examined by the light of this method. Their theory has been explained, and by constant practice the pupils have been led to acquire expertness in them, as well as to pursue the common principles on which they rest, and to ascertain the practical range within which each rule ought to be employed. The ordinary lessons on mental arithmetic have taken their place in the course of instruction separately from the peculiar rules which belong to Pestalozzi's series.

These lessons also prepared the pupils for proceeding at an early period in a similar manner with the elements of algebra, and with practical lessons in mensuration and land surveying.

These last subjects were considered of peculiar importance, as comprising one of the most useful industrial developments of a knowledge of the laws of number. Unless, in elementary schools, the instruction proceed beyond the knowledge of abstract rules, to their actual application to the practical necessities of life, the scholar will have little interest in his studies, because he will not perceive their importance, and, moreover, when he leaves the school, they will be of little use, because he has not learned to apply his knowledge to any purpose. On this account boys, who have been educated in common elementary schools, are frequently found, in a few years after they have left, to have forgotten the greater part even of the slender amount of knowledge they had acquired.

The use of arithmetic to the carpenter, the builder, the labourer, and artisan, ought to be developed by teaching mensuration and land surveying in elementary schools. If the scholars do not remain long enough to attain so high a range, the same principle should be applied to every step of their progress. The practical application of the simplest rules should be shown by familiar examples. As soon as the child can count, he should be made to count objects, such as money, the figures on the face of a clock, &c. When he can add, he should have before him shop-bills, accounts of the expenditure of earnings, accounts of wages. In every arithmetical rule similar useful exercises are a part of the art of a teacher, whose sincere desire is to fit his pupil for the application of his knowledge to the duties of life, the preparation for which should be always suggested to the pupil's mind as a powerful incentive to action . . .

The recently rapid development of the industry and commerce of this country by machinery creates a want for well-instructed mechanics, which in the present state of education it will be difficult adequately to supply. The steam-engines which drain our coal-fields and mineral veins and beds, which whirl along every railroad, which toil on the surface of every river, and issue from every estuary, are committed to the charge of men of some practical skill, but of mean education. The mental resources of the classes who are practically intrusted with the guidance of this great development of national power should not be left uncultivated. This new force has grown rapidly, in consequence of the genius of the people, and the natural resources of this island, and in spite of their ignorance. But our supremacy at sea, and our manufacturing and commercial prosperity (inseparable elements) depend on the successful progress of those arts by which our present position has been attained.

On this account we have deemed inseparable from the education of a schoolmaster a knowledge of the *elements of mechanics* and of the laws of heat, sufficient to enable him to explain the structure of the various kinds of steam-engines in use in this country. This instruction has proved one of the chief features even of the preparatory course, as we feared that some of the young men might leave the establishment as soon as they had obtained the certificates of candidates, and we were unwilling that they should go forth without some knowledge at least of one of the chief elements of our national prosperity, or altogether without power to make the working man acquainted with the great agent, which has had more influence on the destiny of the working classes than any other single fact in our history, and which is probably destined to work still greater changes.

Knowledge and national prosperity are here in strict alliance. Not only do the arts of peace – the success of our trade – our power to compete with foreign rivals – our safety on our railways and in our steam-ships – depend on the spread of this knowledge, but the future defence of this country from foreign aggression can only result from our being superior to every nation in those arts. The schoolmaster is an agent despised at present, but whose importance for the attainment of this end will, by the results of a few years, be placed in bold relief before the public.

The tutor to whom the duty of communicating to the pupils a knowledge of the laws of motion, of the mechanical powers and contrivances, and of the laws of heat, was committed, was selected because he was a self-educated man, and was willing to avail himself of the more popular methods of demonstration, and to postpone the application of his valuable and extensive mathematical acquirements. By his assistance, the pupils and students have been led through a series of demonstrations of mechanical combinations, until they were prepared to consider the several parts of the steam-engine, first separately, and in their successive developments and applications, and they are at present acquainted with the more complex combinations in the steam-engines now in use, and with the principles involved in their construction and action (pp. 341–5).

Besides throwing light on the mathematical syllabus, this extract also, of course, demonstrates the way in which an educated populace was becoming regarded as essential to further industrial, economic and imperial advance.

This, however, is but one of a number of writings which combine to present a most detailed account of Tate's work at Battersea. The annual *Minutes of the Committee of Council* contain detailed reports of visits by the inspectorate, copies of examination papers set by them {38} and even detailed descriptions of individual students, their achievements and failures. When reading them one is left in no doubt of the high regard in which Tate, his books and his students' attainments were held for:

in addition to great mathematical attainments, Mr Tate possesses a rare facility in communicating his stores of practical knowledge, . . . a mere glance at the questions proposed by Mr Tate to his pupils . . . must convince those acquainted with our mining and manufacturing districts, that the practical principles embraced in his teaching possess the highest commercial value . . .; whilst to those who regard simply the intellectual training of the labourer, it must be obvious that he will be more readily led to exercise thought and judgement upon the matters before him, by finding the principles of knowledge acquired at school directly applicable to the employments of his working day life (39).

Battersea was soon to become noted among training institutions for 'the care which it bestows in forming its students *as teachers*' (40). In the same year (1846), St Mark's, Chelsea, was criticised on the grounds that there was little to distinguish it 'from the schools of the upper and middle class or as a place for the education of teachers rather than any other class of persons' (41). Latin and Greek rather than poultry farming figured in its curriculum {42}. Providing a suitable balance (adjusted to changing circumstances) between academic and vocational training has continued to be a major problem for teacher-training institutions {43}.

Certainly, if one compares the examination papers set in the new training institutions with those set fifty years earlier at Cambridge University (see, for example, pp. 212 and 217 below), one can only wonder at the ambitious mathematics and mechanics teaching of Tate and his fellow tutors. Moreover, the mathematical standards of the students at Battersea were probably higher than those of their contemporaries at Oxford University. In 1832, Baden Powell [44], Savilian Professor of Geometry, complained that not more than two or three Oxford degree candidates could add vulgar fractions, tell the cause of day or night, or the principle of the pump (45). Even in 1850, Responsions, an examination held between the third and seventh terms of the undergraduate's residence, demanded only Euclid, Books 1 and 2, or algebra to simple equations; whilst Moderations, an intermediate examination taken between Responsions and the final degree examination, asked for the first three books of Euclid and 'the first part' of algebra or, as an alternative to mathematics, logic. 'Pure mathematics' was only for those few (cf. p. 145 below) who read for mathematical honours.

Yet the training-school curriculum proved too ambitious. As has been demonstrated on many occasions since, what proved possible to highly-motivated innovators such as Tate could not be emulated by less-gifted colleagues. Criticism of academic work superficially treated began to mount.

> Let us see that the trained master possesses the knowledge which he will be called upon to communicate; and more, let us lay in his mind a sound scientific foundation for every part of his knowledge to rest steadily upon; so that the structure may have connection, unity and completeness, as far as it extends. If we send forth the teacher . . . with, in most cases, only lowly attainment, let us provide that he has acquired such a readiness on all that concerns the art of teaching as will render his knowledge at once available (46).

From 1856 the syllabuses were restricted so as to emphasise the study of those subjects taught in elementary schools. However, arithmetic, geometry and mechanics were taught to all in the first year; algebra being twinned as an option with Latin. In the second year, higher mathematics was taught, and this was a specialist option in the third year.

A general survey of the Church training colleges carried out in that year by the Reverend Frederick Temple [47] pointed out particular strengths and weaknesses of the system as it existed. He drew attention to the insufficiently well recognised point that the schoolmaster's business was 'not so much to teach as to make the children learn'. Yet too many college tutors were not themselves good teachers, and the lectures the students received on teaching contained too much abstract, meaningless verbiage which did not prepare them to deal with classroom problems, in particular 'Mental Science [psychology] is in general too abstract, too removed from all practical applications to be of much real value to a normal master'. Nevertheless, it was in Temple's view essential that the demands made on students should not be relaxed: for an ill-informed, ill-educated person never made a good teacher. In a very short time, then, the training colleges were rushed from infancy to adulthood, and faced with what continue to be difficult problems: the recruitment of suitable staff, balancing the claims of academic education and vocational training, and reconciling the theoretical and the practical within educational studies. Yet they were to receive considerable support from the governmental 'Newcastle' commission into elementary education (see section 5), which found: 'As a class [the trained teachers] are marked, men and women, by a quickness of ear and eye, a quiet energy, a facility of command, and a patient self-control, which, with rare exceptions, are not observed in the private instructors of the poor' (48), a welcome evaluation indeed.

4. TATE'S 'PHILOSOPHY OF EDUCATION'

During the early 1850s, when he had moved from Battersea to Kneller Hall (London) {49}, Tate produced a book on the principles and practice of education. In this he sought to provide a balance between the 'purely speculative' and the 'exclusively empirical' – an amalgam of theory and practice. The book, *The Philosophy of Education* {50}, is, as we shall see in extracts quoted below, of considerable interest for not only does it demonstrate Tate's debt to Francis Bacon, Pestalozzi and, above all, Locke, but it admirably illustrates Victorian attitudes to education and the methods on which the more far-sighted innovators proposed to base their teaching.

Few paragraphs are of more moment than that in which Tate sets out the purposes of elementary education:

1. To develope the intellectual and moral faculties; or, in other words, to develope the faculties of the perfect man:
2. To communicate to the pupil that sort of knowledge which is most likely to be useful to him in the sphere of life which Providence has assigned to him (p. 3).

The first aim hints at Tate's beliefs concerning the psychology of education: that clearcut taxonomies could be constructed of the various faculties of the mind. Thus, for example, the intellectual faculties had four stages of development – perceptive, conceptive, understanding and reasoning. Within each of these characteristic classes, individual faculties could be identified: for example, the 'cogniative' faculties associated with the 'knowing faculties', or 'faculties of understanding', were abstraction; classification; generalisation; explicit comparison, composition and analysis; judgment, etc. (pp. 63–4). These faculties could be 'cultivated by being *properly* exercised on appropriate subjects' (p. 53) and so it was the task of the curriculum builder to match suitable subject content with desirable faculty outcomes. For this reason mathematics had an important place in elementary education for:

Mental Arithmetic cultivates the memory and the powers of conception and reasoning. It also especially fosters the habit of promptitude, presence of mind, and mental activity.

Arithmetic cultivates the reasoning powers and induces habits of exactness and order.

Mathematics and Natural Philosophy cultivate the reasoning powers chiefly in relation to the acquisition of necessary truths; they also cultivate habits of abstraction (pp. 248–9).

Mathematical reasoning was particularly valuable since it was:

simple and free from all uncertainty . . .

1. Nothing is taken for granted or on mere authority; for its principles or reasoning are axioms of self-evident truths.
2. Its proper objects are the relations of numbers, lines, and spaces, things which are cognisable by our senses, and which can be defined and measured, with a precision of which the objects or [of?] no other kinds of reasoning are susceptible (p. 212).

Yet not only must a student's faculties, i.e. his intellectual processes, be developed, but he must also be provided with knowledge appropriate to his station in life. The concern with providing education suited to the student's social class, as distinct from his abilities or aptitudes was an over-riding trait of mid-nineteenth-century education. Few suggested that it might be a purpose of education to remove class distinctions. Tate, whilst acknowledging class barriers, was eager to develop the practical powers and well-being of the working class and also to ensure that the class's contribution to society was appreciated and rewarded. Accordingly, a prime goal for elementary education was the production of artisans who would help increase England's prosperity and influence. Curriculum construction was to be governed, therefore, by two other keywords, utility and progress.

Now [Francis] Bacon taught, that geometry, as well as all the other branches of mathematics, was valuable as a branch of education only so far as it contributed to supply the wants of society; and that such practical applications, so far from detracting from the discipline which it gave the mind, in reality made that discipline more forcible and complete. He viewed mathematics as an instrument for the extension of art and science, and considered that it should be studied not as an *end*, but as a *means* to an *end* without which the study would be, in a great measure fruitless (p. 26).

Not surprisingly, therefore, Tate was out of sympathy with the way in which mathematics was taught in the schools of the 'middle and higher' classes. Boys, in his opinion, wanted to be 'taught in matters relating to the employments which they will soon have to follow' (p. 30). The engines, factories, chemical works and various workshops surrounding the schools provided a 'mine of intellectual wealth' to be drawn upon for examples, theories and motivation. If only this were done, then might 'these hives of industry [be filled] with a far more intelligent and skilful class of operatives', which would 'not only advance the interests of the operatives themselves, but contribute to the productive resources of [their] country' (p. 30).

Such principles, then, were to govern the content of instruction; but what could be said about method? Tate firmly believed that 'there is method in Education' (p. 13).

When the Committee of Council on Education published their Minutes of 1846, they virtually announced to the world that there was METHOD IN EDUCATION, and that no man could become a truly useful teacher without a knowledge of that method (p. 21).

Yet, Tate was not to adopt the term 'method' in any facile sense: he did not conceive of any foolproof system of teaching: for:

1. Although the same powers and affections are found in every human being, yet these powers and affections exist in differing degrees and states of development in different individuals. Hence it follows, that a system of instruction which is adapted to one class of pupils may not be suitable to another.
2. Different causes may, and no doubt often do, produce the same or similar effects . . .
3. Teachers differ much in their capabilities and acquirements; and they rarely restrict themselves to the use of any special system of instruction (p. 35).

Tate believed, however, that principles of method could be derived which would overcome these obstacles. Again, this did not mean unthinking adherence to any system for:

no system of teaching can be efficient unless the master possesses all those qualities which the system presupposes. If a teacher is wanting in any of [these] . . . it would probably be better for him to modify the system . . . [Moreover, the particular needs of the pupils may also necessitate modifications.] The blind unreasoning attachment of teachers to systems has often brought ridicule upon themselves and discredit upon the systems which they professed to follow (p. 43).

These failures led to such nihilistic arguments as 'each individual ought to have either his own system or no system at all'. In Tate's view, however, there were principles governing method: the first being that 'a good teacher will vary . . . his methods of instruction . . .: his judgement must be exercised in selecting those methods which are most suited to the existing conditions of his school' (p. 43). Determining methods appropriate to particular circumstances would not, however, be a matter for theoretical debate, but rather for experiment: controlled experiments might well result in the confirmation or overthrowing of 'many of our general theories and systems of instruction' (p. 50). Tate, then, was an early advocate of educational research.

In reading Tate one is constantly impressed by the way in which his practical experience leads him to foreshadow many later educational theories. Thus when discussing the problems of teaching children he finds it helpful to distinguish five stages in their development (pp. 79–81). Up to the age of four 'our instruction should be entirely of a desultory character; we should wait for the spontaneous development of the faculties'. Between four and seven we get 'the first glimmerings of reason': 'at the early part of this period instruction should be identified with amusement, and all technical learning should be carefully excluded', mental arithmetic and later the foundations of common arithmetic are appropriate. The third period, from seven to ten, sees the 'dawn of reason and imagination . . . The abstract terms . . . of arithmetic . . . should be taught in connection with their concrete forms.' From ten to fourteen 'understanding and reason attain a certain degree of strength . . . the mind is now capable of more sustained exertion . . . yet due regard [should be paid] to the imperfect nature of the reflective faculties'. It is only in the post-fourteen period that 'all the faculties of our nature attain their full development . . . competitive examinations and rewards now become appropriate'.

Tate's principles of method are based on this interpretation of child development. Other principles concerned the need to foster self-development and self-instruction, the need to offer a wide curriculum, to proceed from 'the simple to the complex' and to teach 'the concrete before the abstract'. Children should 'learn nothing which they may have afterwards to unlearn'. Tate, goes on to discuss his principles and methods in detail and to provide examples of specimen lessons. However, to end this section it would seem most appropriate to quote him, as a pioneer of teacher education, 'on the qualifications of the schoolmaster in relation to his professional duties' (pp. 275–8). He sees three aspects to these qualifications: his attainments, his capabilities and his character.

The following attainments were considered essential for all teachers:

1. [A thorough acquaintance with] the leading doctrines and narratives of Scripture; mental and common arithmetic; reading, writing and spelling; English history; and the principles of teaching.
2. [A fair knowledge of] drawing, mensuration and practical geometry; geography and astronomy; elementary grammar, composition, and general history;

elementary algebra, to the end of quadratic equations, together with a little demonstrative geometry; industrial mechanics, and some simple course of experimental philosophy.

It is highly desirable that his mind should be well stored with general knowledge, that he should have a ready command of language, and that he should be able to express his ideas, with fluency, clearness, and precision, upon any subject within the range of his knowledge. Profound attainments, in any technical subject of knowledge are scarcely of any value to him, as an elementary teacher. His knowledge should be varied rather than profound. An acquaintance with . . . the higher branches of mathematics . . . would rather interfere with his usefulness, as an elementary teacher. At the same time, it is necessary to bear in mind that a schoolmaster should know a good deal more than he has to teach . . . [However,] all that technical knowledge which leads the mind of the teacher away from the subjects of elementary education, tends most undoubtedly to compromise his usefulness. . . . It is true, people talk much about the discipline which such subjects give to his mind, as if the knowledge, which is essential to his duties, as a teacher, did not sufficiently exercise, discipline, and task his intellectual energies. Would it not be better to raise our standard of his knowledge in physical science and in the principles and art of education . . . ? . . .

Capabilities and Character

A teacher should be a pious, conscientious man; his talents should be, at least, respectable; and he should have a decided predilection and aptitude for teaching. . . .

Aptitude for teaching! what is it? There is no mistaking it when we see it . . . Is it a natural or an acquired gift? . . . Does it grow spontaneously . . . ? . . .

The aptitude for teaching must undoubtedly be a qualification resulting from the development of certain intellectual and moral faculties of our nature . . .

It will be instructive, not only to ascertain what such a man MUST be, but also what he MAY not be.

1. A man having an aptitude for teaching MAY not be
   (1) . . . a man of great technical attainments,
   (2) . . . a man of comprehensive mind, or possessing great reasoning powers,
   (3) . . . a man of robust frame.
2. A man having a great aptitude for teaching MUST
   (1) . . . have a love for children, and a knowledge of their tastes, habits, and capabilities,
   (2) . . . be a man of a kind and benevolent disposition,
   (3) . . . love knowledge, and feel a pleasure in communicating it,
   (4) . . . be a man of fervid imagination, and of great enthusiasm, decision, and force of character,
   (5) . . . be a man of respectable general attainments,
   (6) . . . have considerable fluency of speech, and powers of illustration and exposition,
   (7) . . . have faith in the efficacy of instruction as a means of ameliorating the condition of society,
   (8) . . . be a man of quick and observing habits, and must be in the constant habit of reflecting and reasoning upon the various methods by which knowledge must be communicated to teaching.

Now as all those qualities . . . admit of cultivation, it necessarily follows that the aptitude for teaching also admits of cultivation.

They are interesting lists which cause one to consider what form modern equivalents would take. Attention is not specifically drawn to the teacher's need continually to further his personal knowledge; the concept of initial training had, however, to be more fully accepted before that of in-service education could emerge.

5. RETIREMENT AND THE NEWCASTLE COMMISSION

The years between 1840 and 1856 had been very full ones indeed for Tate. He wrote and had published over twenty textbooks, founded a periodical, *The Educational Expositor* {51}, and also became President of the newly created United Association of Schoolmasters. The books covered all branches of mathematics, Euclid's *Elements, Algebra made Easy, Astronomy and the use of the Globes, Drawing for Schools, Exercises on Mechanics and Natural Philosophy, Hydrostatics, Hydraulics and Pneumatics, The New Coinage considered in Relation to our School Arithmetics* (cf. p. 94 above), *Practical Geometry, The Principles of Geometry, Mensuration, Trigonometry, Land Surveying and Levelling, The Principles of Mechanical Philosophy, The Principles of the Differential and Integral Calculus, A Treatise on the First Principles of Arithmetic, after the Method of Pestalozzi* (which sold over 100,000 copies), etc., and also chemistry, electricity, magnetism, electrodynamics and much of science. As if to emphasise Tate's versatility there was also a monograph on pure mathematics, *A Treatise on Factorial Analysis, with the Summation of Series* (Longman, 1845), which included some original work he had previously communicated to the Royal Society. For, although remembered mainly as a textbook writer, he had a creative mind. In April 1860, for example, he published details in the *Philosophical Magazine* of a new instrument he had invented for trisecting the angle.

During his years at Battersea, Tate was asked by the noted Victorian engineer, (Sir) William Fairbairn, to provide him with mathematical assistance. Fairbairn was 'not a deep mathematician' and relied upon Tate for help (52). It was Tate then who furnished many of the mathematical calculations required in the planning and construction of the famous Britannia and Conway tubular bridges. Later Tate and Fairbairn were to publish several joint papers on the construction of boilers and to present these to the Royal Society and the British Association. Several instruments and mechanisms were developed by him including 'Tate's double piston air pump'.

Most of this work was, however, to take place after 1855, the year when Kneller Hall closed. At that time Tate's health was in a poor state. In March 1856, Temple wrote to the Committee of Council concerning Tate's past work and his future:

Both as an author and as a teacher he must be considered to stand quite at the head of his profession. Most of his books are either in universal use, or are only displaced by others which owe their existence to imitations of his method. His pupils have been eminently successful, and would ascribe most of their success to his instructions . . . It would be difficult to find any man who in proportion to his opportunities has done so much for education in this country.

His health has suffered from his unremitting labours, and even if Kneller Hall had been maintained, he could not have retained his position there much longer. He is not now capable of undertaking any duties which would require that close and continuous attention which is absolutely necessary for success in teaching (53).

Temple's plea that Tate's merits should be recognised by a special retiring pension was seconded by Canon Moseley who felt that such an act:

would not only be a just reward to a very faithful servant, but would be accepted with gratitude by the very great body of the elementary schoolmasters of this country, as a tribute to the honesty, the ability, and the fidelity of one whom they are accustomed to place at the head of their profession. Of the numerous useful works on education published by Mr Tate, there are two the influence of which on elementary teaching has been remarkable. The one is his arithmetic, in which he first, of all the authors who have attempted it in England, succeeded in making arithmetic the 'Logic of the People', by giving for the rules of arithmetic such reasons as it was possible to teach to poor children: and the other is his 'Exercises in Mechanics', in which he first taught, under a popular form, how by the common rules of arithmetic, and by geometrical construction with the scale and compasses, calculations in mechanics of great practical value may be made by persons who have no knowledge of mathematics. I have known him since he first became the lecturer in a training school, and I cannot but feel how immense the influence has been, not only of his writings or his labours as a teacher, but, of his character as a self-dedicated man, who has always thought first of his work and last of himself, and who by his fidelity, and his zeal and industry, offered to those under his care the model of that character of the teacher which it was his object to form in them (54).

The Government acted swiftly and generously. Less than three weeks after Temple wrote his letter, Tate was informed that he had been awarded a pension of £120 p.a.

It was, effectively, the end of Tate's career as an educator. For although he lived for a further thirty years, his death occurring in 1888, from that time onwards he devoted himself, almost entirely, 'to literary and scientific pursuits'. One casualty of this transition from educator to civil engineer was Tate's planned *Mathematics for Working-Men*, for only the first part of this was published (1856). The loss was considerable, for what appeared revealed not only Tate's concern for the lucid and vivid presentation of material, but his intimate knowledge of the concerns of his target audience. To illustrate the former one can quote, for example, Tate's use of figures (dots arranged in rectangles) to illustrate the commutative and distributive laws. The care for the welfare of the working man is demonstrated by Tate's observations on the principles of estimation:

The custom of estimating and contracting for work by the 'lump', is little better than a modified form of gambling, by which the contractor may either make himself a fortune, to the injury of his employer, or render himself penniless, with all the moral evils resulting from hopeless and irretrievable pecuniary embarrassment. Teachers of elementary schools and mechanics' institutes would confer a great benefit on society, by teaching the simple and fundamental principles of estimation, rather than waste the time of their pupils in giving 'sums', which are only calculated to amuse the curious, without conveying any satisfactory information to the man who is engaged in the business of life. It is true that some of these 'sums' may exercise the mind; but it is an important fact, which should never be lost sight of, that those investigations which have the greatest practical bearing invariably form the most healthful and instructive exercise to the intellectual powers (p. 47).

Unfortunately, teachers of elementary schools were soon to be forced to pay more rather than less attention to 'sums'. Indeed, it is possible that Tate's move away from education was encouraged by some of the events of the 1860s, and it is to these we now turn.

The deficiencies of the country's elementary schools, pointed out in the 1830s by Roebuck and others, had not been removed by the establishment of the Committee of Council. Indeed, the newly-formed inspectorate's surveys only served to highlight the 'system's' failings. In 1858, therefore, a public commission, chaired by the Duke of Newcastle, was established 'to inquire into the State of Popular Education in England and to consider and report what measures if any, are required for the extension of sound and cheap elementary instruction to all classes of people'.

The picture of mathematics education revealed in the Commission's report is bleak indeed. Clearly, little mathematics was taught in the majority of elementary schools. Of the children attending the 1,824 public weekday schools visited, only 69.3 per cent were taught arithmetic, 0.6 per cent mechanics, 0.8 per cent algebra and 0.8 per cent Euclid. The corresponding figures for the 3,495 private schools run by individuals were 33.8 per cent, 1.29 per cent, 1.35 per cent and 1.15 per cent respectively. Moreover, as the assistant commissioners' reports make clear, such arithmetic teaching as did take place was frequently ineffective:

In working sums expressly stated, the children were often successful enough, but they were usually quite ignorant of anything that required the simplest knowledge of a principle. 'With respect to arithmetic,' says Mr Cumin, 'I tried the test of dictating a sum in addition or subtraction. The knowledge of notation was singularly defective; but wherever I found a good knowledge of notation, I found the class thoroughly instructed in the elements of arithmetic. It was by no means uncommon to find boys and girls in fractions and compound division who could not write down sums up to 10,000 or even 1,000. It seemed to be the opinion of some that that notation was hardly a fair test to apply.' . . . [Another assistant commissioner reported that] you will find a knowledge of the multiplication and pence tables at as early an age as you could reasonably expect, and familiarity with practice, rule of three, and even fractions, at years not beyond the average of boys of all classes grappling with such calculations; but, if you take the boys back, it may come out that some of them have made more haste than good speed. I confess that it had not occurred to me to try even the younger boys in numeration until Christmas, when one of my colleagues . . . suggested it . . . I subsequently acted upon the hint; and, to the chagrin of masters and astonishment of scholars, many boys of the first [senior] class in their respective schools, able to solve questions in advanced rules with ease and accuracy, were found utterly at fault in a simple addition sum, when consisting of five or six lines of seven figures, slowly and repeatedly dictated, but testingly interspersed with numerous o's. This was a palpable demonstration, that 'the *proper* degree of attention' had not been bestowed upon arithmetic (p. 259).

The main recommendations of the Commission were that the Committee of Council should extend its operations. However, it was not suggested that the grip on education of the denominational bodies should be eased in any way, although a change in the way grants were paid to schools was suggested; that they should be related 'to the attainment of a certain degree of knowledge by the children in the school'. This principle, of 'payment by results' was introduced in the *Revised Code of Regulations* presented to Parliament in 1862 by Robert Lowe, Vice-President of the Council and head of the education department. The grant to a school now depended upon the results obtained by its pupils in tests of reading, writing and arithmetic. Thus, in arithmetic, every scholar had to be examined according to one of the following standards:

Standard I	Form on black-board or slate, from dictation, figures up to 20; name at sight figures up to 20; add and subtract figures up to 10, orally, from examples on blackboard.
Standard II	A sum in simple addition or subtraction, and the multiplication table.
Standard III	A sum in any simple rule as far as short division (inclusive).
Standard IV	A sum in compound rules (money).
Standard V	A sum in compound rules (common weights and measures).
Standard VI	A sum in practice or bills of parcels.

The *Revised Code* was always the subject of considerable controversy. Malcolm Arnold, the leading inspector of schools, thought 'it gives a mechanical turn to the school teaching . . . and must be trying to the intellectual life of the school'. Moreover, in 'the game of mechanical contrivances the teacher will in the end beat us . . . by [getting] children through the . . . examination . . . without their really knowing [anything] of these matters'. Even so, Arnold found the rate of failure in arithmetic 'considerable' for although the child 'sedulously practised [sums] all the year round' he was not taught arithmetical principles 'or introduced into the science of arithmetic' (55).

Nevertheless, the system did offer some advantages. The *Revised Code* ensured that all pupils were introduced to rudimentary arithmetic, and its regulations could be, and were, readily changed. The first modifications were made in 1867 and others followed, including the addition of further examinable subjects, before in 1897 the principle of payment by results was abandoned. In 1871, when it looked as if Britain was at last to adopt the metric system, an addition to the regulations was made to the effect that 'in all schools the children in Standards V and VI should know the principles of the Metric system, and be able to explain the advantages to be gained from uniformity in the method of forming multiples and submultiples of the unit'. In the event Parliament could not be persuaded to acknowledge the merits of the metric system and in 1874 this reference was deleted from the *Code*.

By 1871 the arithmetical standards had been changed considerably. What previously covered two standards was now to be covered in Standard I, and the work previously set for Standards III to VI was accordingly moved forward a year. The new Standard VI was 'Proportion and vulgar or decimal fractions'. It was also set down that in Standards V and VI arithmetical questions

> should correspond with the wants of working men; wages, interest on money put into the savings bank, accounts of work, division of payment for work among a gang according to the time each has given to it, rates on small tenements, profit and loss on joint ventures, freight charges by rail or sea, etc., will serve as instances.

These suggestions, coming as they did after the 1870 Act which finally brought state elementary education to England (see p. 169 below), would, one believes, have had but lukewarm support from Tate. They reflected the 'wants' of working men as seen in Whitehall, and not their day-to-day 'needs'. Moreover, there was a world of difference in learning to do sums on 'profit and loss on joint ventures', and, as Tate had done in his evening classes in York, in bringing science and mathematics to bear on problems from the child's environment. The *Revised Code*, although elevating the standards of the worst schools regrettably served to inhibit excellence and initiative amongst the better teachers. Even so, great advances had been made in the provision of elementary education and in teacher-training during Tate's lifetime, to many of which he contributed. Regrettably, the challenge he set in relating mathematics and science teaching to the pupil's interests, to the world about him, and to the careers he might conceivably follow has still successfully to be resolved.

7 · *James Wilson*

It is not perhaps surprising that James Maurice Wilson became a 'schoolmaster' and 'divine', for that was a family tradition; that the *Dictionary of National Biography* should describe him also as scholar, mathematician, astronomer and antiquary gives an even greater indication of his remarkable all-round ability. As he himself put it, 'I had no special aim or bent in life; but if circumstances showed me that something ought to be done, and no one else came forward to do it, I thought that I ought to try' (1).

1. SCHOOLING

Wilson was born, one of twins, at King William's College in the Isle of Man in November 1836. His father, the first principal of the College, an Anglican institution established in 1833 {2}, was soon to leave teaching to become a parish priest, but it was to King William's that Wilson returned to begin his formal education in 1848. They were not happy times; the school was cheap and attracted a mixed clientèle. Those of 'good birth' were 'swamped in a very rough lot' and there was a considerable amount of bullying and cruelty {3}. Neither could young Wilson find much consolation in his lessons, for 'no one on the staff was a scholar, and no one even a tolerable mathematician'. When he, with the aid of a textbook, learned some algebra, his solutions of ordinary quadratics 'were kept *and copied* by the Principal, in case anyone in the future should reach such a high-water mark'.

To his relief, in 1853 Wilson moved from King William's to the old, endowed grammar school at Sedbergh in the Yorkshire Dales. He later described what he found there: the single schoolroom with the headmaster, Evans, at one end, the classical assistant opposite, and midway down the room the mathematical assistant. The higher classes were grouped near the head and when it was their turn to be taught the pupils were called to sit on the benches surrounding the head's desk at which Evans remained during all the school hours, 7 to 8 a.m., 9.30 to 12 and (on four days in the week) 2.30 to 5. Evans taught the senior boys both classics and mathematics. There were three mathematics lessons each week and the topics taught included Euclid, algebra, geometrical conics and trigonometry. 'In the evenings once a week Evans taught two or three of us a little Statics, and the very beginning of Algebraic Geometry' {4}.

Although Wilson wrote somewhat disparagingly of his mathematical education at Sedbergh, it was probably as good as, or better than, that which he would have received in most contemporary grammar schools {5}. Mathematics was now becoming more safely ensconced in the curriculum of the grammar and public schools, but the amount taught differed considerably depending upon the interests of the headteacher. The external examination system which was to bring with it uniformity had yet to be created. Indeed, charting the development of mathematics teaching in the first half of the nineteenth century is not easy. All that can be done is to consider the histories of individual schools and, again, the position is confused because often mathematics was taught only temporarily according to the whim of the master {6}. As we have seen, in the eighteenth century most mathematics and science teaching took place in private schools and academies, although a number of grammar schools taught the subject (cf. p. 225 below) {7}. Moreover, the Eldon ruling (p. 64 above) served to buttress the classical curriculum and to keep mathematics as an optional extra. That is, pupils could attend classes given by an usher or a writing master outside normal school hours and on payment of an additional fee. Again, the efficiency of this extra mathematics teaching varied widely. For example, Bower, the writing master at Winchester in the 1820s, 'had not pursued his studies beyond the Fourth Book of Euclid' (8) and did little more than 'mend pens and make up the weekly account' {9}. In 1834, however, Winchester was to acquire its first mathematics master, J. D. Walford. This was a decade when the proprietary schools (p. 87 above) began to challenge the established order, and it also marked the first appointment of competent mathematics masters to the major public schools, for example Eton (Hawtrey, 1834) and Harrow (Colenso, 1838).

The presence of a mathematics master did not, however, ensure a firm place for the subject in the curriculum. Mathematics continued to be an optional subject at Winchester for some time after Walford's arrival, and even when it became compulsory, marks obtained in it were not taken into consideration for promotion, etc. The mathematics teacher, too, was left in no doubt that he was an inferior being {10}: the battle for parity of esteem with classics and the classics teacher had still to be won.

It is unlikely, then, bearing in mind the later reports of the governmental commissions on grammar and public-school education, that Wilson found himself at any mathematical disadvantage, *vis-à-vis* his fellow students, when he entered Cambridge in 1855.

2. UNIVERSITY

That Wilson should enter St John's College was to be expected, for it had strong links with Sedbergh School: St John's appointed the headmaster of the school, and three fellowships and ten or so scholarships were reserved at the college for ex-Sedbergh pupils {11}. Yet, surprisingly, Wilson does not recall any of his Sedbergh contemporaries being at St John's when he went up there.

Lectures soon began, and it gradually dawned on me that none of us were in the least degree individualised or directed. It mattered to nobody what I did, whether I read classics or mathematics, or nothing at all. There were lectures in Algebra given to all the freshmen together, about 90 of us, in the College Hall, by Reyner. He would give out about 18 or 20 questions, then he went round the Hall looking at the papers we brought on his last lecture. Of course it was useless. The help we got from him was infinitesimal. Later on the mathematical lectures were very good; Ben Horne's lectures on Optics and Dynamics were as good as lectures could be (12).

Wilson settled down well at Cambridge and in 1856 was bracketed second for the Bell University Scholarship. The omens for a 'good first class and a profitable fellowship' were bright. In the hope of bolstering his family's finances, Wilson also began to prepare for the college scholarship examinations and went to Parkinson, one of the leading university mathematical coaches, for tuition in advanced trigonometry, algebra, geometry and coordinate geometry.

The mathematical coach was a man to be reckoned with in nineteenth-century Cambridge. He, with all the past Tripos papers at his fingertips, would prepare students privately for the great examination contests. The two best known, Hopkins and Routh, produced remarkable results. Hopkins, a contemporary of De Morgan, coached 17 Senior Wranglers ('first of the firsts') in twenty-two years including Routh, Stokes, Thomson, Tait, Todhunter and Maxwell. Routh had an even more impressive list of honours (although his career, in all fifty-four years, was much longer): he produced 27 Senior Wranglers and 41 Smith's Prizemen, and he coached 480 of the 990 wranglers in his time. The success which attended his efforts to turn the Tripos examination into a formalised game was eventually to lead to its reform.

Parkinson, too, achieved his goal with Wilson who came an easy first in mathematics in the college examination, a result which caused his tutor to urge him to continue with mathematics. Somewhat half-heartedly Wilson agreed, and returned to Parkinson for more coaching during the long vacation of 1856.

In the Long Vacation . . . I read Analytical Geometry, Todhunter and Salmon, the Differential and Integral Calculus, both in Todhunter, supplementing it by De

Morgan's great book, which fascinated me. Then Differential Equations, a long intricate subject, and Geometry of Three Dimensions.

His day included 'a good 7 hours of Mathematics and 2 hours of Classics' with, in the afternoon, two hours of rowing, swimming, fives and 'now and then on a velocipede', and, in the evening, perhaps, half an hour's backgammon.

Eventually the fourth, Tripos, year came round. Under Parkinson's guidance he revised and worked through specimen papers. His reading of De Morgan's book on 'double algebra', however, was censured by Parkinson – it was outside the syllabus. For a similar reason, Wilson was forbidden to attend Stokes' lectures on light. The Tripos held pride of place over any interest in mathematics.

At last came the Tripos (then held in the Lent term): it consisted of three days for papers, then an interval of a week or so; then those who had come up to the standard for honours had five days more – ten papers . . . The interval was spent in revision of . . . rigid dynamics, Lunar and Planetary Theory, Astronomy, Calculus of Functions and some of the parts I had omitted in Theory of Equations, Finite Differences, etc.

To Wilson's great joy – and also that of the College's porters who had gambled their money on him – he was named Senior Wrangler. Jubilation was soon dramatically cut short for Wilson had a nervous breakdown. He went to convalesce in the Isle of Wight and then made the striking discovery that everything he had learned at Cambridge had disappeared from his memory: 'I could not differentiate or integrate; I had forgotten . . . all Lunar Theory and Dynamics; nearly the whole of Trigonometry and Conic Sections was a blank . . . Happily Algebra and Euclid were safe.'

Thus, only three months after becoming Senior Wrangler, Wilson had to begin to relearn calculus and other topics. He was elected to a Fellowship of St John's the following year on the strength of his classics but was 'unable to touch the Mathematical papers'.

Wilson retained his Fellowship until he married in 1868, but he never resided in Cambridge {13}. As a Senior Wrangler (even with an impaired memory) many doors were open to him. He was offered a post as private tutor at 'four or five hundred a year with all expenses'; he was pressed to go to the Bar (as Cayley, Sylvester and others had done); and Kay-Shuttleworth, ever on the lookout for new talent, offered him, as he had the young Temple, a post in the Education Department. The most lucrative offer, however, was to become science master at Rugby School and it was this he accepted {14}. That he knew no science was, he thought, a handicap of sorts, but this he attempted to alleviate by working during the summer in the science laboratories of University College, London. It was also agreed that if he failed to teach science satisfactorily he would become a form master (teaching classics) or, failing that, a mathematics teacher.

Although 'successful' in Cambridge, Wilson was disillusioned by the education he received there:

Some of us were eager to read great mathematical works like Laplace's *Mécanique Céleste*: others to prosecute some original work. But all such work was prohibited by the private tutors: it would not pay in the Tripos . . . The examinations demanded much ingenuity in solving problems, and great memory for bookwork; but seemed to us critical and rebellious youngsters to be unworthy.

Despite Wilson's strictures, mathematics at Cambridge was in relatively good shape. Since the late eighteenth century the subject had held an important place there, and had not been neglected as at Oxford. Notwithstanding its deficiencies, the Tripos system encouraged an interest in mathematics and presumably played some part in Cambridge's producing between 1815 and 1850 a group of mathematicians that would readily bear comparison with the output of any overseas university. Yet the university had not come fully to terms with the times; religious and other discrimination still existed and modern disciplines were insufficiently well represented in the curriculum.

Oxford was in a worse state still, and indeed in 1849 Members of the two universities and Fellows of the Royal Society petitioned the Prime Minister to complain that 'the system of the ancient English universities had not advanced and was not calculated to advance the interests of religious and useful learning to an extent commensurable with the great resources and high position of those bodies' (15). In spite of some vigorous opposition the Government established a Royal Commission to inquire into the affairs of the two universities. The Cambridge Commission had relatively few criticisms to make: college fellowships were made open to free competition (and so, for example, the Sedbergh Fellowships at St John's ceased), and the requirement that all candidates for degrees and fellowships should subscribe to the Church of England was relaxed. New chairs in chemistry, zoology, engineering and other subjects were proposed – as well as one in descriptive geometry (the Sadleirian Chair, first filled in 1863 by Cayley).

The criticisms of Oxford were more sweeping and the relations between the university and the Commission, which included Baden Powell (p. 262 below), more acrimonious. In general, though, the outcomes, greater religious toleration, less Church control and an updated curriculum, were the same.

Many of the old links between the endowed grammar schools and the colleges were weakened or removed: only where vested interests were particularly powerful were they maintained as, for example, at Winchester and Eton {16}.

However, although certain types of discrimination were removed, others were created. With their updated curricula the two ancient universities now became more attractive to the prosperous middle class many of whom were no longer debarred on religious grounds. Opening scholarships to all meant that they became the property of the reformed public schools and their new rivals, such as Marlborough and Clifton. From being bulwarks of the Church of England, the two universities became citadels of the middle and upper classes. The percentage of students entering them from the endowed grammar schools decreased rapidly (17), and by the end of the century the 'new' public school system supplied 82 per cent of entrants. Even today (early 1980s), at Cambridge students from the independent schools outnumber those from the state system, while at Oxford the latter have only recently formed the majority.

3. RUGBY AND THE CLARENDON COMMISSION

Despite his appointment as science master, Wilson soon found himself teaching mathematics at Rugby:

> I took the second Mathematical set of the Sixth Form, for three hours a week. As far as I recollect no one in the set wished to learn mathematics . . . I was contented with very little, and tried to teach and inspire, and to be on the alert for any sign of ability. I remember astonishing [two pupils] by telling them that they could get mathematical scholarships if they liked, and would work. They did . . . {18}.
>
> In the Fifth Form, Monday and Friday mornings were given to Mathematics . . . I had the two bottom sets, about twenty or twenty-five in each, mainly big fellows who were regarded as hopeless in mathematics. The classroom was the Cloak Room . . . It was completely unsuited for class teaching: one broad table was in the middle of the room, two narrow ones against the side walls. The class was dull, bored, uninterested, but not ill-behaved . . . I set some work to be brought on Friday . . . and about half the fellows in the set that came . . . had prepared no work, the others very little. [The reason given was] that it was the privilege of the two lowest sets to do no work out of school if they behaved well in school.

Wilson refused to accept this 'immemorial custom' as a valid one, and soon had a strike on his hands: some of the students 'flatly refused to touch their work'.

This was but one of a number of crises which Wilson had to face. As a mathematics master he could not expect deference either for himself or his subject. Yet changes were afoot, and these were accelerated when in 1857, and following the lead provided by the Indian Civil Service, the Royal Military Academies at Woolwich and Sandhurst instituted an entrance examination for would-be entrants. The importance of mathematics was stressed and the subject loomed large in the entrance requirements (19). The effects of these changes formed the basis for a series of questions to Wilson when he gave evidence to the Clarendon Commission established in 1861 to inquire into the state of the nine great public schools.

The 1850s had seen the Royal Commission on the two ancient universities and the establishment of the Newcastle Commission (pp. 120f) on elementary education; now attention was directed at the old-established public schools. So far as the number of their pupils was concerned, the nine schools were of minor importance – in total they had fewer than 3,000 students and of these over 60 per cent were at Eton (850), Rugby (500) or Harrow (475). The other six schools were small, the largest being Merchant Taylors' (260). However, since the governing classes were recruited almost entirely from these schools, their efficiency was a matter of national importance. There was a danger that the new proprietary schools would place the 'upper classes in a state of inferiority to the middle and lower' {20}.

The curriculum of the public schools was thought by many outside observers to be old-fashioned and to be geared entirely to the needs of those few (three out of ten) who aspired to a university education. Something had to be done.

The evidence provided to the Commissioners revealed many defects in the teaching of mathematics and certainly that at Rugby did not seem to have reached a high standard. Despite the headmaster's (Temple's) interest in the subject, it was not accorded the weight given to it at Merchant Taylors', nor was much provision made for the mathematically able. In 1862 only two sixth-formers were tackling the differential calculus, no one attempted conic sections, and although some elementary hydrostatics was taught, it was admitted that the boys did not get on to the applications of mathematics in any 'difficult' subject. Such a mathematical education could hardly match that which Wilson had received at Sedbergh and would not suffice for those aspiring to enter the Royal Military Academy at Woolwich. Some new boarding schools met the needs of the latter students by the creation of 'modern' departments in which mathematics and science were emphasised, and Wilson believed that a similar department at Rugby would in a couple of years attract about half the pupils. Ten of his sixth-form class of twenty-six wished to enter Woolwich but they would have to leave Rugby and attend a mathematical 'crammer' prior to taking the entrance examination. Other pressures were now being exerted on the schools by the two ancient universities through their open scholarship examinations. These were now forcing schools to introduce the calculus – although at the university (p. 220 below) this was second-year work – and, in general, distorting sixth-form curricula.

Not surprisingly, the Commissioners recommended that every boy should be taught arithmetic and the elements of geometry, algebra and plane trigonometry. They also urged (as did the Dainton Report of 1968) that the study of mathematics should continue throughout the pupils' stay at school. More advanced students should be introduced to applied mathematics, especially mechanics, and, so as to facilitate its teaching, mathematics classes should be setted by ability independently of normal class divisions.

Yet, despite these concessions to modernity, it was argued that classical languages and history should still hold the principal place in the curriculum and, with divinity, should account for most of the time-tabled hours {21}.

The proposals concerning mathematics were scarcely far-reaching. Moreover, the twin pressures to prepare boys for the new selection and scholarship examinations and to fight off the challenge of their newly established rivals, would by themselves have ensured a place for mathematics in the curriculum. The form that mathematics took and the resulting disquiet and action are the subjects of the next section.

4. ELEMENTARY GEOMETRY

Soon after I began to teach Euclid I became dissatisfied with it . . . It was not an educational process. It did not bring out originality of thought, and was plainly disliked by the boys: and the amount of knowledge attained was very small (22).

That Euclid should have become the accepted textbook for school geometry is perhaps surprising. There was, of course, no long tradition of teaching mathematics, let alone Euclid, in English public schools; neither was it the case that our Continental rivals used Euclid in theirs {23}. From Recorde's time onwards there were English geometries which departed from Euclid in their order and method of presentation. Yet such geometries came to be associated with those who practised mathematics for a living. Those who learned it as part of an 'education', at university, did not use such utilitarian works, but read Euclid: 'not as an instrument (for the solution of today's mathematical problems) but as an exercise of the intellectual powers; that is, not for their results, but for the intellectual habits which they generate' {24}.

So argued the leading Cambridge don, Whewell. Some sixty years earlier, an Oxford counterpart, Williamson had contemptuously dismissed an alternative approach to geometry proposed by Clairaut (25), who had sought not only to demonstrate the applicability of the subject but to use 'real-life' problems as motivation for geometrical arguments and results. Clairaut's book attracted few immediate disciples. Legendre's *Eléments de*

Géométrie (1794), which attempted a 'modern' axiomatic development, proved more successful. It went through many editions in French, was widely used in translation in the United States and appeared in Britain in translation in 1824 where it failed, however, to emulate the impact of the earlier translation of Lacroix's *Calculus*. The universities refused to desert Euclid, and so helped strengthen its hold in the schools. Moreover, as we have seen, Euclid was not only used in the public and grammar schools, but also in teacher-training institutions and in some exceptional elementary schools (p. 104 above).

The way in which the teaching of Euclid gradually grew in the early nineteenth century can be gauged by the number of new editions: 52 in the period 1801–20, 36 in the next decade, 53 between 1831 and 1840 and 73 in the last ten years of the half-century (26)! Euclid was not only a text, it became synonymous with geometry.

Many of these editions were, in fact, Euclid pure and simple; additional notes were often included, but no exercises, for the student was expected to memorise not to act. Gradually, however, the custom of including exercises grew and, for example, Potts' edition (27), intended 'for the use of the higher forms in public schools and students in the universities', contained a selection of questions from past Tripos and college examination papers. Potts, in his preface, rehearses the familiar arguments that Euclid serves as a mental discipline and is the best exemplar of the deductive method. More interestingly he identified a pedagogical conflict between Euclid and algebra:

> If we consider the nature of Geometrical and Algebraical reasoning, it will be evident that there is a marked distinction between them. To comprehend the one, the whole process must be kept in view from the commencement to the conclusion – while in Algebraical reasonings . . . the attention is altogether withdrawn from the things signified, and confined to the symbols, with the performance of certain mechanical operations, according to rules of which the rationale may or may not be comprehended by the student.

Despite Potts' arguments, the unsuitability of Euclid as a school textbook was becoming increasingly apparent and the 1868 report of the Endowed Schools (Taunton) Commission drew public attention to that fact in a section almost certainly written by Temple. Indeed, well before the publication of the report, Temple had called Wilson's attention 'to the very backward state of Geometry in English schools compared with those of the Continent' and had asked him to 'go into the matter'. Wilson collected and read the major textbooks used in Europe and the United States and, with Temple's encouragement, wrote his own text on material corresponding to the first two books of Euclid (28).

The book first appeared in 1868 and precipitated a long-drawn-out battle of words. In his preface, Wilson set out his objections to Euclid: his artificiality (the manner in which he 'sacrificed . . . simplicity and naturalness' in order to show 'on how few axioms and postulates the whole could be made to depend'), the invariably syllogistic form of his reasoning (which made the subject 'unnecessarily stiff, obscure, tedious and barren'), the length of his demonstrations (for 'great length . . . exercises the memory more than the intelligence') and his unsuggestiveness ('all . . . theorems and problems [are placed] on a level, without giving prominence to the master-theorems, or clearly indicating the master-methods') {29}. Moreover, teaching via Euclid did not help boys to solve problems, 'the true test of geometrical knowledge'.

Wilson's argument, presumably advanced to forestall his critics, that the abandonment of Euclid would not lead to anarchy in the examination hall was, alas, to prove over-optimistic. He, too, felt that 'algebra fails as a discipline because it speedily degenerates into symbols' and so intended to retain pure geometry and reasoning. In fact, Wilson's book was still nearer to Euclid in spirit than to Clairaut.

Reform, however, demanded two essentials: new materials for use in schools, and a change of heart by university and government examiners. The former could readily be supplied {30}, the latter proved more difficult to effect.

Euclid still had some powerful advocates and, as frequently happens, those who desire and seek reform themselves can, out of sheer perspicuity, be the most damning critics of other reformers. Wilson happened to chance on De Morgan in one of his facetious and unsympathetic moods and the result was a withering review in the *Athenaeum*. Wilson's remarks on syllogistic arguments led to a recommendation that he should 'revise his notions of logic' whilst his treatment of angle, which took the notion of direction as self-evident, was 'slip-slop' and '*non est geometria*': 'We feel confidence that no such system [as Mr Wilson's] . . . will replace Euclid in this country. The old geometry is a very English subject.' What De Morgan sought was an 'amended' Euclid, free from its logical faults {31}.

De Morgan's opposition to Wilson's book was echoed by other mathematicians including Todhunter (32) and Dodgson (Lewis Carroll) in his *Euclid and his Modern Rivals* (1879). The latter, whilst approving of Legendre (but only for advanced students), had nothing but scorn for Wilson, Wright (p. 264 below) and the American Benjamin Pierce (*sic*). Wilson, in particular, was heavily criticised and the fact that he was a Senior Wrangler (and so, presumably, should have known better) remarked upon (33). Todhunter's essay is a critique of Wilson's preface (see above) and is characteristically sober and serious. His defence of Euclid rested on the premises that:

(a) ordinary students did learn something from its study – indeed, more than they were likely to learn from its rivals;
(b) those who wished to retain Euclid were more competent to form an opinion on the subject than the reformers, that is, the majority of Cambridge mathematicians opposed change;
(c) the opponents to Euclid were only united in opposition, there was no consensus concerning a replacement;
(d) Euclid's deficiencies could be overcome by 'good teaching';
(e) Wilson underestimated the examining problems which would arise if Euclid were abandoned;
(f) 'English mathematicians . . . trained in Euclid, are unrivalled for their ingenuity and fertility in the construction and solution of problems'.

In the following century the arguments were to acquire a familiar ring; the world does look different when viewed from a Cambridge college. Certainly, Todhunter and Wilson would have differed on what was meant by 'an ordinary student' and offered different estimates of the percentage of students who could demonstrate 'ingenuity' in problem solving. The question of who is the best judge of what is and what is not suitable for inclusion in the school curriculum is still debated, whilst the argument that 'good teaching' will overcome difficulties is used to justify not trying to alleviate them. Yet Todhunter correctly foresaw the difficulties that examiners would face and the fact that those opposing Euclid were not agreed on what should replace him.

Wilson's book generated considerable interest, for he was a Senior Wrangler, a master at a leading public school, a member of the British Association's Committee on Scientific Instruction, and had recently made an outstanding contribution to *Essays on a Liberal Education*. He was invited to address the London Mathematical Society (34) and to speak as far afield as Edinburgh {35}.

Yet what was the value of such a book if the examining bodies still insisted on the letter of Euclid? This point was put in the periodical *Nature* in March 1870 by Robert Tucker, Secretary of the London Mathematical Society and mathematics master at University College School. A second letter, in the issue of 26 May 1870, written by Rawdon Levett, a young teacher at King Edward's School, Birmingham, went further by suggesting that mathematics teachers should unite to form an 'Anti-Euclid Association' which would seek examination reforms. Gradually, the opponents of Euclid began to come together. In June 1870 Tucker, Levett and Wormell invited those readers of *Nature* who objected to Euclid to write to Levett. A meeting was held at Wilson's house to plan further action and it was determined to meet at University College, London, on 17

January 1871 formally to establish an association. In the meantime, the Headmasters' Conference resolved to communicate with the Government, the universities and other examining bodies to seek greater latitude in the use of geometrical textbooks.

5. THE ASSOCIATION FOR THE IMPROVEMENT OF GEOMETRICAL TEACHING

The meeting in London was chaired by Professor T. A. Hirst [36], De Morgan's successor at University College, and was attended by twenty-six 'members' plus a 'few gentlemen' who declined to join the new association. After some initial disagreement the title of the Association for the Improvement of Geometrical Teaching (AIGT) was adopted, Hirst was elected first President, Wilson and Joshua Jones (Headmaster of Wilson's old school, King William's College) Vice-Presidents and Levett and MacCarthy (Second Master at King Edward's, Birmingham) Secretaries. It was resolved

> That this Association aims to promote the general improvement of geometrical teaching, and that, as a necessary preliminary, it will use all its efforts to induce conductors of examinations . . . to frame their questions independently of any particular textbook.

Within a short time the AIGT boasted sixty-one members. Of these about fifty came from the public schools and the leading grammar and proprietary schools; the remainder were mainly not linked with educational institutions or were university men. An interesting exception was J. F. Iselin, Inspector of Science and Art Schools (see below).

As a result of Wilson's urging it was agreed that the first task of the AIGT would be to draw up a syllabus which could be the framework around which future alternatives to Euclid could be written and examined and a committee was appointed to undertake this task (37).

In fact, Wilson was already a member of a committee inquiring into the teaching of geometry, one established in 1869 by the British Association and including, from the universities, Cayley, Clifford, Hirst, H. J. S. Smith, Salmon and Sylvester and, from the schools, Hayward (Harrow) and Wilson. In his old age Wilson (38) was to forget that he served on this committee or that it ever reported! It did, in fact, report twice. In 1873 it felt that nothing so far produced 'is fit to succeed Euclid', nor was it thought that such a book would emerge from cooperative action. Instead, individual authors should be encouraged to write texts around a syllabus authorised by the British Association. The committee agreed with the AIGT that practical work should precede theoretical – a principle often reiterated but not always acted upon – but objected to the introduction on pedagogical grounds of logically redundant axioms. The AIGT syllabus was found 'good so far as it goes' (Euclid, Books 1–3), but further comment was deferred (39).

The AIGT's aims were, in fact, remarkably modest, for most members had a very limited conception of what geometry might mean: proofs would be changed, axioms adjusted, theorems reordered, but the basis would remain unaltered. Two of the members, Laverty of Queen's College, Oxford, and Iselin, the inspector from Kensington who represented an alternative view of education (see chapter 8, section 3), proposed more radical reforms. They wished to see arithmetical and algebraical matter introduced into geometry teaching and consideration given to technical geometry. Iselin had been brought into contact with 'architects, surveyors, carpenters, mechanics and others engaged in manufacturing trades; and . . . had seen the difficulty under which they laboured in obtaining such a knowledge of the principles of geometry as they could apply to the practical work in which they were engaged . . . He . . . hoped that the Association would keep in view the needs of this large and increasing class' (40). Hirst, who had begun life as a surveyor, was unsympathetic and could not hold out any hope of the Association's undertaking such work.

The AIGT, then, made no attempt to reunite the two streams of mathematics education in Britain, the academic and the vocational. It persevered with an élitist, non-utilitarian syllabus which bore little relation to the more practically-oriented geometry syllabus examined by the Department of Science and Art. Yet, what chance would there have been of gaining Cambridge's agreement to a technically-biased syllabus?

Here, the extent to which Cambridge, Cambridge-trained men, and Cambridge-aimed pupils dominated mathematical thought in the secondary schools and the universities must be stressed (cf. pp. 144–5 and 267–8 below).

As it was, progress was proving terribly slow. Membership had passed the hundred mark by 1875, and now included a woman, Miss Beale (pp. 172–3 below), but a survey of schools showed that although alternatives to Euclid might be being used in the lower forms, pupils reverted to Euclid as the examinations loomed. The syllabus of the new Oxford and Cambridge Schools Examination Board (1874; see p. 161 below) was greeted with enthusiasm for it merely defined content in terms of Euclid and did not imply that Euclid's 'particular exposition of the subject' was essential preparation. Alas, the examination paper retained the 'old Euclidic character' and the propositions demanded were 'pure Euclid'. Moreover, in the part of the paper required for a pass, geometry was reduced to bookwork; the universities were 'willing to grant certificates of knowledge of elementary and additional Mathematics to Candidates who [demonstrated not] the slightest amount of original thought' (41). The news in 1877 was even gloomier; the mathematical lecturers at Oxford were in favour of retaining Euclid, Cambridge had turned down the AIGT syllabus as an examination framework, and Durham and the Government were unwilling to express any opinions. Some schools were adopting the syllabus, but, as at the Notting Hill High School for Girls in London, the students 'take up the syllabus first for the sake of their intelligence, and will then take up Euclid for the sake of the Cambridge examination' (42).

Meanwhile, the British Association committee had reported for a second and last time (43). It gave the AIGT its muted support and asked the universities to consider whether or not they wished to adopt the AIGT syllabus. The only unequivocal statement, however, came from Cayley who left no one in any doubt concerning his wish to see Euclid retained.

Seven years of 'reform' had, then, produced little. In retrospect this may seem surprising for, as we have seen, the objectives of the reformers were limited and, moreover, both adequately represented the views of the mathematics staffs of the most esteemed schools in the country and had the backing of the influential Headmasters' Conference. The AIGT could also point to the pronouncements of the Taunton Commission. That such forces could be successfully resisted gives us some idea of the influence, inertia and power of the two ancient universities.

The year 1878 saw a change in the presidency of the AIGT. Hirst resigned, to be succeeded by Hayward. Although a good geometer, Hirst was ill-equipped to lead what was effectively a curriculum development project. He was not an Oxbridge graduate and lacked the necessary foothold and influence within the old universities. Again, although he could provide scholarly surveys (44), he was not a fiery speaker nor an activist. One can only wonder what would have happened had Wilson and Levett approached Sylvester rather than Hirst. Hayward, disregarding the misgivings concerning 'committee' texts, established a group to provide a textbook to accompany the syllabus. Moreover, realising the key rôles played by Oxford and Cambridge, Hayward concentrated attention on them and set up a local group at Cambridge. The Association, too, ceased to be preoccupied with geometry and began to consider other aspects of mathematics teaching.

Again, there were no immediate rewards. By 1890 (45) only London and Edinburgh Universities had accepted the AIGT's recommendations; Cambridge had still to make concessions. Now, however, other forces began to be felt in the world of mathematics education and the pressures for change were gradually building up and would soon sweep away the barriers of restraint.

6. SCIENCE, CLIFTON COLLEGE AND AFTER

Although we have written of Wilson as a mathematics educator, it is arguable that his greatest claim to educational fame is for his work as a scientist.

He was, of course, appointed to Rugby School to take charge of science teaching although he had never studied the subject experimentally at the time of his appointment. Despite this, and the fact that he taught more hours of mathematics than science, his mere presence at Rugby was a milestone in science teaching. For a public school to have a science master was remarkable. Such science education as had previously existed was usually restricted to *ad hoc* lectures and demonstrations given by itinerant scientists who would spend about three weeks in a place lecturing to the local intelligentsia and when possible duplicating these lectures in a school (46). As a step forward Eton had in 1849 instituted a weekly lecture series by visiting experts, such as Baden Powell and Tyndall [47]. Attendance was voluntary and cost 2 s (10 p) a week. A similar system was established at Winchester in 1857, but there science lessons were compulsory; their timetable position, Saturday afternoon, was that traditionally occupied by newcomers to the curriculum! Even so there were critics of science teaching to be answered and Wilson in his contribution to *Essays on a Liberal Education* (48) had to defend the subject against such critics as Dr Moberly, the Headmaster of Winchester, who had already said 'all that may be said on the worthlessness of science as a means of education in schools'.

Wilson, however, not only became a spokesman for science education, but also established a reputation as teacher and scientist. He was particularly interested in astronomy and carried out, together with his pupils, research work on double stars. He interpreted school 'science' as initially physical geography, including elementary geology and astronomy, followed by botany and experimental physics. He did not think chemistry, geology and physiology good school subjects, the first two being 'frightfully crammable'. In 1867 Wilson was invited to become a member of a Special Committee set up by the British Association to consider scientific instruction; other members included Huxley, Tyndall, Farrar and Payne (who in 1872 became the holder of the first chair in England of 'the science and art of education'). Outside the school system, Wilson spent considerable time, effort and money supporting the Rugby Mechanics' Institute. There he presented lectures on many aspects of astronomy, but his work, and that of his colleagues, went largely unrewarded. For, despite the fact that Rugby was a rapidly-growing industrial centre, the attendances were very low (49).

The late 1860s and early 1870s, therefore, saw Wilson very much occupied with the politics and practice of mathematics and science education. It is noticeable, however, that he appeared to view these as distinct concerns; problems resulting from the scientists' mathematical demands or from demarcation issues, such as who should teach mechanics, still lay ahead. It was also a time of domestic change for Wilson, for in 1868 he married. There were also changes at Rugby School when in 1869 Temple, the headmaster, accepted the bishopric of Exeter. A most unhappy chapter in the history of Rugby School ensued. The staff's hostility to the newly appointed headmaster, Hayman, was made apparent even before he took up his appointment and his time at Rugby was one of continued unrest. Eventually, the governing body was provided with the opportunity for which it had waited and Hayman was dismissed. Wilson was an active participant in an affair which brought credit to none. For science teaching too, Hayman's years at Rugby were unfortunate, for he held that though science might have a place in the education of the lower orders, it was beneath the dignity of his pupils. Moreover, Wilson received little encouragement from Hayman's successor, Jex-Blake, whom he also found uncongenial.

In 1878 Wilson's wife died leaving three young children and shortly afterwards he took the decision to leave teaching. The following year he was ordained as a priest of the Church of England; however, this was not to enable him to take up parish work, but the better to qualify him for his new post, Headmaster of Clifton College, Bristol.

Clifton College, a proprietary school which opened in 1862, had already established a tradition of recruiting headteachers from Rugby. The first of these departed to King Edward's School, Birmingham, before a boy was admitted. The second, John Percival, arrived less than four weeks before the school was due to open. He was a man of whom many stories are told {50}, yet, whatever his oddities, he proved a fine headmaster and an outstanding judge of science and mathematics teachers. In 1871 he appointed as physics and second mathematics master the young John Perry who, although he was not to remain long in school teaching, was greatly to influence mathematics education throughout the world (see pp. 147f below) [51]. Before Perry's departure in 1874, H. S. Hall, an old Cliftonian who was to become the most successful textbook author of his age, joined the staff. When Wilson arrived at Clifton he found mathematics being taught by Watson, Hall and Stevens, three teachers who taught him 'what first-rate mathematical teaching was'. They helped Wilson to realise how much could be achieved with brighter boys, although he reckoned that he himself had always been 'very successful with average and dull boys'. There was no immediate need for Wilson to teach science either, for he was fortunate enough to have several future FRSs on his staff [52]! Accordingly, he left well alone and turned to teaching classics.

He did, however, still retain an interest in mathematics and, in particular, in that art fostered by the Tripos examination, problem solving. Once a term he set 'challenge problems' to the school. One of these may, to present-day eyes, seem unfair: it was the four-colour problem, thought to have been proved by Kempe. Wilson received one proof which satisfied him – from Temple, then Bishop of London, who had composed it whilst 'attending' a church meeting (53).

Mathematics, however, occupied less and less of Wilson's time. As a headmaster he had other cares and duties. For example, he introduced a summer camp for working-class and Clifton boys and, to astonish those who believe that violence in school is a recent phenomenon, survived an attempted assassination by one of his pupils {54}.

In 1890 Wilson, now aged fifty-three, remarried and with a second family, resigned the headmastership of Clifton to become Vicar of Rochdale. This was a surprisingly 'lowly' Church post for such a high-ranking educator, but we are told that Wilson had been offered and refused a bishopric {55}.

There were even fewer opportunities for Wilson to practise his mathematics in Rochdale, although significantly his path crossed that of the young E. T. Whittaker {56}. In his late sixties, Wilson moved from Rochdale to become a Canon at Worcester Cathedral. There, in addition to his clerical and civic work, he established a new reputation as an antiquary. In 1908 he acted for a time as temporary headmaster of the King's School, Worcester, and in 1921, in his mid-eighties, he was elected President of the Mathematical Association, as the AIGT had been renamed, in its golden jubilee year. Gradually, however, old age crept up on him and in his ninetieth year he was forced to resign his canonry. He was to live for another five years before his death in 1931. By then the reforms he had sought had been achieved and indeed had become a new orthodoxy.

Wilson's was in many ways a remarkable life: he achieved and was capable of so much and yet an unwillingness to concentrate his interests in any one area meant that in none of the fields to which he contributed did he attain the very top rank. Within mathematics education he serves well to illustrate an age and an attitude. His view of school mathematics was limited, as is to be expected of one who taught only a social and academic élite. He lacked, for example, the experience of his mentor, Temple, who for a period had ventured outside the public-school system. Yet he was a genuine curriculum developer in a section of the curriculum which made him particularly susceptible to criticism; and he uncomplainingly accepted the ridicule and abuse which followed. The association he helped to found, although initially limited in membership, scope and success, was, as the first professional subject association, to provide a model for teachers in many disciplines in Britain and elsewhere. By helping to create this first subject association, Wilson made an outstanding contribution to the continuing work of improving mathematics education and, in particular, to curriculum development.

8 · *Charles Godfrey*

In the previous chapter we described how mathematics established itself within the curriculum of the older endowed schools and how the AIGT came to be established. Meanwhile, state intervention in education had sought to ensure that a uniform brand of arithmetic was taught in those elementary schools which received public money, and the State had also, somewhat unwillingly, become involved in a form of secondary education. During the period which we now describe, secondary education, too, was to be accepted as a state responsibility and mathematics teaching within that sector was to be reformed. Moreover, not only was the *practice* of mathematics education to be reviewed, but considerable attention was also to be directed at the *study* of mathematics education. Whereas, for example, J. M. Wilson was largely happy to practise as a mathematics teacher, Godfrey and his contemporaries, such as Branford, Carson and Nunn, attempted to found a discipline of mathematics education.

1. SCHOOL AND RAWDON LEVETT

Many, if not all, mathematicians are greatly indebted to their first mathematics teacher and this was certainly true of Charles Godfrey. His master at King Edward's School, Birmingham, was Rawdon Levett, one of the founder members of the AIGT (pp. 134f above). Levett, later described as 'probably the best schoolmaster I ever knew' (1), was clearly the kind of teacher profoundly to influence a young pupil. He was a Cambridge mathematician {2} who had come to King Edward's in 1869 after four years teaching at Rossall School. King Edward's was an old endowed grammar school refounded in 1552. It had boasted a mathematics teacher as early as 1743 (3) but, as elsewhere, there had been no continuous tradition of mathematics teaching and when Levett arrived he was the only mathematician in a school of 400 boys. All the boys were taught arithmetic, but only sixty or so learned any more mathematics and even these had only the merest 'elements' of geometry and algebra – those of the former being, of course, Euclid's (4).

It was his fight to replace Euclid which made Levett known nationally, and it was on him that much of the writing of the AIGT's textbook (p. 136 above) fell. The book, *Elements of Plane Geometry,* illustrates the strengths and weaknesses of Levett and the AIGT. In order to establish their mathematical respectability, Levett and his colleagues resorted to a demonstration of scholarship intended to win over Cayley and other opponents of reform, but which rendered the book useless in the hands of schoolboys. As Godfrey said, it suggested 'that geometry must have been a dull subject to learn' (5), although in Levett's hands it was not. Yet, dull or not, geometry for Levett was a theoretical subject and he does not appear to have allowed practical work or drawing {6}.

Levett remained at King Edward's School until his retirement in 1903. By then advanced mathematics was being taught to a large part of the school which had acquired a national reputation for its mathematics teaching.

Some indication of what it was in Levett's teaching methods which appealed to, and helped create, young mathematicians is given by another of his pupils, Arthur Siddons, who was later to form a renowned textbook-writing partnership with Godfrey.

> As a teacher he must have been rather unique: he made us think for ourselves and had wonderful discretion in leaving us to fight out our own battles. The one thing he abhorred was cramming: he never let a boy sit down to a scholarship paper. If ever a boy ventured to ask whether the work in hand was of any help for his scholarship examination, Levett's invariable reply would be: 'It is not my business to win scholarships for you, I have to make you love beautiful series.' . . .
>
> With our ordinary work we went on quietly but he always kept some outside subject going; sometimes it would be Astronomy – eclipses, the Harvest Moon, or the precession of the equinoxes. I remember vividly a term in which we worked at falling chains and eventually solved problems by differential equations – I remember still that he never gave us any rules for solving them, and terms later, when at Cambridge I had learnt more about differential equations, I asked him why he had not given us one particular rule that would have been so helpful, he merely replied: 'It was so good for you to flounder.'

Later, in 1902, when the changes in the mathematical curriculum had begun to get under way, Levett wrote to Siddons about the principles that should govern reform:

> I trust that in the new reforms you will aim at giving freedom to the teacher; my present position with regard to reform might perhaps be summed up in the following resolutions:
>
> (1) That it is desirable that every schoolmaster should be intelligent, zealous and stimulating.
> (2) That good schoolmasters should be well paid in cash and in repute, and that bad ones should be requested to find other occupation.
> (3) That a good schoolmaster should be allowed to do what he liked.
> (4) That examiners should go to the devil (7).

From such a schoolmaster, Godfrey obviously learned much more than the mathematics which enabled him to win a major scholarship to Trinity College, Cambridge.

2. UNIVERSITY

Godfrey entered Trinity in 1892, a time when, in the words of G. H. Hardy, 'mathematics meant to me primarily a Fellowship at Trinity' (8). Russell was a student there with Godfrey, and the Fellows included Cayley, Forsyth, Whitehead and Rouse Ball. Even more importantly, Godfrey showed, in the only way in which Cambridge then recognised, his right to be classed with these. For in 1895 Godfrey was ranked Fourth Wrangler after Bromwich, Grace and Whittaker, the position, incidentally, that De Morgan, Whitehead and, later, Hardy occupied. Although Godfrey's position was not so elevated as Wilson's, the competition was of a different order: six of the students in his year were to become Fellows of the Royal Society, an honour which now had to be earned!

The mathematics, too, had changed since Wilson's day (see p. 126 above) and no longer were students 'expected to know any lemma [of Newton's *Principia*] by its number alone, as if it were one of the commandments' (9). The influence of the French analysts had, however, yet to be fully felt. A. R. Forsyth [10], who was himself to introduce many new ideas, tells of a college examination question set about that time: 'Define a function and prove that every function has a differential coefficient' (11). Hardy was later to speak of the astonishment with which he read Jordan's *Cours d'Analyse* and 'learned for the first time . . . what mathematics really meant' (12).

The reign, or tyranny, of the coach was, however, not yet over: the person who 'knew all the obstacles, all the tricks of the examiners, and was sublimely uninterested in the subject itself' (13). His days were, however, numbered as the opposition to the Tripos grew. Early in 1906 a Special Board made recommendations for reform. It recommended splitting the examination into two parts, Part I to be taken by abler students after one year and by others at the end of their second year, and the abolition of the order of merit. The ablest students, relieved of the pressure of competing for places, would then be free to devote more time to advanced work (14). Weaker students, too, would also gain, for in the Board's opinion the performance of at least half the candidates was unsatisfactory; they demonstrated no grasp of principles and occupied themselves too much with the mere manipulation of symbols. Controversy arose and, as the Senate's vote neared, the columns of *The Times* carried arguments from reformers and those who, like Routh (p. 125 above), wished to see the old system retained. One particularly telling point was made by Joseph Larmor (15) who argued that:

> The vast expansion of pure mathematics and the specialization thereby necessitated . . . have made it impossible to frame a schedule for three years' undergraduate work, the

same for all students which would meet the legitimate requirements of all branches of that study.

Differentiated courses and options had become necessary and 'a satisfactory order of merit can be drawn up, if at all, only when all the candidates are examined in the same subjects' (16).

The reforms did not have the backing of the majority of college lecturers in mathematics; nevertheless the recommendations were agreed, albeit by a narrow margin {17}. They marked, in Hassé's words, the move from the 'problem age' with its emphasis on a highly developed technique, to the 'example age' in which understanding was required. To some critics, though, the reforms encouraged students 'to read mathematics instead of learn mathematics'. Hardy, however, viewed the reforms as largely cosmetic (18). The Tripos still distorted teaching, and not only at Cambridge. As we have noted before, Cambridge exerted an enormous influence (see Figure 3).

The study of Higher Mathematics in the British Empire is now practically concentrated at Cambridge. Thither come graduates from the Scotch Universities and all the best men trained in the provincial universities of England and Wales {19}. Practically all the professorial chairs of mathematics and mathematical physics in London and the provincial Universities of England, in the Universities of Scotland, the Colonies and India are filled by Cambridge men (20).

The modern universities, led by Cambridge men, adopted 'honours' systems based on the Cambridge Tripos and the education they provided was similarly distorted. 'Reforms' of such systems would never, in Hardy's view, provide a cure:

The only satisfactory reform of the Tripos was to abolish it, on the grounds that the only harmless examinations, without cramping effect on teaching are tests of competence with a low passing standard and no classes (21).

Hardy's views have yet to be accepted in England.

The reform of the Tripos was not, however, the first significant decision to be taken at Cambridge in the new century. It is to consideration of the changes of 1903 that we now turn.

3. ALTERNATIVES

In the previous chapter we had cause to mention the Department of Science and Art and the way in which alternative forms of secondary education were developing. We now consider these, and their consequences, in more detail.

Partly as a result of the Great Exhibition of 1851, the Government established in 1853 a Department of Science and Art (DSA) charged with

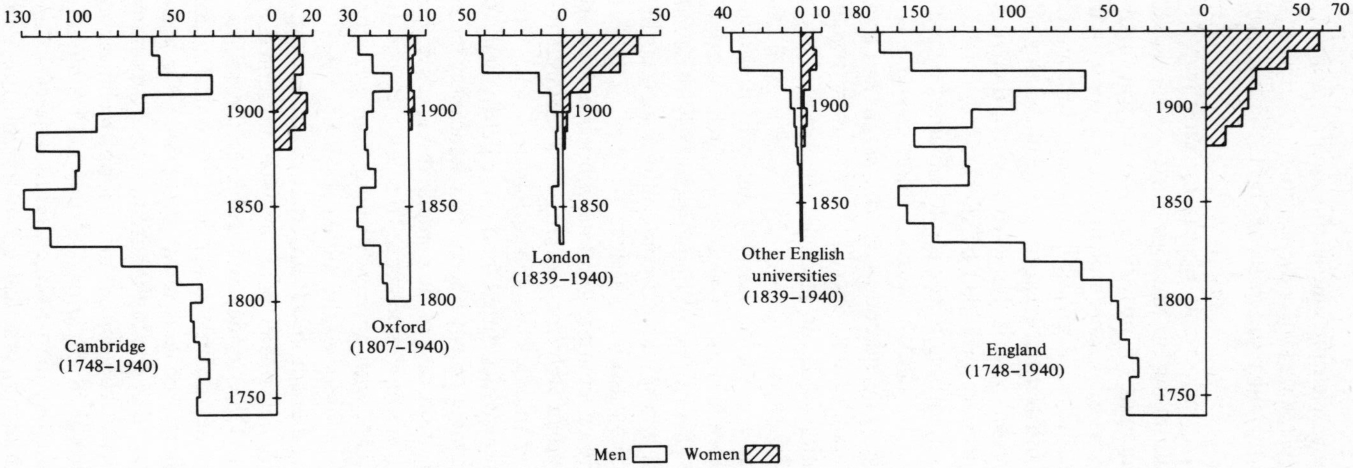

Figure 3. Annual average numbers of honours graduates in mathematics in each decade for the universities of England. (Women 'graduates' include those passing final examinations but debarred from obtaining degrees.) From Chapman, S. 'University training of mathematicians', *Math. Gazette*, 30 (1946), 61–70.

the task of improving scientific and technical education. It was soon to come, like the Education Department which exercised control over elementary education, under the aegis of the Committee of Council on Education.

The DSA sought to achieve its objects in many ways: by providing museums, by administering institutions such as the Royal School of Mines, and by encouraging the provision of science teaching which would 'assist the industrial classes . . . in supplying themselves with instruction in the rudiments of Geometry, Mechanical Drawing, Building Construction, Physics, Chemistry and Natural History' (22). Any existing school or class could apply for approval and, if this were granted, then a grant ensued. Annual examinations were provided by the DSA and teachers received additional payments according to the results obtained by pupils {23}. So, for example, in 1860 the DSA subsidised thirty classes and 1,340 candidates, mainly from private and endowed schools. By 1873 there were 1,182 classes and 24,674 candidates. Special courses were mounted for the teachers in the vacations with no fees, all expenses paid and a bonus for attending {24}.

Soon, however, as a result of the 1870 Act, which established a state system of elementary education, a new type of school came into being. This was the higher-elementary, or higher-grade school which catered for the more able pupil whose parents could afford to keep him at school beyond the age of compulsory attendance.

These schools differed from the grammar schools in several ways. First, they catered more adequately for working-class children (although the children of the lower middle-class still dominated) {25}. Secondly, they provided an education which fitted 'end-on' to elementary school. This was not the current view of what secondary education meant: 'Primary and Secondary Education cannot be compared respectively to the lower and upper storeys of a single tenement. They are rather to be figured as two adjacent tenements with an easy passage from near the top of the lower to the mezzanine floor of the higher' (26).

Yet the schools' legal position was uncertain. Certainly the 1870 Act did not permit public money to be spent on what amounted to secondary education. However, DSA money was available for science classes and from 1890 further money became available for technical education which included those subjects recognised by the DSA. The result of these financial measures was that the higher-grade schools, along with the less well-endowed grammar schools, began to adopt technical and scientific curricula. However, the Regulations of the DSA required a 'science school' to devote not less than thirteen hours a week to an obligatory course of science, drawing, practical geometry and not more than five hours of mathematics; of the remaining ten hours of the working week two might be given to manual work and two to mathematics or art. This left only six hours for English and other languages, history, geography and other general subjects (27). The curriculum of a science school, then, was as educationally unbalanced as that of the classical grammar school. The effects of this were to be considerable, but as was pointed out in 1900 'numbers of the small grammar schools are at present confronted with a choice between ruin and a transformation into schools of science and technology' (28).

Alongside this development in the schools, the DSA also encouraged the growth of evening classes and institutes. Technical education was developing apace and in 1881 the City and Guilds Institute established the first English technical college at Finsbury. A year later John Perry (p. 138 above) joined its staff and, with the encouragement of H. E. Armstrong and W. E. Ayrton, proceeded to develop a new mathematics course more suited to the needs of engineers and scientists than the academic course traditionally offered (29). The syllabus of 'practical mathematics' which Perry developed at Finsbury was in 1899 adopted by the DSA as an alternative to its conventional one. The syllabus (see p. 222) proved most popular and by 1909 it attracted more candidates than the traditional one. Perry rarely doubted the solutions he offered, and in 1887 he had informed the London School Board that 'I consider that the problem of Middle Class Technical Education is completely solved' (30).

If, however, the syllabus worked at Finsbury and at the Royal College of Science, to which Perry moved in 1896, and if it could be followed in the 'science schools' of the DSA, might it not have value in all secondary schools? The Mathematical Association (as the AIGT had then become) sought minor changes to the geometry syllabus, but

> Why not . . . let a boy jump over all the Euclidian philosophy of geometry and assume even the forty-seventh proposition of the first book of Euclid [Pythagoras' Theorem] to be true? Why not let him replace the second and fifth books of Euclid by a page of simple algebra, and give him much of the sixth book as axiomatic? . . . The present rules of the game are really a little too absurd. A difficult vector subject like geometry must be studied before algebra. Simple exercises on squared paper . . . must not be approached until one has wasted years on higher algebra and trigonometry and geometrical conics, because they belong to the subject of coordinate geometry. It is assumed that it is not until after coordinate geometry is thoroughly studied that a man can take in the idea which underlies the calculus, an idea which is possessed by every young boy with absolute accuracy . . .
>
> The young applier of physics, the engineer, needs a teaching of mathematics which will make his mathematical knowledge part of his mental machinery, which he shall use . . . readily and certainly . . .
>
> [This] method is one which may be adopted in every school in the country, and

adopted even with the one or two boys in a thousand who are likely to become able mathematicians (31).

Perry's criticisms of school mathematics teaching were to be echoed by other engineers, scientists and mathematicians. Action was demanded.

4. REFORM

The problems of the poorly-endowed grammar school were not an immediate concern of Godfrey. For, after spending the years 1896–9 at Cardiff University College, coaching at Cambridge and assisting Levett at his old school, he was in September 1899 appointed Senior Mathematical Master at Winchester College, Hardy's old school and 'then and for long afterwards the best mathematical school in England' {32}. His headmaster's confidential reports reveal how quickly Godfrey settled to his new post, impressed his colleagues and began to make changes in the way in which mathematics was taught, in particular, in relating the earlier stages of mathematics and science teaching (33). By 1900 a revised scheme of work had been drawn up and brought into use (34) which included some of the newer topics to be found in Perry's 'Practical Mathematics' syllabus; in particular the first term's geometry was 'experimental work with ruler, compass, protractor, and set squares; paper cutting' and there was also emphasis on graphical work, but, of course, later geometry meant Euclid, for Cambridge had still to yield.

The pressures for change, though, were mounting and when Perry was invited to give an address on 'The teaching of mathematics' in September 1901 to the new 'Education' section of the British Association, Forsyth, Cayley's successor to the Sadleirian Chair at Cambridge, was forced into action. It was clear to Forsyth that Perry's denunciation of school mathematics as it existed would receive widespread support and he was also aware that the British Association intended to establish a committee to report on possible improvements in mathematics teaching. Perry's syllabus would clearly have to be given serious consideration, but the syllabus and, one suspects, Perry himself were anathema to many pure mathematicians. Might not an alternative to Perry's proposals be found?

With this in mind, Forsyth approached Godfrey some months before the September meeting of the British Association and suggested that Godfrey should prepare a letter for the prospective committee outlining desirable reforms (35). Presumably, Godfrey was chosen as a recent Trinity product who had shown promise as a mathematician and was already beginning to demonstrate a shrewd but vigorous approach to the problems of mathematics education {36}.

Perry's attack on contemporary mathematics teaching proved as vehement as, no doubt, Forsyth had expected. It had repercussions in many countries of the world {37}, and attracted the support of many well-known scientists including Lord Kelvin, Oliver Heaviside, and Sylvanus Thompson {38}. Criticisms were, however, to come from some quarters; Hudson thought Perry gave insufficient emphasis to 'reasoning'; Horace Lamb felt that from secondary school upwards mathematics 'should be more or less deductive in form' and feared that though Perry's method might suit Perry, 'in other hands it is likely to become as mechanical and wooden as any other'; Larmor, too, pointed to the practical difficulties of implementation; and Forsyth did not conceal his overall hostility.

As expected, a committee to report upon improvements in teaching was established with Forsyth as Chairman and Perry as Secretary. Godfrey, meanwhile, had been collecting signatures in support of his suggestions and these were published on 16 January 1902 in *Nature*.

The proposals, which appeared in a letter signed by twenty-two masters drawn from the country's leading public schools, surprisingly, did not call for Euclid's abolition; instead, it was stated that 'It may be felt convenient to retain Euclid.' Thirty years without success had left the schoolteachers completely dispirited and Siddons, when commenting on Perry's Glasgow speech, dejectedly wrote: 'The only possible chance of any new order [of geometrical propositions] being agreed to is its adoption by the great examining bodies, such as the Universities and the Civil Service Commissioners, and the most hopeful of reformers would never hope for that' {39}. Langley, however, was more optimistic and felt that the Mathematical Association might well approach the universities and other examining bodies again (40). Indeed, the Association was to awake from its slumbers and realise that, thanks to the interest aroused by Perry, reforms might be possible. At its Annual Meeting in January 1902, Alfred Lodge, a member of the British Association Committee, introduced a discussion on the reform of mathematics teaching, as a result of which it was decided to establish a Teaching Committee which would:

(i) suggest reforms in mathematical teaching,
(ii) persuade examining bodies to revise their syllabuses and papers so that the reforms could be made.

It was also agreed 'that the wise course was to suggest moderate reforms, in the hope that we could get many mathematical teachers to approve our proposals and that with a large body of support we should be able to bring much pressure to bear on examining bodies' (41). The Committee began work enthusiastically; a (conservative) report on geometry appeared in May 1902 (42), a second, on algebra and arithmetic, two months later.

The report of the British Association's Committee, drawn up by

Forsyth, was presented at Belfast in September 1902. It was a carefully worded document which 'for the most part [suggested] only broad lines of change' (43). On geometry it was equivocal 'it is not necessary that one (and only one) textbook should be placed in the position of authority'. Privately, however, Forsyth was more forthcoming: he told Godfrey and Siddons that they were to be invited by Cambridge University Press to write a textbook on elementary geometry for use in schools. Moreover, he advised them to ignore Euclid's order for he felt that he could now persuade Cambridge to revise its examination requirements.

The final overthrow of Euclid was soon accomplished. A Cambridge Syndicate was appointed to consider the requirements in mathematics for the Pass Degree Examination and its proposal that 'Any proof of a Proposition shall be accepted which appears to the Examiners to form part of a systematic treatment of the subject' was accepted by the Cambridge Senate in June 1903 {44}.

Thus, in eighteen months, what had appeared to be a hopeless cause had, largely through Perry's initiative, been won.

Godfrey and Siddons were now encouraged to finish their text. They worked at Cambridge until mid-August – a messenger from the Press came to Siddons' rooms in Jesus College each morning to collect any manuscript that was ready for the printer – and the book was published in September. Those, at least for authors, were the days! Within ten months 13,000 copies of the complete book and 9,000 copies of vol. 1 had been sold, and it soon established itself as the most popular geometry text {45}.

As E. M. Langley, reviewing the book in the *Mathematical Gazette*, wrote:

> The work well deserves the attention it will naturally receive from the prominent part taken by the authors in the movement for securing reform in geometrical teaching. It is an excellent treatise and shows throughout its pages the marks of the enthusiastic teacher who loves his subject. An experimental portion (55 pages) precedes the theoretical (300 pages). In arrangement the authors follow the schedule of geometry recently adopted at Cambridge . . . They have given a profusion of concrete examples well calculated to lead the young student to recognise that there is geometry everywhere, and not merely between the covers of a school-book. Among these we notice the forms of constellations given as an exercise on graphs, and the motion of a point on the connecting rod of a steam engine as an exercise on loci. But while the many and varied illustrations give the book a very different aspect from the pages of the Potts and Todhunter of our youth, there is no decline from Euclidean rigour in the formal proofs, though a well-chosen notation and use of symbols enables these to be given concisely (46).

Not only did Cambridge relax its rules so far as Euclid was concerned, but there were other changes: for example, graphs and the use of four-figure (as opposed to seven-figure) logarithm tables were introduced into the Previous Examination.

Such reforms, however, took place within the context of an even greater reform. The legal standing of the publicly-financed higher-grade schools was dubious, and led in 1899 to an engineered test-case in which a district auditor, Cockerton, disallowed London School Board expenditure on higher-elementary classes. This precipitated the establishment of a state system of secondary education, a move effected by means of the Education Act of 1902 (47). The nucleus of the new system included many endowed grammar schools and the majority of the old higher-grade schools. Initially, the Regulations permitted the coexistence of two types of schools: Division A schools (formed from the old 'science schools') which devoted most of their time to science, and Division B schools which were only required to devote a third of their time to science, mathematics and manual subjects. Now, however, the effects of the unbalanced DSA curriculum came to be felt. There was a reaction to such a scientifically biased course of study and fears were expressed concerning the future of literary education. The Regulations were revised in 1904 and laid down a compromise course (48) in which 'not less than 4½ hours per week must be allotted to English, Geography and History; not less than 3½ hours to the Language where only one is taken or less than 6 hours where two are taken; and not less than 7½ hours to Science and Mathematics, of which at least three must be for Science'. In girls' schools the time given to science and mathematics could be reduced to one-third of the total number of hours (if less than 22) provided that at least 3 hours were devoted to science.

Thus, concurrently with the reform of the mathematics syllabus, a vital reconstitution of secondary education was taking place. (We note that this was to happen again in the 1960s.) For these reasons it is difficult to present a balance sheet for the period indicating debits and credits. As we shall see in the next section, Perry's views were to be translated into action in some quarters of the secondary-school system. However, as Armstrong was later to write (49), Perry's methods made special demands on the intelligence of the teacher and few really understood him. One suspects that, taking into account the switch in bias of the higher-grade schools as they were absorbed into the new state secondary system, it was the reforms acceptable to the academic Forsyth rather than the technical Perry which gained more general acceptance {50}. Concessions were made to the scientist and engineer, but these were limited. More importantly, few concessions were made to the boys and girls who would end their mathematical education at the age of sixteen. Yet this latter group was now to attain considerable significance {51}.

5. CHAOS AND REACTION

Godfrey, writing in 1920 (52), compared pre-1903 to an ice age of geometry teaching. In 1903 came the break up of the ice.

> The effect seemed to us at the time to be very great; but I suppose that as a matter of fact a great majority of schools went on teaching in the old way, and no doubt many to this day have been quite unaffected by the new spirit. But in those schools which moved with the times the immediate effect was probably rather chaotic. This was natural and quite inevitable . . . We did not at once realise how to make the best use of our liberty. One mistaken development was the great abuse of drawing and measurement. These methods have their proper and permanent place in the teaching of geometry, but many of us did not understand at the outset their proper limitations.

Godfrey and Siddons' book was by no means the only geometry text to be published in the early 1900s. Wolff (53) lists a number which he classifies as 'after Euclid', for example, Hall and Stevens; in 'the sense of Perry', for example, Sanderson and Brewster; and 'compromise', for example, Godfrey and Siddons and, that which Wolff admired most, but which attracted few adherents, Fletcher (54).

Such an abundance of alternative courses proved perplexing for teachers. Where were they to turn for guidance? How were they to cope with new types of examination questions and how were examiners to set questions which did not depend upon the order and spirit in which the pupils had learned the subject? Wouldn't it be better if all schools agreed to use one fixed order?

There was a reaction against the new freedom. As Siddons later wrote: 'being free from one set of fetters some people could shackle themselves with another . . . just for the sake of simplifying examinations' (55).

Hobson, too, was to comment in 1910 on this reaction and on the way in which teachers were allowing the curriculum to become dominated by the examination system:

> From the point of view of the examiner it is without doubt an enormous simplification if all the students have learned the subject in the same order, and have studied the same textbook, but, admitting this fact, ought decisive weight to be allowed to it? I am decidedly of opinion that it ought not. I think the convenience of the examiner, and even precision in the results of examiners, ought unhesitatingly to be sacrificed when they are in conflict – as I believe they are in this case – with the vastly more important interests of education. Of the many evils which our examination system has inflicted upon us, the central one has consisted in forcing our school and university teaching into moulds determined not by the true interests of education, but by the mechanical exigencies of the examination syllabus. The examiner has thus exercised a potent influence in discouraging initiative and individuality of method on the part of the teacher; he has robbed the teacher of that freedom which is essential for any high degree of efficiency (56).

Some guidance for teachers was contained in a circular issued by the

Board of Education in 1909 entitled *The Teaching of Geometry and Graphic Algebra*. Briefly, it suggested that teachers should recognise three stages in the teaching of geometry, and that deductive geometry in the style of Euclid should not be attempted until the third of those stages. The three stages were to be:

(1) Introductory work concerned with the fundamental concepts, and not primarily designed to give facility in using instruments.

(2) Discovery of the fundamental facts of geometry by experiment and intuition. The solving of easy riders based on these fundamental facts. This stage was intended to make the pupil familiar with basic geometrical facts and with the accurate use of instruments, and to introduce him to logical argument as used in strict theoretical geometry.

(3) A deductive development based on the propositions dealt with by experiment and intuition in the second stage {57}.

It was a most impressive document which also served to illustrate the enormous changes in attitude and expectation that had taken place in less than a decade {58}.

Not surprisingly, Godfrey was to express his support for the 1909 circular, and, indeed, in that year he and Siddons published their *Geometry for Beginners* which echoed much that was to be found in the circular.

Geometry was not the only field, however, in which Godfrey interested himself. Although his view of geometry teaching did not tally entirely with Perry's, for Godfrey wished to retain greater emphasis on deductive work, he was with Perry in wishing to make mathematics more laboratory-based. He, too, was firmly of the opinion that stress should be given to the utility of mathematics.

> I hold that Mathematics should be taught as a tool. If regarded in this light, elementary mathematics gains rather than loses in dignity. The advanced pure mathematician has a position in the world of study to which everyone will defer; the elementary pure mathematician is neither useful nor impressive (59).

Godfrey established a laboratory for practical mathematics at Winchester College and in 1905, together with a colleague G. M. Bell, published *A Note-Book of Experimental Mathematics* (Edward Arnold) which contained instructions for carrying out a number of simple quantitative experiments in mechanics.

By 1912, 'practical mathematics' was flourishing in several public schools (60). Here again, though, there was a reaction and in the 1920s mathematics masters closed their laboratories and gave up 'pandering to the spirit of mere utility in education' (61). What Godfrey feared happened: mechanics became divided into two parts, theoretical, belonging to the mathematician, and practical, belonging to the physicist.

The year 1905 also saw Godfrey's departure from Winchester on his appointment (at the age of thirty-two) to be Headmaster of the Royal Naval College, Osborne (Isle of Wight). This was an esteemed post, for the Royal Navy then played a most important rôle in Britain's Empire, and indeed one of Godfrey's first pupils was the future King George VI. At Osborne, Godfrey, free from the constraints of the external examination system was able to devise a curriculum which was modern in outlook and met the particular needs of his pupils (62).

His writing partnership with Siddons continued and they combined to write a textbook on algebra (63) with 'ideas rather than manipulation as its main aim'. With E. A. Price he was also to write an *Arithmetic* (64). However, it was with Siddons that he compiled the best-selling publication of all, their *Four-Figure Tables*. By 1913 four-figure tables had almost entirely supplanted the seven-figure ones, but those in use tended to be badly set out and printed, and inadequately bound. Cambridge University Press, at the prompting of Godfrey and Siddons, produced tables which were better spaced out, easier to use and better sewn. They were rewarded by sales which by 1980 totalled over six million.

Godfrey, somewhat surprisingly, was not to write a 'Calculus', although that subject is introduced in the second volume of *Elementary Algebra*. This was, of course, a topic which became established in the curriculum of the sixth-form during the early 1900s, and one to which we shall return in later sections. Here, however, it will be used to throw light on one of the major barriers to successful reform, namely, the poverty of the teaching in the secondary schools and the lack of qualifications of the teachers.

Many of the hopes of the reformers came to be dashed when the new materials reached the classroom. Thus, for example, Godfrey and Siddons were to complain that teachers treated the practical and theoretical geometry of their *Elementary Geometry* as two distinct and unrelated entities, the former to be despatched before the latter was introduced. Moreover, a sample of those teaching mathematics in secondary schools showed that of those teaching mathematics only 47 per cent of the men and 52 per cent of the women had read any mathematics at a university; and that the proportions of those who had studied calculus were 44 and 33 per cent respectively. As W. C. Fletcher, the Chief Inspector of Secondary Schools, remarked

> The efficiency of individual teachers [cannot] . . . be measured by . . . academic qualifications . . . None the less when the question is not of an individual or of a small group but of a large number, it remains true that the lack of good qualifications must seriously limit the efficiency of teaching (65).

The question of whether the projected reforms could conceivably have been carried out by the existing teaching force was rarely put. It might indeed be argued that if this were made a condition for the initiation of reform, then development would never take place. Yet these particular reforms were attempted at a time when there was very little provision for in-service education. It is true that the Mathematical Association made some attempt to adjust to the new demands; local branches were opened and from 1907 or so onwards the *Gazette* began to concern itself more with pedagogical questions. (The Association was rewarded by a very significant increase in membership; from about 230 in 1900, to 375 in 1905, 570 in 1910 and 750 by 1914.) In general, though, help for teachers was limited. No wonder then that Barnard, a fellow public-school master and author, was to write in 1912 concerning Godfrey's proposals on the reform of the teaching of algebra:

> Like many other reformers, Mr Godfrey is at no pains to consider whether his scheme is practicable under existing conditions. He will hardly deny that a much higher degree of competence on the part of teachers would be required than at present. Is he sure of a sufficient supply of competent teachers (66)?

It was, then, a period which not only saw considerable attempts at reform, but which demonstrated clearly some of the problems of curriculum development: that many schools were unaffected by change and 'went on teaching in the old way'; that in many cases the intention of the reform 'was quickly forgotten or not understood'; that an absence of 'an accepted guiding principle' led to 'various fantastic and ill-balanced developments'; that an ill-equipped teaching force is likely to adhere 'slavishly to the textbook'; that the 'backwash' of examinations can distort aims and methods; and that

> It is not sufficient to take the ordinary teachers . . . and by a short course of lectures arouse in them a temporary enthusiasm for new methods. I have myself tried that plan. The enthusiasm was there and all seemed intelligent and interested until I followed the teachers into the schools and heard the reproduction and extension of what I taught. In many ways the patching of the new cloth into the old garment was a lesson in more senses than one (67).

6. A GOLDEN ERA

Writing in the *Mathematical Gazette* in a tribute to that journal's fiftieth anniversary (68), T. A. A. Broadbent described the period 1900–14 as 'perhaps the golden age of the *Gazette*'; indeed, that period was a golden time for mathematics education in England. In preceding sections we have observed the significant changes which took place in mathematics teaching. Equally impressive were the attempts to supplement what Nunn referred to as Perry's 'pedagogical pragmatism' (69) with deeper studies of possible principles, theories and guidelines.

Leaders in this movement were Branford [70], Carson [71], Nunn himself, and, of course, Godfrey. Branford's *A Study of Mathematical Education* (Oxford University Press, 1908), which was translated into German and Russian, was described by Nunn as 'the most important and original of recent English contributions to the pedagogy of mathematics'. Less encyclopaedic, but equally worthy of respect, was Carson's *Essays on Mathematical Education* (Ginn, 1913). In these Carson goes beyond the glib and investigates the possible meanings of so many current catchwords. Thus in one essay he considers what is meant by 'real', 'useful' and 'concrete' and, for example, concludes that

the essence of reality is thus found in definite recognisable precepts or concepts, and is therefore a function of the individual and the time; what is real to me is not necessarily real to another, and much that was real to me in childhood is no longer so. It is for the teacher to determine the realities of his pupils and exemplify mathematical principles by as many as are suitable for the purpose. He will also find it necessary to enlarge their spheres of reality . . . [A common policy] confuses the many worlds of reality, different for each individual, with some absolute world of reality supposed to be common to all. This absolute world is usually based on those applications of mathematics which have some commercial or scientific utility, such utility being considered to involve reality for the pupil (pp. 36–7).

As Carson shows, 'utility' and 'reality' are far from synonymous – a point which should be borne in mind whenever that modern catchword 'relevance' is used.

Carson was also, in cooperation with the American D. E. Smith, to write textbooks on algebra and geometry. Another significant contribution to mathematics education was his initiative in 1911 in forming a teachers' association intended to have a wider appeal than the Mathematical Association. As we have seen, the latter's membership had grown rapidly in the first decade of the century, yet there was criticism that the *Gazette* could not be read by the ordinary teacher; it and the Association's thinking were geared too much to the male public-school master with a good Cambridge degree. Carson, Head of Mathematics at Tonbridge School, Kent, established the new body – the Association of Teachers of Mathematics for the South-Eastern Part of England

to facilitate interchange of experience and opinions, and promote common action, among teachers of Mathematics, including Arithmetic, in schools of every type . . . There is not a Mathematics of the Elementry School, another of the Secondary School, and so on; the subject is one and indivisible. Differences of treatment there are or should be . . . But these are merely different aspects of a single structure, and since a structure can only be comprehended by viewing it in its various aspects, it follows that those whose knowledge of mathematical teaching is confined to experience in, and the traditions of, schools of one type, are debarred from a comprehensive view of the subject in all its possibilities (72).

The first number of the Association's *Journal* appeared in 1911; it listed forty-six members, mainly from secondary schools, but, somewhat surprisingly, with the majority female {73}. The contributions to the *Journal* were weighted in favour of secondary-school mathematics but Carson ensured that the elementary school was not neglected. The Association was, however, to suffer a great blow when, in 1914, Carson, its Chairman, editor and initiator, left Tonbridge for Liverpool University. A new editor for the *Journal* was found, but he produced only one issue, No. 8 (July 1914), before an even greater blow, the commencement in August 1914 of the First World War, effectively brought the Association's activities to an end. The need which it had attempted to meet, was to remain.

Nunn's contributions to education extended far beyond the confines of mathematics. As Director of the Institute of Education, London University, as author of the popular *Education: its Data and First Principles* (Edward Arnold, 1920) and as a consultant to many bodies, he exerted influence in all parts of the school curriculum. In addition to this, however, he made gigantic contributions to mathematics education. He was to wield considerable influence within the Mathematical Association, and to write numerous reports and articles; his thinking pervades the sections on mathematics in the Spens Report (see pp. 186–7 below); his *Relativity and Gravitation* (University of London Press, 1923) demonstrates his ability to popularise advanced mathematics; and his three volumes on the teaching of algebra remain unrivalled as a detailed translation of theory into practice within one area of the mathematics curriculum. Notwithstanding all this, his greatest contribution to mathematics education probably resulted from his teaching and the influence this had on future teachers and teacher-trainers such as C. T. Daltry and Elizabeth Williams (see chapter 9).

Godfrey's contributions to the study of mathematics education were quantitatively more modest, although his early articles in the *Mathematical Gazette* with their emphasis on principles and theories rather than mathematical detail were for that journal innovative. His best-known work, the preliminary chapters to Godfrey and Siddons' *The Teaching of Elementary Mathematics* (Cambridge University Press, 1931), the source of the unattributed quotations below, appeared after his death. These were written about 1911 and found later among his papers.

Like Tate (pp. 113f above), Godfrey wished to build a curriculum upon a psychological basis, but no longer was it faculty psychology. This he rejected:

> But when it is said that mathematics develops the memory, the logical and reasoning faculty, the power of generalisation, develops all these powers as applied not only to mathematics but also general activities – well. I hope that it may all be true, but I have not met with a proof (p. 38).

This rejection of faculty psychology was crucial for it also meant that no longer could Godfrey argue for teaching mathematics as a 'mental discipline'. Some of his contemporaries, however, were unwilling to set such a powerful (even if discredited) argument aside {74}. In place of faculty psychology, Godfrey favoured that derived from Herbart {75}. Although, this could not offer a justification for teaching mathematics, it did provide guidelines for method:

The Herbartian allows the mental content to grow by laying hold of ideas that will hook on to the old. He will not be much moved by the contention that this or that bit of work will strengthen this or that intellectual muscle, for he does not view the mind as a muscular system. He will rather enquire what affinities there are between the new and the old. And the old will include not only old knowledge but old experience; he does not, in fact, discriminate between knowledge and the results of experience, for he holds that knowledge and experience really form a single whole (p. 15).

The principal justification for teaching mathematics was for Godfrey, as for Perry (although they may have differed slightly in their reasons for reaching the same conclusion), the subject's utility:

In England we have a ruling class whose interests are sporting, athletic and literary. They do not know, or if they know do not realise, that this western civilisation on which they are parasitic is based on applied mathematics. This defect will lead to difficulties, it is curable and the place for curing it is school (p. 35).

The syllabus should be constructed to give boys a 'mathematical outlook'; calculus, then, was essential, for without it much of chemistry, biology, economics and statistics would remain unintelligible.

The world has moved on and left school mathematics in a backwater. Geometry stands as a venerable monument of antiquity . . . Formal and demonstrative geometry is not going to help us very much on the side of outlook . . . Now consider school arithmetic . . . mainly it is a complete collection of methods for solving all problems that have been set by all examiners since the invention of printing, a snowball that still grows, a burden to boyhood, a nightmare to mathematicians . . . The invention of logarithms has left much of arithmetic on the scrap-heap, a shrine where it is still worshipped . . . Algebra is perhaps the mathematical subject which gives the smallest return for the amount of time spent on it . . . the main preoccupation of teachers is to impart to their pupils a higher degree of mechanical manipulative dexterity in handling algebraical expressions . . . They want boys to *understand in order to manipulate correctly*, whereas their ideal should be just reversed. The ultimate aim should be, not manipulation, but understanding and outlook.

English education is dominated by examinations. Examiners cannot test outlook and they can test understanding only by testing manipulation; teachers have to supply what examiners demand; it is not surprising then that many teachers have mistaken the means for the end.

If time can be saved from algebra, there is no difficulty in using it. Why should a 3-dimensional boy be tied down to a 2-dimensional geometry? . . . Now that geometry has become numerical and mathematical tables no longer a luxury, trigonometry cannot be kept out of the non-specialist course . . . with trigonometry, we begin to get

> 'outlook' . . . We now come to mechanics . . . I am all in favour of amalgamating the mathematics and physics staff. The physics or engineering man, if he is a competent mathematician, will often be a better mathematical teacher than the pure mathematician; the latter will generally be disinclined to regard his own subject as a tool: he may treat it as self-contained, insular . . . The more mathematical teaching looks outwards, the better it will be for schools, the place for self-centred mathematics is the university (pp. 40–2).

Godfrey gradually appeared to recognise that his thoughts were running in advance of what could possibly be implemented in schools. The practicalities of which Barnard had written (p. 155 above) were unlikely to have passed him by. Perhaps it was for these reasons his writings lay unpublished.

Indeed, the era had spawned so many new ideas that the school system suddenly appeared to become saturated and to refuse to contemplate anything further. Many innovations or suggestions were completely ignored. Thus David Mair's *A School Course of Mathematics* (Oxford University Press, 1907), a unified approach based on Branford's theories, and utilising the problem as a motivation for mathematical activity, had little impact. The subsequent text nearest to it in spirit was School Mathematics Project (SMP) *Book 1*, published in 1965. That same SMP series also utilised an idea advanced by Carson; distinguishing, say, 'minus one' from 'negative one' {76}. Yet another fore-runner of SMP mathematics was W. J. Dobbs' *A School Course of Geometry* (1913) which appeared in the same Longman's Modern Mathematical Series as Nunn's *Algebra* and Boon's *A Companion to Elementary School Mathematics* (1924). Dobbs' course was 'founded on the elementary notions of (a) motion of rotation of one plane on another, (b) motion of translation of one plane on another, and (c) folding'. According to Dobbs, such a treatment 'runs parallel to the intuitive apprehension of the pupil; . . . leads more rapidly to the acquisition of knowledge; [and] . . . serves as a more efficient training of the geometric sense'. Again, Dobbs' book had little impact, for it led to no obvious examination: yet his thesis that 'boys are still brought up to believe that the subject of congruent triangles is at the root of all geometry . . . instead of being . . . of relatively small importance' was one to be advanced over forty years later by Dieudonné {77}. The alternative proposed by the latter was, however, radically different!

The period from 1900–14 could be seen, then, as one in which mathematics educators laid up a stock of capital which if not utilised at the time, could be drawn upon later. In fact, however, the ideas of the 1960s were largely developed *ab initio*. A shared tradition led to common solutions being offered; that many of these had been advanced before was never realised.

7. THE EXAMINATION SYSTEM

We have already seen how the examination system served first to inhibit and later to distort the objectives of reform. The effect of examinations on the teacher and on the curriculum is indeed, a recurring theme in Godfrey's writings (78). It is of interest then briefly to survey the growth of the external examination system up to the outbreak of the First World War and the subsequent efforts made to effect changes in it.

The virtues of examinations as a means for stimulating exertion and healthy rivalry and for combatting nepotism and patronage were advanced by many radicals in the first half of the nineteenth century. The latter argument led in 1853 to the establishment of a competitive examination for entrance to the Indian Civil Service, and soon afterwards the Military Academies (p. 253 below). In the same decade the first external examinations for schools were also held; for, in 1850 the College of Preceptors, an institution concerned with raising the professional standing of teachers, examined certain pupils of a school in Nottingham (79). This pilot scheme was developed and by 1870 there were some 1571 candidates for the College's pupils' certificate. By that time, however, more powerful rivals had arisen. In 1856 a competition had been held in Exeter for young men aged between eighteen and twenty-three. The subjects examined were English, the history and geography of the Empire and practical mathematics, and the prize £20. From this emerged the idea of a school examination for schools in the Exeter district, and in 1857 106 candidates entered. This time the examination was a 'pass–fail' one: 16 out of 34 seniors (up to eighteen) and 22 of 72 juniors were adjudged to have passed. The examiner was Temple and it was he who in February 1857 wrote to the Master of Balliol College, Oxford, to ask if the university would appoint a board to oversee and encourage such school examinations: 'If Oxford began, Cambridge would soon follow. In this way the universities would give guidance to those schools which is sadly needed' (80).

Oxford was persuaded to establish a Delegacy to conduct examinations in that same year and, as Temple predicted, Cambridge was quick to follow (1858). The regulations for the Oxford examination of 1858 stipulated that all candidates should satisfy the examiners in arithmetic. After that juniors might offer mathematics as an option (Euclid Books 1 and 2, arithmetic and algebra to simple equations, with additional questions on Euclid Books 3, 4 and 6, quadratic equations, progressions and proportion, plane trigonometry, not beyond the solution of triangles, the use of logarithms, mensuration and practical geometry). Again, mathematics was an option at senior level and to pass it was sufficient to show a knowledge of algebra to quadratic equations and Euclid Books 1–4 (i.e. the core for the senior was slightly greater than that for the junior); additional questions introduced problems from applied mathematics. In 1858 examinations were held at eleven centres; 750 junior and 401 senior candidates entered and 280 and 150 respectively were awarded certificates. 575 juniors and 308 seniors opted for mathematics. The seniors were considered to have performed more satisfactorily at pure mathematics than at any other subject. However, it was felt by the Cambridge Syndicate that the mathematics of their juniors showed imperfect preparation and suggested defects in the methods of teaching and the textbooks followed (81). The examination and its syllabuses had Temple's hall-mark, namely, the desire to impart a liberal education.

Yet Temple's involvement did not mean that when at Rugby he entered his pupils for the examination. The 'strong' schools found that their recommendations proved more valuable to leaving pupils than a certificate from the Oxford Delegacy. It was only the threat of possible state intervention in the examining process which in 1873 led the leading schools to establish their own board, The Oxford and Cambridge Schools Examination Board. Its syllabuses differed from those of the Oxford and Cambridge Locals and, in particular, appealed more to Miss Beale (p. 173 below) who urged girls' schools to adopt the new system. As a result, by 1903 the Oxford and Cambridge Board examined 100 boys' schools and 92 girls' schools (cf. p. 173 below) {82}.

It will be noted that the early examinations had a high failure rate, something which continued and which caused great concern.

The involvement of universities in examining external students was, of course, not new. The London University Matriculation examination (see p. 214 below) was taken as a qualification by many who never intended to read for a London University degree, and, as a result of the Oxford Delegacy's establishment, the London Committee of Senate agreed to make it possible for its examination to be taken in centres outside London. Soon new universities were created and with them new examining boards. By the first decade of the twentieth century the school examination system had grown to frightening proportions: a survey conducted in twenty-seven secondary schools in Lancashire revealed that twenty-six different examinations were being taken in a single year and that each pupil might take up to seven external examinations in the course of his or her school life. These took various forms: some were openly competitive to select the best students for particular posts or scholarships, some attempted to classify those students who were worthy of distinction or distinctly above average in any subject, others aimed to certify that students had been reasonably well educated (83). Unfortunately, these aims had become confused and examiners attempted to cover several aims in the same paper.

The results of the Oxford and Cambridge Board exemplified this problem. The School Certificate examination 'intended as a test of general education' was in 1910 passed by only 41.3 per cent of the candidates (38.2 per cent were successful in elementary mathematics and 50.3 per cent in additional mathematics). Not unexpectedly there was criticism of such a high failure rate, as there was of the way in which the Oxford and Cambridge scholarship examinations were forcing the school to become 'the ante-room to the university' {84}. The question of whether or not examinations imposed too high a price on education was raised but the response was that on balance they performed a useful purpose {85} and that rather than being removed they should be brought under control and directed so as better to aid educational ends and less to militate against their achievement. The first major official step led to the creation in 1917 of the Secondary Schools Examination Council (SSEC) to act as a coordinating body. Existing boards agreed at that time to modify their examinations and recast them into the First (or Lower) and the Second (or Higher) School Certificate Examinations.

An often forgotten feature of the examination system is the way in which it has stimulated work in statistics and educational research. For example, in 1888 and 1890 F. Y. Edgeworth published papers in the *Journal of the Royal Statistical Society* on problems of errors of marking and the significance of the demarcation lines drawn for the purposes of awarding distinctions. Several years passed before Edgeworth's pioneering work was developed further (86); however, the connection between examinations (and, later, more general methods of assessment) and educational research has remained strong.

Godfrey was always aware of the enormous influence of examinations on mathematics education in the secondary schools and, not content to criticise the examination system from outside, sought to reform it from within {87}. In 1912 he was appointed one of the two Examiners in Mathematics for the School Certificate of the Oxford and Cambridge Board. Perhaps he had attracted the Board's attention when attending a meeting the previous autumn to consider a possible response to the Mathematical Association's report *The Teaching of Elementary Algebra and Numerical Trigonometry*. The meeting proposed that trigonometry should be substituted for certain algebraic topics on the Pass Higher Certificate papers and Godfrey was asked to criticise sample papers. Later, in 1918, Godfrey (cf. p. 167) was to press for reconsideration of the Board's geometry syllabuses; this time, however, he had no immediate success. Shortly afterwards, though, he was able to make his major contribution to the process of examining mathematics. As an Awarder for Mathematics he participated in a 1919 conference which recommended that 'instead of papers on the subjects Arithmetic, Algebra and Geometry separately, three mixed papers should be set'.

A new examination pattern was proposed, therefore, which would encourage schools to treat, and consider, elementary mathematics as a unified subject and not the union of three disjoint strands. The Board invited Godfrey and Mair (whose pioneering 'unified' course had appeared in 1907) to prepare new syllabuses for the School Certificate and these were approved in July 1919. The new 'unified' papers first appeared in 1921. The dying down of the reform movement and the slow rate of diffusion of ideas in the non-centralised English educational system, are indicated by the way in which the case for 'unified' papers had to be argued strongly by the Jeffery Committee in 1944 (88), that until 1962 the majority of candidates for the Joint Matriculation Board examinations took the old 'separate' papers (89), and that these papers were still offered until the mid-1970s. The decentralised system offered freedom to the confident innovator, but it also provided alternative escape routes for the 'autonomous conservators' {90}.

8. THE INTERNATIONAL SCENE

England was not the only country in which educational changes were taking place or were being advocated. In the USA, E. H. Moore had shown that a leading mathematician could demonstrate concern about mathematics teaching in schools (91), D. E. Smith was attempting to establish the foundations of a discipline of mathematical education, and G. B. Halsted produced a school geometry based, not on Euclid, but on Hilbert (92). In France, a government decree of July 1905 invited 'the teachers to follow a method entirely new in geometry' (93) – that the 'entirely new' was basically a book published thirty years earlier by Charles Méray (94), need neither worry nor surprise us! In Germany, Klein was giving the lectures which were later to be published as *Elementary Mathematics from an Advanced Standpoint*.

The need to exchange information and experience was strongly felt and this manifested itself in an increased interest in the work of the Education Section of the International Congress of Mathematicians (ICM).

At the 1908 meeting held in Rome, papers were presented on mathematical reform in many countries; for example, by Gutzmer (Germany), Smith (USA), Fehr (Switzerland) and Godfrey (England) (95). These papers indicated the level of activity in the various countries, but could not answer questions concerning details. For a proper understanding of each country's problems and attempted solutions, something more was needed.

The idea of an International Commission to enquire into mathematical education had been floated by Smith in 1905 in *L'Enseignement Mathématique*; now he was able to submit it formally to the ICM. As a result, the 1908 Congress passed a resolution empowering Klein, Fehr and Sir George Greenhill to form a commission which would compare methods and syllabuses for mathematics education in the secondary schools of the various countries. This commission, known as the 'Commission Internationale de l'Enseignement Mathématique' (CIEM – ICMI in its anglicised form), had Klein as its President, and met for the first time in Cologne in September 1908. One of the three British delegates was Charles Godfrey. The Commission had several other meetings before the 1912 ICM held at Cambridge (England). At some of these meetings general topics were discussed. At others, for example that in Milan in 1911, attention was concentrated on specific questions – in that particular case on 'What mathematics should be taught to those students studying the physical and natural sciences?', 'What is the place of rigour in mathematics teaching?' and 'How can the teaching of the different branches of mathematics best be integrated?'(96); three questions that are equally relevant today!

These meetings enabled Godfrey better to appreciate other approaches to mathematics education. For example, reporting back to the Mathematical Association after the Milan meeting, he spoke (97) of the different ways in which countries tackled the problem of rigour. To Italians rigour meant not Euclid, but Peano and Hilbert, whereas 'in Germany the freest use is made of intuition and the sphere for rigorous mathematics is admitted to be the university'. As he saw it, there was a 'rigour spectrum' as far as mathematics education was concerned and the order was Italy, France, Britain, Germany. He also observed that Continental educators appeared to possess a greater vocabulary for describing features of mathematics education {98}.

The main work of the CIEM was, however, preparation of a vast report on teaching practices in member countries. Each country appointed a subcommittee to prepare national reports, often in many volumes, and the result was outstanding in terms of quantity and quality {99}. Certainly, nothing on the same scale had been attempted before, or has been attempted since.

In addition to receiving the various reports, the 1912 Congress paid special attention to two problems, namely 'The mathematical training of the physicist' and 'Intuition and experiment in mathematical teaching in the secondary schools'. Godfrey was responsible for preparing the English contribution on the latter subject and his report (100) is of interest in the way in which it shows the extent of the reforms of the past ten years and the proportion of schools that had adopted new methods.

It shows, for example, that about half the 370 schools {101} which replied to the questionnaire undertook out-of-doors practical work in surveying and mensuration using such instruments as the theodolite and the plane table. Descriptive geometry (in the sense of Monge) was now being treated as mathematics rather than art, and was taught by over half the schools.

Elementary statistical work – the diagrammatic representation of data – was a common feature {102}, the graphical representation of functions was now taught in almost all schools, and (somewhat more surprisingly) about half the schools claimed to use vectors in connection with the teaching of mechanics and/or complex numbers. Graphical methods were used in statistics and the battle for the adoption of four-figure tables had been won. The first decade of the century had seen a 'marked advance' in mathematical education in England.

Yet another modern movement was 'calculus for the average boy'. For twenty years or more it had been the custom for those boys intending to compete for university scholarships in mathematics to study the calculus at school. More recently, those preparing for entrance to the Military Academies had begun to study it, as had those hoping to become engineers. Now attention turned to the boy aged sixteen to eighteen and of average ability {103} and, indeed, to pupils under the age of sixteen.

The CIEM decided that a study of the position then occupied by the calculus in secondary schools would form a suitable subject for discussion at its 1914 meeting to be held in Paris. Once again Godfrey was entrusted with the preparation of the English report. The other subject for discussion was to be 'The place of mathematics in higher technical education' and the speakers were to include Klein and Borel {104}.

The Sub-Commission on Calculus sought information from the member countries on, for example, what proportion of schools considered functions of several variables, discussed the remainder term in Taylor's Theorem, introduced the integral as the limit of a summation or as a primitive function, called attention to non-differentiable functions and dealt with irrationals logically and systematically (which is not the same as rigorously).

Godfrey was able to report {105} on the success which attempts to teach calculus in England had achieved and of the support provided by physicists and engineers. Such opposition as there was came from those who felt that teaching the calculus would mean less emphasis on the manipulation and formal side of algebra. The biggest obstacle to change had been '*vis inertiae* – the most powerful force in educational matters'.

An interesting paragraph in Godfrey's report concerns rigour and demonstrates the great changes that had occurred in university mathematics in the past twenty years, mainly as a result of such books as Jordan's

Cours d'Analyse (p. 143 above) and its English offspring Hardy's *Pure Mathematics* (Cambridge University Press, 1908) {106}.

Twenty years ago, when I was up at Cambridge, the teaching given to ordinary students at that University was not conducted on rigorous lines. A movement towards rigour of mathematical treatment was then just beginning to make itself felt, and since that date has, I believe, gone to considerable lengths. Many of the best prepared students turned out at that date and earlier are unfamiliar with the modern standard of rigour, and presumably do not employ it in the teaching they now impart to others. Belonging to the non-rigorous period myself, I do not feel myself entirely competent to criticise the standard of rigour adopted in various textbooks. But books of this type do not pretend to be exhaustive in their discussion of fundamentals. Their ideal appears to be this – state nothing but the truth, but do not necessarily state the whole truth {107}.

The English will recognise yet another similarity between the present day and seventy years ago in Sir George Greenhill's statement that 'An application to our Board of Education for patronage [in connection with the ICM planned for 1916] met a very curt refusal – No funds' {108}.

9. A PERIOD OF CONSOLIDATION AND STAGNATION

Godfrey (p. 152 above) compared the happenings of 1903 with the breaking up of the ice. For a variety of reasons the 1920s and 30s were to see a 'drop in the temperature' within the grammar-school system and a slowing down in the movements for reform {109}.

Of course, the First World War had many effects, particularly on the international scene, for not until the 1960s did CIEM's work reach a comparable level. Another victim of the war – and the changing arts of warfare – was the Royal Naval College, Osborne, which closed in 1921. Godfrey found new employment as Professor of Mathematics at the Royal Naval College, Greenwich. Unlike Osborne, RNC Greenwich corresponded to a university college and so Godfrey had now to concentrate more on higher-level work; however, his interests in school matters and, in particular, the teaching of geometry continued. As we saw earlier, the freedom allowed to teachers had resulted in a variety of approaches, not all of which were successful. Pressure was renewed for guidance on a best possible sequence and advice was sought from the Mathematical Association. Godfrey was opposed to the imposition of a fixed sequence on teachers and felt that 'it was undesirable that the Association should attempt to crystallise teaching'. Writing in 1920, he felt that the 'movement of reform' was not yet over. He, himself, was already looking forward to the next advance in geometry teaching. Indeed, that year saw the publication of his and Siddons' *Practical Geometry* intended to 'be especially useful to those students whose school education ends at 14 to 16'. In this book the two authors attempted to remove that 'abrupt plunge' from the intuitive methods of Stage B to the Euclidean methods of Stage C. Instead, they introduced a Stage C in which a mixture of methods, formal and informal, and a variety of techniques including symmetry, were used, followed by a stage D 'a recapitulation stage, in which the attention is directed to a formal chain of propositions and the interest is logical rather than geometrical' (110). That deductive geometry should be taught to all secondary-school children aspiring to the School Certificate disturbed him: 'a boy of 13 is not ripe for the full rigour of the game: I am sure he is not ready . . . before . . . 15, and I expect that even 15 is too early' (111).

This wish to replace formal proofs based on congruence by more informal ones making use of symmetry was not shared by many contemporaries, although it was, of course, to be met with again in the 1960s.

Yet despite Godfrey's pleas he found himself in 1922 on an MA Committee established 'to consider the effects of a diversity of sequences on the teaching of geometry, and the desirability and feasibility of a return to . . . uniformity' (112). The Committee, when it reported, rejected such a fixed sequence but, as Godfrey had predicted (113), its recommendations were directed at the average teacher and were more in accord with his views of 1911 (*A Shorter Geometry*) than those of 1920 (*Practical Geometry*). Besides being lukewarm in its approach to methods other than those used by Euclid, the report reinforced the viewpoint that deductive geometry could profitably be taught to pupils under the age of sixteen although this was not the opinion of two of the most influential committee members, Godfrey and Nunn.

The reasons for this conservatism are not hard to find. First, the external examination system was still an enormous obstacle to change. As Godfrey pointed out:

> until the examining bodies have been converted – that is the next campaign – it may be necessary to attack . . . [strict deductive geometry] a year or so before the School Certificate is taken.

Secondly, there was the problem of the teachers. The report had to provide for the average teacher – his experience and his capabilities – and despite Barnard's misgivings (see p. 155 above), it is unlikely that Godfrey would have proposed changes which he thought the average teacher was incapable of implementing.

For these reasons, therefore, the 1923 report ushered in a period of consolidation in the secondary grammar schools. Yet, clearly, Godfrey was eager to proceed with what he saw as the next step in geometry teaching, but this was not to be, for early in 1924 he died of bronchial pneumonia. Siddons wrote in his obituary that:

> it is only those who knew him intimately who can appreciate how much of the reform of mathematical teaching in the last twenty years is due to his influence . . ., he seemed likely to be a dominant influence on the mathematical teaching of the country for many years to come [114].

One can only speculate on what Godfrey would have achieved, but there is no doubt that his death marked the end of an era of mathematical reform in England. The 1923 'geometry report' of the Mathematical Association had exactly the effect Godfrey had prophesied – it crystallised geometry teaching. Its excellence (although it was not without faults (115)) and the way in which it was allowed to stand unchallenged through its many reprints and several editions (116) made it a restraining influence on the development of geometry.

To use Godfrey's metaphor, it marked the return of the ice, and of the start of a second ice age which, like the first, came to an end in a period of great excitement and also of great confusion: a time of too frenzied activity which is unsettling for teacher and pupil alike.

Yet Godfrey's metaphor although appealing is, nevertheless, somewhat misleading, for it is the view of a person primarily concerned with one type of secondary education. As we have seen the period could also be seen as marking the confluence of two streams, that representing the academic, grammar-school tradition, and the newer, 'faster-running' flow of the higher-grade schools and the DSA science classes. Some turbulence then was to be expected; the breaking of a dam in the former stream merely added to the resulting turmoil. It was, with the possible exception of the 1960s, the most creative period in English mathematics education. New opportunities were seen to exist and were taken. Moreover, unlike the contemporaries of Wilson whose outlook even within the limited sphere in which they taught was blinkered and who, perhaps through no fault of their own, never developed significantly as mathematics educators, the reformers of the early twentieth century thrived on the experiences they were allowed to gain and grew professionally. As a mathematics educator, the Godfrey of 1920 was significantly different from the Godfrey of 1900. It was he who came closest in his time to resolving the tensions arising between the academic tradition and the Perry movement, those same tensions to which Vives had pointed. With Siddons, one can only regret that England was denied the Godfrey of 1930.

9 · *Elizabeth Williams*

The most obvious difference between the subject of this final biographical essay and those of previous chapters is doubtless that of gender. By the twentieth century it had become possible for a woman to be ranked alongside her male colleagues as both a mathematician and a mathematics educator. There are, however, other significant differences: for the first time the study is of someone who trained professionally as a teacher, who studied at a university other than Oxford or Cambridge, and who undertook postgraduate work in mathematics education. Most importantly of all, Elizabeth Williams' interests were to span all sectors of school education. For, during her lifetime, the concept of two independent elementary and secondary systems existing side-by-side was first weakened and later replaced by a unified scheme in which all children received first primary and then secondary education.

1. PRIMARY SCHOOL EDUCATION

Elizabeth May Larby was born in January 1895, the second in the family of four children. Her childhood was to be divided between the country where her forbears had farmed for generations, and the suburbs of London in which the family lived during term-time until the children's education had been secured [1].

So it was that at the age of four Elizabeth began school at St Michael's Church of England Infants' School, Chelsea. She attended that school for two years – each Monday paying her 'fees' of 3d (slightly over 1 p) – before transferring to Shaftesbury Road Elementary School in Forest Gate. The move from a church to a 'board' school illustrates a general movement in English elementary education. As a result of the Education Act of 1870, school boards had been established throughout England to provide elementary education for children up to the age of thirteen. Where denominational schools did not exist it was the duty of the school board, using public money, to build them. The boards were also authorised, but not compelled, to frame by-laws making attendance at school compulsory for children between the ages of five and thirteen {2}. An Act of 1880 turned this power into an obligation and for the first time England had compulsory (but, as we have seen, not necessarily free) education for all {3}. The dame schools which had survived in large numbers until 1870 now disappeared {4}. The denominational schools had still, however, a large part to play and at the time Elizabeth Larby first enrolled at her infants' school the Church was still educating roughly the same number of children as attended the newer 'board' schools.

We have already referred (p. 103 above) to the creation in 1824 of an Infant School Society, and the need to provide separate schools for young children was to be specifically recognised by the Newcastle Commission in its 1861 Report, and by the newly-established, and highly influential London School Board {5}. Nevertheless, in the nineteenth century infant education too often meant dame schools and child-minding. The trained infant teacher with a knowledge of Pestalozzi and Froebel was still a rarity, and, in general, infant education was slow to develop nationally {6}. The early board schools, too, were established in the days of the *Revised Code* and 'Payment by Results'. As the first standard examination was for children of six to seven, the infants' schools were marginally affected by the *Code*'s drastic conditions. They had, however, greater freedom than the schools for older pupils and, once the rigours of the *Revised Code* were removed {7}, the infants' schools were quicker to take advantage of the new opportunities. In 1893 a Board of Education circular recommended teachers to provide 'various occupations . . . such as will relieve the younger children, especially during the afternoon, from the strain of ordinary lessons, and train them to observe and imitate'. It was necessary to recognise and stimulate 'the child's spontaneous activity' and to pay attention to 'the love of movement . . . and to that eager desire of questioning which intelligent children exhibit' (8). By 1900 it was claimed that in no branch of elementary education 'have greater strides been made in the last 30 years than in the methods of teaching infants from 4 to 7 years of age' (9).

Yet, Elizabeth Williams' memories of her infant school days are not particularly rosy. She still recalls 'with horror the crowded "gallery" for the five-year-olds' (10). The classes were smaller, however, and the discipline less harsh than at Shaftesbury Road. One lesson at St Michael's stands out in her memory. 'Very exceptionally we were given squared paper [cf. my pp. 147 and 223] and told to copy patterns made up of squares on a blackboard grid. I was excited by this because I could *count* the squares in the different shapes, and so I could reproduce them. Who thought of this unusual activity . . . I do not know.' That squared paper should have so quickly found its way into an infants' school is, of course, surprising (11). Perhaps it was due to the influence of Mrs Boole who at that time was trying to encourage the early development of mathematical, and, in particular, geometrical, imagination [12]. In general, however, mathematics meant attempts to come to terms with the first few natural numbers and the associated tables.

Shaftesbury Road Elementary School was a bewildering contrast to St Michael's: 'a great three-storey barracks with classes of fifty or so. We learnt arithmetic by oral repetition and much practice of "sums".' Indeed, arithmetic was synonymous with mathematics in many elementary schools {13}. This was not surprising, for a fixed syllabus in arithmetic had only just ceased to be prescribed. Although teachers were now exhorted to take advantage of the newly-granted autonomy {14}, the habits and procedures of thirty years were not to be quickly set aside. Indeed, the textbooks used in most schools were built upon a scheme which appeared annually in the *Code* from 1894 to 1905 (see pp. 224–5 below).

The conditions were far from conducive to good teaching, with overcrowded, ill-equipped and depressing classrooms and often poorly-trained teachers. There were no opportunities for the children to engage in active learning and the only memorable lessons for Elizabeth Williams were given by pupil-teachers {15} engaged on teaching-practice. Their use of diagrams and, in particular, of visual representations for the multiplication of fractions shone out in the prevailing dimness and also serve in retrospect to exemplify changing emphases in teacher-training.

Following the 1902 Act there was, however, increased provision for children to transfer at the age of eleven from the elementary schools to the new secondary schools. Success in the 'scholarship examination' meant a free place at such a school. The examination included general knowledge (history and geography), arithmetic (cf. pp. 229–30 below) and written composition. The pressure to do well was considerable and Elizabeth Larby was placed in a special 'scholarship' class of six girls who spent their time practising 'sums' and 'compositions'; with the desired result in Elizabeth's case. Yet the number of children who made the transition was small, about one in twenty-three in 1911 (16). Opportunities, too, varied considerably throughout the country, and selection was clearly shown to depend more on social factors than academic ability {17}.

As an alternative to the highly regarded secondary schools a new type of school for older pupils arose within the elementary-school sector: the central school, of which over fifty existed in London by 1910. These had a commercial and/or industrial bias which was reflected in their curricula:

> The common programme . . . comprises arithmetic, algebra, drawing to scale, mensuration and experimental geometry. The arithmetic includes the use of graphs, logarithms, and contracted methods of multiplication and division of decimal quantities; and the algebra includes quadratics, square and cube roots, surds, variations and progressions. Peculiar to the commercial syllabus are long and cross tots, foreign exchange, bankruptcy, brokerage, commission and other commercial transactions, and peculiar to the industrial syllabus are simple trigonometry, graphic algebra, kinematics, and the use of such instruments as the slide rule, vernier, micrometer and theodolite (18).

Some schools also taught deductive geometry.

The technically and vocationally biased mathematics curriculum of the central schools differed considerably, therefore, from that of the elementary schools which offered only that arithmetic thought to be socially useful.

After passing the 'scholarship examination', however, Elizabeth Larby could escape from Shaftesbury Road to East Ham Girls' Grammar School. It was a move to a more privileged and esteemed section of education, where one was taught by specialist 'masters' and 'mistresses' rather than 'teachers' {19}. The first rung of the recently-established educational ladder had been scaled.

2. SECONDARY EDUCATION FOR GIRLS

Although I have referred to individuals such as Mrs Bryan and Miss Beale, little has been said in the preceding pages concerning the mathematics education of girls. So far as elementary education was concerned, girls were, in fact, treated like boys; the only marked difference being that for older girls needlework and 'housewifery' were emphasised, often at the expense of arithmetic. More significant differences existed in the other sectors of education.

There were boarding schools for girls even in the early decades of the nineteenth century, but these were often either very expensive, 'finishing' schools which aimed at equipping pupils with first 'accomplishments' – music, dancing, deportment and a foreign language – and then, as a result, a husband; or cheaper, smaller schools with a curriculum consisting of the three R's, grammar, geography, history and a little French and music. Mathematics and, more surprisingly, the classics were not seen as fit subjects for well-to-do girls {20}. The only career open to an unmarried girl from such a school was that of governess – an unenviable position – or as a companion in a rich household. It was to help protect governesses that the Governesses' Benevolent Institution was founded; a body which in 1848 opened its own training establishment, Queen's College, London. The College admitted girls of twelve years and over, but also served as an adult education centre, two of its early students being Dorothea Beale (then in her late teens) and Frances Mary Buss (a schoolmistress in her early twenties). At about the same time as Queen's College was established, Elizabeth Jesser Reid began to hold classes in her own house. In 1849 the classes became known as Bedford Square College; a name changed in 1860 to Bedford College for Women. It was as Bedford College that the institution later became recognised as a school of London University.

The aims of the two institutions were such that neither could provide a model for girls' secondary education. Here it was that Miss Beale and Miss Buss made their great contribution. The latter was proprietor of a private school which in 1850 was re-established as the North London Collegiate School for Ladies {21}. Miss Beale was (like Mrs Boole several years later) to teach at Queen's College before being appointed Principal of the Ladies' College, Cheltenham in 1858. Immediately she began to reform its curriculum: even so she dared not introduce Euclid at first, for 'had I done so, it might have been the death of the college' (22).

Soon other schools modelled on Cheltenham and the North London Collegiate School began to appear: notably those of the Girls' Public Day School Company (or Trust as it became) which was set up in 1872 {23}. In theory the curriculum in each of these schools was a matter for the individual school and its headmistress. In practice, however, the curriculum came to be determined by the new public examinations {24}: as so often happens in education, institutions created to satisfy new demands and to seek new objectives had, in order to gain acceptance, to adopt an existing pattern which guaranteed some measure of comparability. It was a time when women, in order to achieve independence, had to conform to the existing rules: they could not make their own. The mathematics curriculum which they were forced to follow was regrettably one which was increasingly being recognised as unsuitable even in boys' schools.

Miss Buss in 1881 was able to write to Maria Grey (p. 179 below), 'There is now no such thing as a "woman's education question" apart from that of education generally; and the real question . . . is one as old as education itself: how is the child of either sex to be trained to the measure of the stature of the perfect human being' (25). In some respects Miss Buss was correct; the 1902 Act did not differentiate between the education of girls and of boys, and the 1918 Act sought to ensure that girls were given equal opportunities in the newer central schools. However, there were many prejudices still to be overcome and many doubts to be removed. This was particularly the case in the field of mathematics education.

At the time that Queen's College was founded, arguments for teaching mathematics to girls were advanced but these were somewhat muted (26). F. D. Maurice, one of the founders of Queen's, pointed out that the subject could prove of interest as well as utility. The Professor of Mathematics at King's College, T. G. Hall, was even more apologetic: although mathematics might tend 'to unfit the mind for application to the purposes of life' and render its followers 'unfit to enter upon any other train of thought or inquiry' (pp. 323–4), rest assured, the students at Queen's would not do sufficient mathematics to have their mental well-being endangered! Mathematics 'chief advantage' for girls, was as a mental discipline:

> It may be a proud exercise of the intellect to read the language which Newton taught . . . but our task is a more humble one. We must teach, and you learn, the grammar of a science which demands and will repay your attention; diligence and thoughtful patience are the chief requisites to obtain success: and these being given, a reward will certainly follow (pp. 344–5).

The belief that mathematics, or, at least, serious mathematics, was not a suitable subject for 'delicate' girls was for long to be held by many educators. Writing in 1912, at a time when Elizabeth Larby was studying the subject at undergraduate level, Miss M. M. Ford, mathematics mistress at Roedean School, summed up the objections to teaching mathematics to girls as (a) 'the subject has no attraction for [them, it] is dry and uninteresting'; (b) its study 'is of little or no practical use for a girl'; (c) 'Mathematics is too difficult for the average girl, who is only successful at the expense of an undue amount of time and effort' (27).

As ever, the reply to (a) was that any subject can be dry and uninteresting when badly taught. Here Miss Ford pointed to the recent advances made in the teaching of geometry as indicative of the way in which mathematics teaching could be made more interesting. Again, Miss Ford was able to identify instances when school mathematics could have practical worth

> I do not say that school Arithmetic is going to turn out capable business women, but at least girls must be given an opportunity of learning . . . something of business terms and processes . . . Graphical illustrations of statistics, . . . the ideas of variation, are all extraordinarily interesting to the average girl, simply because she thinks them practical and useful for the affairs of everyday life (p. 34).

That girls might be physically unsuited to learning mathematics was yet another objection raised. In 1859 Dr J. S. Howson had argued that girls because of 'their more excitable and sensitive constitutions' should not be exposed to examinations. Speaking in 1911, Sara Burstall, Headmistress of Manchester High School for Girls, claimed that 'even a moderate degree of success in mathematical study . . . can only be attained at an excessive cost, in time, energy and teaching power' and 'occasionally, there is so much strain and effort . . . that, in consideration of her future health, the girl is obliged to give up the idea of going to college' (28). Three years later the same girls were being asked to work long hours in ammunition factories or hospitals!

Yet, writing elsewhere (29), Sara Burstall had some very cogent points to make. The demise of faculty psychology had weakened the case for teaching mathematics as a means of developing the logical facilities 'which are, it is thought, deficient in women'. Moreover, the reforms, instigated by Perry and others, which emphasised utility, had been entirely directed at satisfying the needs of boys: mathematics was now 'definitely directed towards the demands of careers which do not concern girls' {30}.

It is the opinion of many . . . that the amount of Mathematics demanded of girls must be materially reduced, and that it should be no longer compulsory . . . in all examinations of matriculation standard . . . For the ordinary girl, a moderate amount of mathematical study: . . . for those who will never go to College, and whose gifts are practical rather than academic, a very limited course of generalised arithmetic and elementary geometry; for the exceptional girl Mathematics as it is now . . .: such . . . are the lines on which [I] would solve the problem of the place of Mathematics in the education of girls (p. 581).

Existing practices within mathematics education for girls were being questioned then prior to the First World War. However, women were ill-represented in the general reform movement – there were, for example, no significant female textbook authors and no concessions made to the special interests of girls.

Not surprisingly, therefore, Elizabeth Larby found that the mathematics taught at East Ham Girls' Grammar School did not differ markedly from that taught in the local boys' school (which shared the same building). The syllabuses were those of the Oxford Local Examinations, and neither they nor the mathematics teaching had been greatly influenced by the more radical changes of the preceding decade {31}. However, unlike Shaftesbury Road the new school did provide 'humane treatment, free speech and keen intellectual enjoyment'.

Initially, geometry, despite Godfrey's efforts, was 'chiefly a matter of definitions and proofs of obvious properties; algebra meant dealing with symbols and operations on them'. Soon, however, it was time for the external examinations, for the girls were entered for the Oxford Junior examination at the end of their third year and the Senior examination a year later. There were three mathematical sections in the latter: arithmetic, mathematics and higher mathematics. In mathematics, algebra to the binomial theorem (positive integral index) and geometry to the end of the circle theorems sufficed for a pass, but those wishing to obtain a distinction had also to offer trigonometry up to the solution of triangles. In higher mathematics there were three divisions, higher algebra and plane trigonometry, higher geometry including analytical conics, and applied mathematics. A pass could be obtained on a single division, but to obtain a distinction two of the divisions had to be passed.

Elizabeth Larby was to obtain a first class in both the Junior and Senior examinations (in which she took the arithmetic and mathematics papers).

She then decided to try to secure admission to London University, but for this financial aid was essential. To obtain an Essex County Major

Scholarship {32} it was essential to obtain a First Division Pass in the London Matriculation examination and so she remained at school to prepare for this and studied a new topic, trigonometry (33). Since a university place was uncertain, alternative plans were made and she applied to enter the Pupil Teachers' Centre to train as an elementary-school teacher. Unfortunately, she failed the medical examination. However, she succeeded in obtaining the required First Division Pass and applied, though under-age, to enter Bedford College. She was accepted, but what was she to study? A chance meeting with the mathematics master who had recently taught her proved crucial. He surprised her by suggesting that she should read for an honours degree in mathematics. The suggestion appealed and soon Professor Harold Hilton agreed to let her attempt the Mathematics Honours Degree Course.

3. UNIVERSITY

When Miss Buss spoke in 1881 of the disappearance of the 'woman's education question' she was thinking predominantly of secondary education. Battles had still to be fought and won in the universities. As we have seen, Bedford College, to which Elizabeth Larby went, had existed for over sixty years. In its early years it hardly aspired to provide a university education; yet, by distinguishing carefully between its junior and senior sections, it created a foundation on which a university college could arise. Even at the time of the Taunton Commission its studies included pure and applied mathematics.

Bedford College, however, could not by itself satisfy the demands of those who fought for higher education for women. In 1862 a 'Committee for obtaining the admission of Women to University Examinations' was established, and five years later 'The Committee for establishing a New College for Women'. Both enterprises were soon to bear fruit: in 1868 London University instituted a General Examination for women and the following year University College opened its classes to them. That same year, 1869, a 'new' college opened at Hitchin, Hertfordshire. Six students were admitted to it to study for the courses set for the Previous examination and the ordinary degree of Cambridge University. In 1870 the University refused formally to admit these students to the examination, but raised no objection to individual examiners allowing the women students to take their papers. Five students, therefore, sat the examination which included Euclid (Books 1–4) and arithmetic, with additional papers in algebra, trigonometry and mechanics. Four were 'awarded' first class passes and one a second, but, of course, their names did not appear in the class list. Under the same informal arrangement, one student in 1872 passed the examination for the Mathematical Tripos. In 1872 the 'college' was incorporated as Girton College and the following year it moved from Hitchin to the outskirts of Cambridge. In the meantime, a Women's Examination (soon to become the Higher Local Examination) had been established in Cambridge and about seventy students attended a variety of lectures, provided by the 'Lectures Association', including some by Cayley on algebra and arithmetic. This low-level fare did not satisfy those at Girton who had set their sights on the Cambridge degree course. The Association was, however, soon to present students for the Tripos examinations under the same informal arrangements as Girton; and it also prompted the establishment in 1876 of Newnham Hall which sought to offer academic education to women. Again, considerable support was given by many senior members of the university and students were admitted first to professorial and then to college lectures.

In 1881, Miss C. A. Scott of Girton was judged as standing 'equal in proficiency to the eighth wrangler'. Nine years later, Philippa Fawcett of Newnham was placed above the Senior Wrangler. No more effective evidence of the rise of women in mathematics could have been given. Nevertheless, attempts to open Oxford and Cambridge degrees to women continued to fail. Elsewhere, matters were different. In 1878 the University of London obtained power to confer degrees upon women, and the Victoria University (Manchester and, later, Liverpool and Leeds) from its establishment in 1880 did not discriminate between the sexes. Slowly, the opposition was eroded at Oxford and women were admitted to degrees in 1920. At Cambridge the fight took even longer; 'titular' degrees were approved in 1921 but it was not until 1948, nearly seventy years after Philippa Fawcett's victory in the 'wranglership' race, that women were admitted to Cambridge degrees as equals.

There was no obvious discrimination against women students, however, at London University which Elizabeth Larby entered in 1911. Mathematics was not a subject studied by many at that time and in order to provide a sufficiently large audience there was a combined programme of lectures for students from Bedford, King's, University and Westfield colleges. Even so the class was barely twenty strong. Yet there were almost as many women as men. Moreover, the women 'had no sense of inferiority', but they were 'sorely handicapped by [their] lack of knowledge of applied mathematics'. None of the women had taken any public examination in applied mathematics and their knowledge of physics was very limited: at East Ham Girls' Grammar School only heat was taught – no mechanics, electricity, sound or light {34}. In the degree course, however, dynamics, statics and other applied topics were compulsory. Elizabeth Larby was also disadvantaged in that she had never studied the calculus; a subject which most of her fellow female students, who came from independent schools, had already met.

The inter-collegiate system of lecturing meant that the students came into contact with a wide range of teachers. At that time there were within London University several women lecturers, but few who taught mathematics. One of those who did had an international reputation, 'but she was markedly eccentric and students did not take her very seriously'. Elizabeth Larby had only one woman lecturer, who gave a 'sound but dull' course on spherical trigonometry. The male lecturers varied from the competent to, in the cases of Hilton and A. N. Whitehead, the brilliant. Hilton taught the calculus and differential geometry and impressed with the way in which he engaged the interest of his audience and made them feel 'obliged' to understand and appreciate his message. Whitehead lectured in a different manner: his eyes were focused on infinity and he 'declared his thoughts like an Old Testament prophet revealing the great truths of God'. At that time between spells as a Lecturer in Applied Mathematics and the Chair of Applied Mathematics at Imperial College, Whitehead was a Reader in Geometry; and it was on projective geometry and complex numbers that he lectured to Elizabeth Larby. Whitehead, a man of so many parts [35], was to have a significant effect on mathematics education. He was President of the London Branch of the Mathematical Association, and later of the Association itself, as well as of its new rival, the Association of Teachers of Mathematics (p. 156 above)! The titles of his presidential addresses, 'The place of mathematics in education', 'The aims of education – a plea for reform' and 'Technical education and its relation to science and literature' (36), reflect Whitehead's insistence on viewing mathematics education within the context of education in general. He was also to produce, in 1911, *An Introduction to Mathematics* (Oxford University Press), which attempted, in a novel and elementary way, to explain not how to do mathematics, but rather what mathematics is about. Clearly, such a man would exert considerable influence on those with whom he came into contact. It was to be many years, however, before the full influence was felt, when, after Whitehead's death, a reform movement in primary mathematics teaching was led by Edith Biggs and Elizabeth Williams using as a slogan Whitehead's dictum 'From the very beginning of his education, the child should experience the joy of discovery' (37).

Despite the deficiencies in her mathematical background and the fact that university requirements meant that in the first year she had also to study English and Latin, Elizabeth Larby coped well with her undergraduate work and her tutor felt that, given a year's extra study to compensate for her youth, she would easily gain that sought-for 'first'. Essex, however, refused an application to extend her scholarship and so she had to take her final degree examinations in October 1914. The result, an upper second, was a remarkable achievement for a nineteen-year-old. It set the seal on a memorable year, for, just before taking the examinations, she had become engaged to a postgraduate student at King's College – a young man whose response to the recent declaration of war was to enlist as an infantry officer. As with so many of her generation, joy was quickly to change to sorrow: within the year her fiancé had been killed on active service. By that time, however, Elizabeth Larby had embarked on what was to prove a lifelong task, the teaching of mathematics.

4. PROFESSIONAL TRAINING

The principle that secondary-school teachers should have professional, in addition to academic, training does not have a very long history. In the mid-nineteenth century the College of Preceptors in London provided what were in effect in-service courses of lectures on topics such as 'The teaching of Euclid', 'The best means of registering the progress of pupils' and 'Public examinations'. Miss Beale, too, established a 'training department' at Cheltenham, but this provided little more than pupil-teachership. It was, however, girls' education which prompted the establishment of the first significant institutions for the training of secondary-school teachers, Bishopgate (Maria Grey) Training College (1878) and the Cambridge Training College for Women (1885). The former was in the training-college tradition and its students studied for examinations such as the Cambridge Higher Local; the latter, however, recruited a number of students who had already attended a university (38). An attempt to create a training college for men proved a disaster, for the Finsbury Training College (1883) attracted few students and lasted for only three years.

The seeds of yet another type of training had, however, by then been sown. In 1876 the universities of Edinburgh and St Andrews established chairs of education, thus recognising education as a university subject {39}. The next year the Headmasters' Conference approached the universities of Oxford and Cambridge concerning teacher training. A proposal to establish a training department at Oxford under a professor was narrowly defeated. Cambridge, however, responded by setting up a Teachers' Training Syndicate to arrange lectures in education and to award certificates to those passing an examination and successfully completing teaching experience. In 1879–80 three series of lectures were given, on the theory of education, its history and its practice (40). The experiment was not very successful and the lectures were later abandoned {41}, although the Syndicate continued to act as an examining and inspecting body.

The early history of professional training for men was, then, discouraging. Neither was the cause helped by such pronouncements as Bowen's (Headmaster of Harrow) to the 1895 Bryce Commission: 'A bad man teaching history well is a far worse thing than a good man teaching history badly' (42). 'Character' was what the public schools valued most in teachers, not professional competence. Significantly, the Headmasters' Conference was not represented on a body which met in March 1897 to make recommendations on professional training. It proposed (43) a 'consecutive' course: first academic training to degree level, and then a year's professional study encompassing psychology, school organisation, method, educational administration, the history of education and practical work. Professional training, it was argued, should take place in universities and university colleges. Embryonic facilities, in fact, existed in several such institutions for, following legislation in 1890, 'day training colleges' for teachers had been established in, for example, Birmingham, Cardiff, London, Manchester and Nottingham {44}. Such colleges offered a concurrent academic/professional course and in some instances (5 per cent in 1897–8) students obtained degrees in addition to teaching certificates. Initially the colleges were intended to train elementary-school teachers, but once education had secured a foothold within the universities, emphasis switched to the production of secondary-school teachers. In 1907 it was insisted that all students studied for a degree and in 1911 concurrent professional training ended, to be replaced by the system familiar to us nowadays, the postgraduate year recommended by the 1897 Joint Committee.

By 1912, therefore, Nunn [45] could write that:

> It is now beginning to be recognised generally that a knowledge of Mathematics is not by itself a sufficient equipment for the business of teaching the subject in a school . . . The change of opinion . . . implies the recognition that 'training' is concerned with something more than the tricks of management needed to keep a class . . . in order on a hot afternoon, and that method is not merely a system of devices for insinuating unwelcome facts into dull minds {46}.

It was from Nunn that Elizabeth Larby was to receive her training in mathematics teaching. She had entered Bedford College with an interest in teaching and this interest was fanned by a fellow student and friend, Frances Pendry (Pen), who was following a four-year teacher-training course which included the honours degree course and a further year's professional training at London Day Training College (later the London Institute of Education). The whole four-year course was supervised by 'London Day' and, in order to emphasise the 'teacher-training' aspect, students received general lectures on education on Saturday mornings during their undergraduate days. 'Pen' discussed the content of these lectures, including Nunn's contributions, with Elizabeth Larby and in so doing persuaded Elizabeth of the value of undertaking professional training. In September 1914, then, she applied to Essex for an extension of her grant to permit her to train for teaching. The grant was awarded and she was offered a place in the Bedford College Education Department from January 1915. In the intervening months she took up a temporary post at Nottingham High School for Boys, replacing a recruit to the army. She was the only woman on the staff and was unsure of what she would find: in fact 'I received the utmost kindness, courtesy and even respect. I was fortified in my belief that women and men can share responsibility without regard to gender.' Although still at Bedford College, Elizabeth Larby went for her main lectures to the London Day Training College where her teachers included Adams, on the principles of education, and Nunn on the teaching of mathematics.

Adams introduced his students to the ideas of Herbart (cf. p. 158 above) and stressed the importance of illustration and anecdote. In general, though, 'the psychology lectures were not very helpful'. However,

> the classes on mathematics in the earlier years at school were surprisingly stimulating, owing much . . . to Froebel and Montessori. We saw one 'demonstration' lesson given to a small class of girls (? 9+) on fractions in which the children actually cut up strips of paper! . . . Nunn was a joy; he believed passionately in the value of algebra as expressing relationships and of graphs as their visual image. I took his ideas into the mathematics classes that I had to teach as part of my training {47}. The pattern of training at that time was dual – 3 days a week in college, the two intervening days in school. I was attached for the whole year to Greycoat Hospital, the girls' school in Westminster, where I was given . . . a large measure of independence in my planning of lessons. My practical examination took place there, with Nunn as one of the two examiners . . . My lesson was an introduction of calculus to the sixth-form mathematics group. I had chosen a logical approach; Nunn disagreed and invited me to a personal discussion with him . . . the following Saturday. This was for me the start of a valued lifelong friendship. It was in such individual encouragement that Nunn's wide influence was spread. Yet it was his educational philosophy that gave depth to our thinking.

Elizabeth Larby, then, gained much from her year of professional training, all the more because of her practical teaching experiences at Nottingham. For, in addition to the lectures and practice-teaching, she had time to read such authors as Perry and Dobbs (see p. 159 above), and was even able to see the latter in action: the group was 'almost scandalised' to observe a geometry lesson in which 'there was none of the usual verbal argument'. She was, of course, privileged to have Nunn as a tutor, and, because of the small number of mathematics students, so much of his attention. The fact that she had opted to 'qualify' as a teacher was, however, still somewhat unusual; for, far from being over as Nunn had hoped, the fight to secure professional training for all mathematics teachers was only just beginning: it has still to be won.

5. TEACHING: THE INTER-WAR YEARS

In addition to her month at Nottingham, Elizabeth Larby gained further teaching experience in the summer of 1915 when she deputised for the senior mathematics mistress at the girls' Christ's Hospital at Hertford. Once more she was called upon to teach the upper school including the university candidates. With these two spells of teaching behind her, she was undaunted at taking up her first permanent post, as the 'senior' of three mathematics specialists at Haberdashers' Aske's Hatcham Girls' School at New Cross, London. 'Of the other two, one had maths in a general pass degree; the other had not gone beyond the Higher Local in maths. The shortage of fully-qualified maths teachers in girls' schools was acute.' Immediately, she was forced to construct curricula. There was a fund of advice from authors such as Godfrey to call upon,

> but one's own attempts at finding fresh approaches, materials and types of organisation were very important in building confidence and in providing new insights and skills. For instance, algebra was introduced from numerical situations in which an unknown could be discovered by the use of a symbol; geometry was an investigation of basic relationships in important shapes in common use, made evident in constructional and measuring activities; many types of pattern were discovered or invented . . . The second-year forms made their own geometry textbooks with a succession of groups of theorems . . . We had a non-examination form which had a new fourth-year course: accounts, maps and house plans, simple statistics, pattern and design.

As we have indicated earlier (p. 167 above), there were few significant changes in the content of secondary mathematics during the inter-war years. It was, however, a time of considerable thought and experiment. In particular, the 1920s produced one teaching innovation of considerable importance. Many schools, particularly those for girls, adopted an American innovation known as the Dalton Plan. This was a move towards individualisation of instruction. However, unlike many such attempts in the 1960s and 1970s, it did not depend upon the 'atomisation' of content and the precise identification of a host of 'sub-goals'. Pupils were given considerable 'assignments' to undertake in the belief that learning to bear responsibility is a prime goal of education (48). Elizabeth Larby visited several schools to see the Dalton Plan in action and found that 'the interest and vigour were splendid, but the pressure on teachers' time and the complexity of organisation made general adoption unlikely'. Enthusiasm for the Dalton Plan soon waned, but the experiment did serve to highlight both the increased motivation which can come from suitably individualised work and also the great individual differences which exist between children; in so doing it acted as a spur for much subsequent research work. Despite the attractions of the Dalton Plan, Elizabeth Larby preferred to use her own style of classroom organisation which involved a considerable amount of working in pairs or in somewhat larger groups.

In addition to teaching secondary-school mathematics, Elizabeth Larby was also responsible for the overall mathematics schemes covering the age range five to eleven. Thus she was soon introduced to mathematics teaching at the primary-school level, a field to which she was to contribute much in the succeeding years. In this work she was greatly influenced by Margaret McMillan. There was an acute shortage of apparatus and materials in schools, and Elizabeth Larby was able to profit from Margaret McMillan's demonstration of how Maria Montessori's methods could be employed using a variety of informal everyday objects. Not only did this solve the immediate problem of the absence of manufactured apparatus, but it also militated against the artificiality and formality of much of the apparatus work that was finding its way into schools. It also served to reinforce Elizabeth Larby's belief in the view, expressed by Comenius three centuries earlier, that direct experience with 'real' objects was an essential part of the education of the young.

Elizabeth Larby's stay at New Cross came to an abrupt end on her marriage in 1922. In those days there was no place in state or endowed schools for married women {49}. Only a private school would offer employment to Elizabeth Williams, as she now was. Accordingly, she and her husband, Richard, became joint principals of a private school in north London for pupils aged from five to eighteen, an arrangement which provided increased opportunities for experimentation. She was now able to introduce applied mathematics to the curriculum and 'whatever topics in pure mathematics seemed to be congenial'. The 'little school' attracted the interest of Nunn and Elizabeth Williams' work there was eventually to lead to her appointment as a teacher-trainer.

In a national context, however, Elizabeth Williams was demonstrating what 'might be' rather than exemplifying what 'was'. The educational system of the country was in a depressed state throughout the 1920s. Yet it was a time when important educational issues were being raised and discussed. The Education Act of 1918 had fixed the school-leaving age at fourteen and abolished all exemptions {50}. It also made provision for many other improvements, for example the establishment of nursery schools, the raising of the school-leaving age as soon as possible to fifteen and the provision of 'continuation schools'. Originally, it was proposed that every boy and girl from fourteen to eighteen (and not in full-time school attendance) should attend a day continuation course for 320 hours a year, but this proposal was fiercely attacked. Eventually a compromise was reached; the introduction of part-time education for sixteen to eighteen-year-olds was postponed for seven years and the number of hours attendance required of fourteen to sixteen-year-olds was reduced. The local authorities, however, chose to ignore their obligations. Only a handful began to operate day continuation schools and when, in 1922, the London County Council closed those of its schools working on a compulsory basis, the battle was irretrievably lost. Only Rugby continued to pursue the scheme throughout the 1920s {51}.

The local education authorities argued, with some justification, that the desperate shortage of teachers and financial resources prevented their building and staffing central and day continuation schools and that their first priority should be to increase the provision of secondary education. The actual extent of this provision and the form that secondary education might take were subjects for much discussion. The Labour Party in 1922 argued for secondary education for all (52). Quite how great a step forward this would represent can be gauged from the fact that even though the numbers attending secondary schools had increased rapidly, by 1920 only one in ten elementary school children was proceeding to secondary education (53). Yet the Labour Party seemed to lack any clear view of what forms such universal secondary education might take:

> There must be local initiative and experiment . . . within the secondary system there must be more than one type of school . . . The demand of Labour for the democratising of secondary education implies no wish to sacrifice the peculiar excellence of particular institutions to a pedantic State-imposed uniformity (54).

In an attempt to define policy and to establish an accepted pattern for post-primary education the Board of Education in 1924 asked its Consultative Committee, chaired by Sir Henry Hadow,

> to consider and report upon the organisation, objective and curriculum of courses of study suitable for children who will remain in full-time attendance at schools, other than Secondary Schools, up to the age of 15, regard being had on the one hand to the requirements of a good general education and the desirability of providing a reasonable variety of curriculum, so far as it is practicable, for children of varying tastes and abilities, and on the other to the probable occupations of the pupil in commerce, industry and agriculture.

The Committee, when reporting in 1926, recommended that all children should be provided with 'some form of post-primary education' and that three types of secondary schools would be required: Grammar Schools, 'Modern Schools', and Technical (Trade) Schools {55}. The report also contained specific suggestions on how mathematics might be taught in modern schools and the senior classes of elementary schools providing post-primary education (56). Attention was drawn to the way that 'arithmetic had been too long dominated by the utilitarian value of the subject', whereas

> the amount of . . . indispensable arithmetical knowledge is in reality comparatively

small and would not in itself justify the time given to the subject . . . On the other hand . . . even for a superficial understanding [of our modern industrial society] all required . . . some knowledge of mathematical principles and their applications . . . It is desirable, therefore, that much of the traditionary arithmetic . . . should be replaced by new material . . . and there should be included in the mathematical training of all normal children suitable parts of mensuration, algebra, geometry and trigonometry, especially such as are necessary for the intelligent comprehension of some of the problems of our everyday life.

New material had, however, to be complemented by new teaching approaches:

there is need . . . for more vivid, more logical and more practical methods in teaching the subject, methods which will cause the pupil to appreciate both the beauty of mathematical truths and their practical applications . . . *Every course therefore should aim at developing in the pupil an appreciation of the meaning and teaching of a coherent system of mathematical ideas and the realisation of the subject as an instrument of scientific, industrial and social progress.*

There was, then, a timeless air to the Committee's recommendations!

Yet the Committee could observe 'great improvements in the teaching of arithmetic in primary schools during recent years'; improved teacher-training was beginning to bear fruit. Now the aim was to bring about similar improvements in the teaching of mathematics at more senior levels; and one way in which this could be done was by giving greater emphasis to practical work for:

not only does it supply a concrete and experimental basis upon which the child may proceed to abstract reasoning, but it vitalises the work for the pupil and stimulates his interest in it. It will also lead to co-ordination with other subjects in the curriculum, especially hand-work, geography and elementary science.

So as to facilitate transfer between the various types of school at 13+ it was suggested that 'for the first two years of the course the work . . . will not vary materially whether in a Grammar School or a Modern School' {57}.

The Cabinet and the Board of Education received the Hadow Report with little enthusiasm; the will and the resources to put its recommendations into practice were non-existent (58). The message which the report contained was not, however, forgotten.

Two years later, Sir Henry's Consultative Committee was asked to inquire and report upon the education of children aged between seven and eleven. Once more the Committee had specific recommendations to make on the teaching of mathematics (59). In particular, the report regretted that

too much time is given to arithmetic . . . and too little attention to the study of simple geometry form . . . this liberality in the matter of time is undoubtedly due to the traditional position given to arithmetic in the public elementary school, but it is often maintained by the importance attached to [it] in . . . scholarship examinations . . . [However] it is essential that these fundamental processes of arithmetic shall become automatic before the child leaves the primary school . . . The aim in the primary school is to secure ability and readiness in using the processes . . . [it is not] reasonable to expect a child . . . to justify the process he employs, say in subtraction or division, this is too hard an exercise of his reasoning powers and should be left to the secondary school . . . From the first, increasing attention should be paid to the applications of arithmetic to matters within the children's environment . . . In dealing with money, the children's home experiences will provide a sound foundation . . . In dealing with shape, size and weight, however, home experiences have to be clarified and arranged before they can be used. This can only be done satisfactorily through practical work . . . In this way children will acquire a knowledge of the simpler properties of spatial figures, plane and solid . . . and also form clear concepts of area and volume through their measurements of the simpler regular figures . . . By the age of ten they may reach the measurement of angles and be ready for 'Boy Scout Geometry'. Nothing in the nature of formal geometry will be attempted . . .

In all this we have had in view the child of ordinary abilities . . . We believe that the primary school can give this nucleus of knowledge and that the chief variation from child to child will lie, not in the certainty of his knowledge, but in the speed with which he can use it. But in many schools there are likely to be children who will be capable . . . of . . . more . . . It would be disastrous both to their keenness and enthusiasm and to their habits of industry if they were allowed to 'mark time'. We think that these children should be allowed to go forward to the new work of which they are capable.

The increased availability of education was already giving rise to questions about the need for differentiated curricula to cater for the wider range of abilities, aptitudes and aspirations (60). The Labour Party continued to favour the notion of separate secondary schools 'varying in type and educational methods . . . through which all children shall pass' (61). However, the Trades Union Congress objected to the segregation of children in separate schools and in 1934 expressed itself in favour of 'multilateral' schools in which 'the Grammar side and the Modern side [would be] housed under the same roof' (62). Such schools were essential if 'real parity in education or equality of opportunity' (63) were to be achieved. The creation of multilateral schools was one possibility considered by the Spens Committee when it reported in 1938 on secondary education in grammar and technical schools (64). The Committee rejected multilateralism on a number of grounds, including the fear that 'the prestige of the academic "side" would prejudice the free development of the Modern school form of secondary education' (p. 291). Instead, it supported the Hadow view of separate schools, while realising the difficulties which would arise in securing parity of esteem and in facilitating transfers between the different types of school. Attention was also drawn to the 'latent prejudice against technical or quasi-vocational studies' to be observed in schools, and to the fact that

The curriculum should be thought of in terms of activity and experience rather than of knowledge to be acquired and facts to be stored. Both the conservative and creative elements ... must be represented in the curriculum ... The studies of schools ... should be brought into closer contact ... with the practical affairs of life ... The School Certificate Examination determines the curriculum unduly. It should follow the curriculum, not determine it.

The Spens Committee had some interesting recommendations to make on mathematics. 'The content of school mathematics should be reduced ... If it be taught (as one of the main lines which the creative spirit of man has followed in its development) it will no longer be necessary to devote the same number of hours to the subject.' The Committee felt that all pupils should study science or mathematics to the age of sixteen, but that the latter could be dropped in the 'third and later' years of study. Mathematics teaching suffered 'from the tendency to stress secondary rather than primary aims'; instead of giving 'broad views' it concentrated too much on tricky problem-solving. 'It is sometimes utilitarian, even crudely so, but it ignores considerable truths in which actual mathematics subserves important activities and adventures of civilised man.' The type and 'rigour' of the logic it presented had 'not been properly adjusted to the natural growth of young minds'. 'These defects are largely due to an imperfect synthesis between the idea that some parts of mathematics are useful to the ordinary citizen or to certain widely followed vocations, and should therefore be taught to everybody, and the old idea that, when mathematics is not directly useful, it has indirect utility in strengthening the process of reasoning or in inducing a general accuracy of mind' (pp. 235–42).

The report of the Spens Committee was debated in Parliament in February 1939, but immediate action was not to follow for the war intervened. Two decades of peace had seen few of the educators' hopes fulfilled; they had been frustrated by social and economic realities. The need for secondary education for all had, however, been generally accepted and the various governmental reports had identified significant failings in the teaching of mathematics at all school levels and had made important suggestions for its improvement. In the post-war years significant changes were to take place.

6. RESEARCH IN MATHEMATICS EDUCATION

In May 1934 the *Mathematical Gazette* published a paper by Elizabeth Williams on 'The geometrical notions of young children' (65). The paper is of significance on several counts. One obvious difference between it and almost every other article published up to that time in the *Gazette* is that it concerns primary-school children. It was to prove the first attempt of many by Mrs Williams to convince the Mathematical Association, then totally preoccupied with the curriculum in grammar and independent schools, that the foundations of mathematics learning were laid prior to the age of eleven and that the needs to establish these foundations satisfactorily and then systematically to build upon them were matters of vital concern to *all* those involved in mathematics education.

The paper is, however, of interest for other reasons, for it indicates both a recognition of the need for research in mathematics education and also serves to exemplify contemporary approaches to educational psychology and to research design.

It is a feature of the 1912 Board of Education *Reports* that they contain no contribution relating to research in mathematics education. True, there is a paper on 'Research as a training for mathematical teachers' (66), but this merely sets out the case for mathematics teachers having experience of creative mathematics. It does not suggest that research might have pedagogical rather than mathematical ends. Today's reader might, however, still be surprised to find that as early as 1912 it was felt that 'research . . . has become a vague and elastic term, . . . sometimes used to cover up rubbish that would be better consigned to the dust-heap' (p. 318). Yet educational research was soon to be given a powerful boost, both as a result of work carried out in the USA by Thorndike, Woodworth and others, and of the First World War which served to publicise mental and scholastic tests. Some idea of the quantitative growth in research in the field of mathematics education is given by the details of dissertations and theses written for higher degrees. Frobisher and Joy (67) suggest that the first such thesis to be written in Britain was W. McClelland's B.Ed. (Edinburgh) thesis on different methods of subtraction (1910); in 1919 came the first thesis to be presented to an English university (a statistical survey of arithmetical ability in a London school); and in 1930 the first Ph.D. thesis, D. Shandarkar's 'An experimental investigation in teaching to solve problems in arithmetic, and the light it throws on the doctrine of formal training'. Postgraduate degrees within the field of mathematics education were, however, rarely awarded in the 1920s; there were only ten or so in all. The 1930s witnessed an expansion in research facilities and in 1934 five higher degrees were awarded, one, an MA, going to Elizabeth Williams for her thesis on 'The geometrical concepts of children from 5–8 years of age'. This was the work on which the *Gazette* article was based {68}.

Mrs Williams' MA thesis was written after she had taken up a new appointment in the Department of Education at King's College, London. It also serves, then, to illustrate yet one more change; that the staff of the university education departments were now seen as having other tasks than merely lecturing on method. It was their duty to pursue and to foster educational research. This point was strongly emphasised by Dover Wilson, the Shakespearean scholar, when he appointed Mrs Williams to King's College: the lecturer's power of passing on insights to students, he asserted, depended on his continuing investigation and experiment.

Yet, what form was that educational research to take? Reference has been made (p. 162 above) to F. Y. Edgeworth's work on examinations, but he was not the first to undertake what could be termed educational research. Earlier work had been carried out by, among others, Francis Galton (1822–1911), and it was Galton's work which was to prove seminal. For over half a century, Galton pursued studies in anthropology, psychology and eugenics and during that time developed many statistical procedures (e.g. ranking, percentiles and correlation) to assist him in his analyses. His application of quantitative procedures to biological research was to provide a model for many educational researchers. At the present time it is commonplace to speak of the 'biological' or 'agricultural' model of educational research. However, as Hamilton has indicated (69), educational researchers have tended successively to use two such biological models: first, that based on Galton, which sought to generate hypotheses inductively on the bases of analyses of natural situations, and secondly those of Fisher, building upon the statistical tests he devised and perfected in the 1920s and 1930s, which were concerned with the deductive testing of hypotheses through controlled experimentation. The methodology of Mrs Williams' work, which concerned the child's notion of symmetry, exemplified the approach of the earlier, 'Galton' school.

A common use of statistical procedures was not, however, the only link between biology and (mathematics) education. The reader of Nunn's *Education* will be immediately struck by the number of times Nunn uses biological analogies and metaphors. Educational psychologists, too, were to make frequent use of animals in their experiments: Pavlov's dog has entered common folk-lore, but Thorndike, for example, carried out experiments on cats, dogs, fish and monkeys. In her paper, Mrs Williams referred to the 'amusing mistake of Köhler's apes'. Such links, and that which bound educational research to psychological research {70}, were, partly as a result of their initial fruitfulness, to prove severe constraints on the development of alternative research procedures within mathematics education.

This, however, is not meant to imply that educational psychology was itself moribund. Wolfgang Köhler, who was mentioned above, was, for instance, one of the Gestalt psychologists offering theories which were effectively to supplant Herbartian ones. Yet Nunn was to write in 1945 (71) that the publications of the Gestalt school did 'not seem to entail any modification of our doctrine or any important addition to it' {72}. Nevertheless, the work of the school was welcomed as confirming the practical beliefs of many British educators.

What was eventually to prove a much more influential contribution to educational practice was the work of Jean Piaget – a name which most of the readers of Elizabeth Williams' *Gazette* article would probably be meeting for the first time. Piaget, in Mrs Williams' words, followed a 'different line of investigation' from that pursued by Burt and other leading British researchers, 'in numerous interviews with children he has studied the way in which they actually reason about matters of their daily experience'.

Yet, despite the interest shown by Mrs Williams and some other workers, and the fact that by 1934 several of his books had appeared in translation (73), and even though he had been invited to lecture in England in the late 1920s, it was to be another twenty years before Piaget's work was to become widely esteemed (often, however, at 'second-hand'). There were, for example, immediate objections to his 'clinical method' from such influential educators as Susan Isaacs (74).

Not surprisingly, therefore, although in its choice of subject Mrs Williams' thesis reflected Piaget's interests and the fact that by then she was mother of three children in the target age-range, the research methods she adopted were those currently fashionable in England. This early paper reflects the methods and views of colleagues such as Nunn and Hamley (fresh back from work in the United States), and of such writers as Burt, Isaacs, Rignano and Spearman {75}, rather than those of Piaget. Even so, the paper and thesis demonstrated clearly that research in mathematics education could mean more than psychometrics and the design and analysis of tests, and that research could hold out the promise of practical assistance for the teacher and curriculum developer.

7. TEACHER-TRAINING

Elizabeth Williams was to spend four years at King's College, London, where she acted as tutor to students in training and also devised and presented to them a course in statistics, then beginning to play a prominent part in educational research work. The students attended Nunn's lectures on the teaching of mathematics, but Mrs Williams was responsible for conducting seminars and supervising teaching practice. That this took place largely in the well-established and esteemed schools was not altogether an advantage, for 'in many of the schools which were held in high regard, opposition to new topics and less formal organisation was very strong. "No models", said one Headmistress, "They fall off the desks"!'

In 1935 she moved from King's College to Goldsmiths' College, a training college administered by the University of London. The appointment was almost accidental, for originally she was 'lent' by King's College to deputise part-time for a member of the mathematics staff who had suffered a breakdown.

> Very soon I realised what a great institution it was . . . The College had its own nursery school, a school of art, a well-known choir, a revolutionary school for modern dance, etc. I was asked to join the team and share in the Education courses and also to be a member of the mathematics team . . . The liveliness of thought, the ready sponsorship of any worthwhile experiment and the co-operation of very different people . . . led the way [in] so many teacher training developments.

It was at Goldsmiths' that Mrs Williams learned to make use of what is now called 'micro-teaching'; she was also able to cooperate in the establishment of a large mathematics workshop with 'the most varied' equipment, and to participate as a tutor at residential in-service courses mounted by various local education authorities. When the war came, the workshop, like the rest of Goldsmiths', was evacuated to Nottingham.

Mrs Williams was to remain in the Midlands when the war ended, for she was appointed the first Principal of the City of Leicester Teacher Training College.

The years at Goldsmiths' had brought Mrs Williams into close contact with the preparation of both elementary- and secondary- (grammar) school teachers, for the College offered both two-year training courses and one leading to a postgraduate teachers' certificate. As we have seen the training requirements and provisions for teachers in elementary and secondary schools had always differed. The teacher-training system had arisen in a piecemeal fashion as various bodies, Church, State and independent, had identified areas of need and attempted to meet them, and, despite the establishment of Church and other training colleges in the mid nineteenth century and of the later day-training colleges, in 1920 less than half the country's elementary-school teachers were college trained. Gradually, however, the percentage of untrained teachers diminished as new local education authority training colleges were opened. The economic recession of the 1930s also had the effect of diverting many graduates into elementary-school teaching {76}. The growth of the education system during the inter-war years was not, however, very rapid and completely new teacher-training measures were called for to deal with the expansion and changes following the 1944 Education Act.

Secondary education for all and the raising in 1947 of the school-leaving age to fifteen necessitated an increase in the number of teachers and in the level of their qualifications. One answer to this need was a programme of one-year crash courses for mature, ex-service students {77}. In a six-year period 35,000 such teachers were trained {78}. Partly to meet the requirements of this programme, and also to meet the demand for a greater number of two-year trained teachers, several new training colleges were opened, the first being the City of Leicester College, that to which Elizabeth Williams was appointed.

The purpose and procedures of the training colleges had been the subject of a governmental report (the McNair Report) in 1944 (79), and disquiet had been expressed both at the overcrowded nature of the two-year course and the fact that students leaving such colleges had not 'by twenty years of age, reached a maturity equal to the responsibility of educating children' (p. 65). As a result, the committee recommended that the course should be extended by a year, a recommendation which was to go unheeded for well over a decade. 'Maturity' was, however, to become an increasing theme in teacher-training.

Yet, despite the misgivings of the McNair Report, the City of Leicester College opened with considerable hopes:

Staff and students were all fired with enthusiasm for programmes related to social needs and fulfilling the demands of the Butler Act. It was too small [a college to offer] main courses in mathematics, but professional courses in basic subjects were carefully planned to offer appropriate information on a subject as well as its educational aims and modes; and so it was for mathematics. [No] single textbook [was] prescribed for the course; there was really nothing quite suitable. We recommended Mathematical Association Reports, Westaway's craftmanship book, Ballard and others on arithmetic, and Carson and Smith's *First Ideas of Geometry*. More important were the learning materials provided and local school contacts showing the experiential approach. We were fortunate in the imaginative qualities of the initiating staff.

The college was also to derive much help and encouragement from the new Leicester Institute of Education and, in the field of mathematics, from W. W. Sawyer who was then working in Leicester.

In 1951 Mrs Williams left Leicester to become Principal of the old-established Whitelands College (p. 106 above).

Its new buildings were wonderful; its old students were devoted; its standards in academic studies were high and showed the trend to degree standards. Alas! Physical Science and Mathematics were not adequately recognised and had no proper accommodation. It was difficult to put things right. We did our best and certainly a start was made . . . It is a sad reflection on what can be done by highly intelligent women in discouraging girls from taking up maths or physics. Fortunately, additional buildings and the admission of men students have changed the whole situation.

Even if Mrs Williams was having only limited success in changing attitudes and courses at Whitelands, her activities outside the college were proving more fruitful. Some of these will be referred to in the next section.

It suffices here to mention her work on the National Advisory Council for the Training and Supply of Teachers (NACTST) and also in the Association of Teachers in Colleges and Departments of Education. She was very much involved in the fight to see the recommendation of the McNair Committee on the extension of the teacher-training course put into practice; efforts which were rewarded in 1957 when it was announced that from 1960 the teacher-training course would be of three years' duration. The NACTST responded in *The Scope and Content of the Three Year Course of Teacher Training* (HMSO, 1957) in which it welcomed the greater time which could now be placed at the students' disposal. The change was, indeed, a most significant one;

> it stimulated more advanced work in specialist subjects, provided fuller studies in the 'curriculum' subjects for primary schools, and permitted much greater opportunities for child study in many forms. All these improvements showed in the various mathematics courses. It was possible to base formal studies on practical experience, observation and experiment both in a student's own learning and in teaching practice.
>
> [Developments in the two-year course had led to] some of the mathematical work [being] of considerable interest and value. It was good enough for many of us to be convinced that the quality of work possible in a 3-year course would justify proposals for the award of a degree.

8. THREE REPORTS

In the two decades following the Second World War, Elizabeth Williams participated in the production of three important reports on mathematics education which had as their fields of interest, primary schools, secondary modern schools and teacher-training. These reports, and a brief study of their backgrounds, do much to illustrate national developments in that era.

Following the 1944 Education Act, secondary education came to be organised on tripartite lines throughout most of England. Only a few authorities at that time opted for 'multilateral' or 'comprehensive' schools {80}. The majority took the recommendations of the Norwood Report {81} to heart and formed grammar, technical and secondary modern schools. Attempts to influence mathematics teaching in the traditional grammar schools were made by the Jeffery Committee (82) which in 1944 recommended that the subject should be taught and examined as a unified whole and not as a number of disjoint components, arithmetic, algebra, . . . (cf. p. 163 above). The immediate response was lukewarm and often cosmetic; 'unified' courses appeared which were merely shuffled collections of chapters from older, 'separate' texts {83}. In general, few major changes took place in grammar-school mathematics until the 1960s. A few enthusiasts and innovators were, however, beginning to experiment with new methods and content during the late 1940s and the 1950s.

Technical schools were, in fact, never to develop as originally envisaged – even by 1961 only 3.1 per cent of pupils in England and Wales attended them (compared with 22.1 per cent in maintained grammar schools, 53.8 per cent in secondary modern and 10.4 per cent in independent and direct grant schools (84)). Their mathematics syllabus was fashioned on that of the grammar school, but tended to emphasise the technical applications of mathematics, and its connections with other courses, for example workshop and general science courses, and, in particular, technical drawing. The traditions and aims of the DSA classes, Perry and the central schools were thus kept alive (85).

It was, however, in the secondary modern schools that the major problems were posed. In these a new curriculum had to be developed almost from scratch. Moreover, there were few teachers in this sector who were well qualified mathematically, for almost all those who had taught in the senior elementary schools prior to 1939 and who possessed good qualifications had either, during the war years, joined the services or transferred to grammar schools (86). In any description of mathematics education in the immediate post-war years special attention must, then, be directed at developments in the secondary modern school.

The 1944 Act by providing secondary education for all ended elementary education and effectively created for the first time a readily identifiable primary-school system. An educational divide was established at 11+, with important repercussions for education in the five to eleven age-range. The abolition of fee-paying in the maintained grammar schools increased the pressures to make success in the 11+ examination, the major criterion of successful primary-school education, even more vital. The disappearance of the senior classes meant that it was no longer possible to deal with the brighter child simply by 'accelerating' him through the various 'standards' {87}. On the other hand, the slower child could not repeat the same standard as in the past, sometimes year after year. New opportunities had been created, but these had brought in their wake new problems.

(a) *'The Teaching of Mathematics in Primary Schools' (Bell, 1955)*

In 1938 the Teaching Committee of the Mathematical Association established a subcommittee 'to consider the teaching of mathematics to children under eleven'. Elizabeth Williams, soon to become one of the two Secretaries of the Association, was a member. Work started on the preparation of a detailed syllabus for Junior Schools (7–11):

Enquiries were made into current practice, and the help of teachers was enlisted. Specimens of arithmetic papers used in the Special Place [Scholarship] Examinations at the age of eleven were collected . . . a detailed programme of arithmetic in half-year units [was drawn up] as a basis for discussion (p. v).

The war brought an end to the committee's work and in 1946 it was a new subcommittee which was charged with reporting on mathematics in primary schools {88}.

[At first] it was expected that we should complete the fairly substantial pre-war work. As we read and discussed it, however, we realised that we could not continue on that basis. We could no longer envisage the organised class lesson, confined to symbols on blackboard and paper. So we decided to go over to *mathematics* in place of *arithmetic* and to concentrate on the ways in which children think and the activities they find satisfying. There were two main influences which inspired our new start: first the freedom that had developed in nursery and infant schools and in art, English, etc. at the junior stage; secondly, the investigations into children's modes of thought which were being conducted by the Piagetian school in Geneva. We knew of Gattegno's lectures and the new ideas he was advancing, having just returned from work with Piaget, and we invited him to come [and join the committee] and give us his views on what we were attempting. We thought we had brought ourselves up to date; he told us that we were entirely old-fashioned, and were ignoring vital new concepts.

The impact of Caleb Gattegno was to prove considerable. That the Association should concern itself with the education of junior-school children was itself remarkable, for it remained essentially a body of public and grammar-school masters which became more representative, and so more conservative, as the years went by. For example, only once in over fifty years had it chosen a President who did not hail from Cambridge – and he, Percy Nunn, could hardly be ignored {89}! The Association contained few primary- and modern-school teachers and had indeed little to offer them, for would-be *Gazette* readers had to be well equipped mathematically; needs which Carson's short-lived association had attempted to satisfy were still unmet. The members of the primary committee were, therefore, neither typical Association members nor representative primary-school teachers. They were innovators and enthusiasts, who had been trained as mathematicians and were familiar with primary-school classrooms. Moreover, since the vast bulk of the Association's membership neither knew nor, one suspects, cared greatly about what went on in primary schools (so long as their more able products reached a satisfactory standard), the committee were effectively free from constraints. The result was an exciting report, more *avant garde* than that typically issued by the Association, which effectively established a new direction for developments in primary education. Its publication, too, was followed by countless meetings held throughout the country which set a new pattern for in-service courses.

Yet, although the enthusiasts on the primary committee had been willing to be told that they were 'entirely old-fashioned' and 'should seek new ways', those whose interests lay more in the field of secondary education were not. Gattegno could not exert the same influence on the bulk of Association members. It was, in any event, a hopeless task. An association created to deal with the interests of a particular sector of education could not readily convert itself into a body satisfying wider and different needs {90}. Such needs could only be met quickly through the establishment of a new body which could cater for the enthusiastic innovator. Gattegno, then, was led to form the Association for Teaching Aids in Mathematics (ATAM); a body which in 1962 adopted the less limiting title, the Association of Teachers of Mathematics (ATM). The Association's journal, *Mathematics Teaching*, first appeared in 1955. ATM was to grow very rapidly indeed during the 1960s and its appeal to primary- and modern-school teachers led to its soon having more members than the older Mathematical Association {91}.

Certainly, however, the primary report would have had the full support of the new association, its principal theme being that primary-school mathematics should not be solely concerned with the art of reckoning but 'with an approach to the relations and ideas . . . fundamental . . . to Mathematics as a whole' (p. vii).

> Understanding must precede drill or formal exercises intended to develop memory, mechanical accuracy or speed . . . It is not because we under-estimate the value of accuracy and the orderly arrangement of computation that we have given comparatively little emphasis to the written arithmetic which is traditional . . . Rather it is because we are convinced that the important thing is to help children to understand mathematical ideas and to recognise the kind of computation or other thought processes which a problem situation demands {92} . . . We plead therefore for attempts to develop mathematical ideas through the study of broad environmental topics and through the investigation of situations and phenomena at first hand (pp. 17, 20). The importance of experience as the necessary foundation for the formation of . . . concepts and patterns of thinking is stressed . . . No amount of practice and rote-learning will take the place of this experience; unless practice and rote-learning are concerned with ideas that have become 'a part of' a child they will never lead to real understanding and mathematical thinking. Practice without the power of mathematical thinking leads nowhere; the power of mathematical thinking without practice is like knowing what to do but not having the skill or tools to do it; but the power of mathematical thinking supported by practice and rote-learning will give the best opportunity for all children to enjoy and pursue Mathematics as far as their individual abilities allow (p. 4).

Alas, the balance for which the report pressed was occasionally forgotten in the excitement of the ensuing reforms.

(b) *'Mathematics in Secondary Modern Schools' (Bell, 1959)*

Considering as a whole the education of all children over 11, emphasis has shifted during recent years from the education of the relatively few children in Grammar Schools to that of a very much larger number outside these schools.

To consider the mathematics education suitable for the latter type of children was the remit that in 1946 the Mathematical Association gave a subcommittee, chaired by Elizabeth Williams. The committee published an interim report in 1949 and it is from this that the above quotation is taken. As we have observed, there were several 'traditional' curricula which the newly created secondary modern schools might copy; the academic one of the grammar schools, the technically and vocationally oriented curricula of the senior (central) schools, and that of the elementary schools which encompassed little more than arithmetic and mensuration. The 1949 report rejected the idea that the new schools should ape the grammar schools: 'for modern school children traditional academic standards and methods, appropriate as they may be to the needs of grammar school children, lead only to failure . . . and distaste for the artificial culture of the schools' (p. 3). Instead, it demanded that new curricula should be devised bearing in mind the three claims which any branch of knowledge had to meet when being considered for a place within the school curriculum:

1. Economic, or 'general utility', the claim to a place among basic knowledge or skills without which the individual cannot easily carry on normal life . . .
2. Community value, or the contribution towards the achievement of a constructive and well-adjusted life for the individual among his fellows . . .
3. Cultural, overlapping with social . . ., but going further in the direction of spiritual values; giving insight and confidence, deepening feelings, engendering delight.

The case for mathematics for the ordinary child, it was felt, had in the past been founded too much on the first two aspects. Moreover, the second claim now demanded even more attention, for some knowledge of, for example, graphical representation, simple statistics and probability would seem essential for all good citizens. Nevertheless, attention should still be focused on the third aspect for 'the possibility of developing in the ordinary child a cultural appreciation of mathematics has been insufficiently realised, but the few who have experimented in this direction have had encouraging results'. It was, however, considered impossible to attempt to draw up a standard syllabus for modern schools which would meet these claims.

Contemporary emphasis on the needs of the individual is common to all aspects of education, and it is unfortunate that existing conditions in schools, enforcing very large classes, tend to make it appear a counsel of unattainable perfection. It is urged, nevertheless, that the teacher should aim consistently at the ideal, and resist the pressure of circumstances towards the fatal expedient of standardised syllabuses and methods.

> The needs of Modern School children, their environmental conditions, and the special interests of individual teachers as well as children, all vary so much that anything approaching standardisation would be detrimental and undesirable even if it were not impossible. In any case, in the present stage of development, any standardisation is obviously premature . . . The task of the teacher of mathematics in a Modern School is as difficult as any that has ever confronted a teacher. Having in mind the mathematical ideas which are to be his long-term aim, he must study the children to discover interests which he can use as starting-points and on the basis of these decide upon and often create the environment and experience which will afford opportunity for the desired development; he must seize upon the child's needs as soon as they become conscious and from them derive the stimulus necessary to carry the child through the voluntary disciplinary process of acquiring skills and techniques; finally, he must integrate the whole into the subject of mathematics, seeking always for generalisation and the opportunity to demonstrate the power which generalisation confers . . . The reader may feel that the present Report emphasises unduly the activity aspect of teaching at the expense of the practice of skills and the direct teaching of the generalised, integrated, subject of mathematics; it is necessary to devote attention mainly to the less familiar aspect, but the ultimate aim remains the endowment of every child with the maximum command of the language, insight, and power of the subject of mathematics (pp. 9–10) {93}.

The principles outlined in the interim report were exemplified in more detail in the 1959 publication. This was a book of over 200 pages, the first third being concerned with 'Purpose, approach and organisation', the rest with 'Mathematical content'.

A review, written by Cyril Hope, appeared in the ATAM's *Mathematics Teaching* (94). Hope began by contrasting the depressing beginnings of secondary-modern-school mathematics teaching with the developments of the mid and late 1950s. Now some schools were entering candidates for O-level mathematics examinations and 'the success of these efforts [has] led to the inclusion of more and more mathematics in the curricula . . . and with freedom from external constraints some secondary modern schools are doing interesting work of high standard and quality, many including calculus in the third year'. Although, 'many' in the last sentence might better have been replaced by 'some'; it was nonetheless true that in certain schools great advances were being made {95}. Hope found 'nothing new in the proposals' in the first part, which was hardly surprising since this only built upon the interim report, however he was 'struck by the need for more thought about the way to select topics and to relate them to mathematics and the other subjects of the school curriculum' {96}. The second part of the book he found to contain 'one of the best treatments of Stage A geometry and algebra which has been published'. The sections concerned were indeed excellent. Yet, in retrospect, the mathematical content was backward-looking; as Hope pointed out 'many teachers will bemoan [the report's] lack of forward vision, which reaches out to a new concept of secondary mathematics, stressing mathematical ideas and principles in a form which enables modern developments to be glimpsed'. Hope was writing, of course, after the famous Royaumont Seminar (97) which launched 'modern mathematics' in Europe; indeed the following year he, himself, was to be part of an international team which produced suggestions for 'modern' secondary-school mathematics syllabuses (98).

Yet in describing the report 'as a valuable source for much B and C stream mathematics', Hope was very much a man of his time. For it was a period when setting and streaming were accepted, indeed, encouraged {99}.

In fact, the 1959 report was soon overtaken by other events. The following year the Beloe Committee proposed the establishment of an external examination, the Certificate of Secondary Education (CSE), directed at pupils in secondary modern schools (100), and soon the spread of comprehensive education added further complications. For a variety of reasons 'the ordinary child' was too often subjected to that watered-down academic curriculum which the MA report had rejected, with, unfortunately, the results the committee had foretold.

(c) *The Supply and Training of Teachers of Mathematics (Bell, 1963)*

In 1958 Elizabeth Williams resigned her post as Principal of Whitelands College. She had passed the minimum retiring age and her husband's health, at the age of eighty, was giving concern. Yet 'retirement' was not to mean the cessation of active participation in mathematics education – indeed, there were now fewer administrative chores to divert her energy. Not surprisingly, therefore, when the Mathematical Association, in conjunction with the Mathematics Section of the Association of Teachers in Training Colleges and Departments of Education (ATCDE), established a committee to consider teacher training, Elizabeth Williams was asked to be its chairman. The committee reported in 1963, just as the first 'projects' were beginning their work in schools and shortly after the switch to the three-year training course had been made.

Its first concern was the acute shortage of mathematics teachers. There had, of course, always been a shortage of well-qualified teachers (cf. pp. 154 and 182 above), but the problem, following the expansion of the educational system, was now worse than ever before. The situation was further complicated by the rapid expansion then taking place in the university sector, which absorbed many of the ablest young mathematical talents, and in the colleges of education, which were recruiting experienced, able teachers, and even more tellingly by a switch in graduate employment patterns. 'In 1938 well over 75 per cent of newly qualified mathematics graduates entered the teaching profession. Largely owing to the recognition of the important part played by professional mathematicians, industry now absorbs a comparable percentage of all mathematics graduates, including about half of those with honours' (p. 3). The growth of the computer industry, the development of operations research and the spread of quantitative methods was simultaneously creating a new demand for mathematicians and threatening the well-being of the system which produced them (101). To some extent the problem was eased by the rapid expansion of university departments of mathematics {102}, but the shortfall in the number of qualified mathematics teachers has continued {103}. Supply is, however, but one side of the problem; what training was it desirable for teachers to have? The late 1950s and early 1960s saw great changes in the way mathematics was approached in the universities and these had implications for teacher-training as well as important repercussions on the schools' curricula. The increasing diversity of university courses and the growth of specialist options militated against the production of good teachers. Thus, for example,

> A prospective teacher requires a mathematical training in breadth rather than depth. The academic pressure at which the usual university course proceeds leaves little time for the study of the history and development of mathematics and its impact on other branches of learning. Yet it is precisely an appreciation of this aspect of the subject which is one of the most important assets of a good teacher in the schools (p. 16).

Much of the new material, too, was being introduced to undergraduates in a manner which not only made comprehension difficult, but gave them no assistance in understanding how such concepts might be introduced at a pre-university level:

> The emphasis on the axiomatic method in modern mathematics has led to the belief that it is impossible to teach the subject at university on the lines of the 'Stages A, B and C approach' familiar in the schools, and that all mathematics at university must begin with Stage C. This belief is exaggerated . . . an axiomatic approach is more reasonably comprehended when accompanied by motivation for the choice of axioms. There is nothing intrinsic in modern mathematics which forbids concrete examples and motivation – in fact, the whole point of the unifying nature of abstraction is missed without knowledge of apparently different examples of the same concept (p. 18).

Some of the new ideas would, however, have to be brought down into the schools and to do this successfully it was essential to have teachers who fully grasped them:

> It is evident that to revitalise school mathematics, many ideas will have to be synthesised: *mathematics is not necessarily bad because it is old*. The making of a new synthesis is the urgent problem facing equally training departments, colleges and groups of teachers (p. 13).

What of professional training? Here the arguments for and against teacher-training had, despite Nunn's claims (cf. p. 180 above), to be repeated. It was still the case that about half the mathematics graduates entering teaching had not completed professional training (p. 2). Yet was the system which had now stood for some fifty years meeting contemporary needs? The committee was dubious and considered alternatives to the 3 + 1 system. One possibility was a four-year university course leading to an honours degree in mathematics and education and to the Postgraduate Certificate in Education (PGCE) {104}. Another was to make the postgraduate training course of two year's duration with responsibility for the training and certification of the student shared between school and university, training being supervised by a new type of school-based teacher-trainer. In the event, the PGCE system was not only to remain essentially undisturbed, but was in the two succeeding decades to gain ground on what the committee saw as a viable alternative, 'four-year concurrent degree and professional courses in training colleges'. For many years a few training colleges had offered courses in which the student worked for a University of London external degree whilst at the same time studying for his Teacher's Certificate. In the mid 1960s, however, what Elizabeth Williams had foreseen when the three-year training programme was established came to pass. College of Education courses were granted degree status and gradually 'certificate' courses were phased out as students aspired to a B.Ed. degree and the teaching profession moved to becoming all graduate. The immediate results of this change were not all beneficial. Academic work, often only half understood, was frequently accorded more weight than professional training (105), and the time allocated in the training of primary-school teachers to aspects of mathematics teaching was insufficient. The 1970s and early 1980s were, however, to see an enormous contraction within the colleges of education and the proportion of student teachers following a concurrent, as opposed to a consecutive, PGCE course declined rapidly. Thus while the recommendations of the 1963 report on consecutive training have still to be effectively implemented, those referring to 'non-graduate' and concurrent training have from an organisational point-of-view become largely out-dated.

The report also drew attention to what was now becoming recognised as an additional, essential form of teacher-training, in-service education. It was able to point to 'a growing demand from experienced teachers for courses which will keep them in touch with developments in educational thought and in . . . mathematics' and to make suggestions as to how that demand might be met. The 1960s were, indeed, to see an enormous increase in opportunities for in-service education and arguments were advanced for making such provisions statutory and nationally organised

{106}. However, such training was, in the event, provided in a haphazard piecemeal fashion by a bewildering variety of agencies {107}.

In retrospect, the report stands up well. In it many of the key problems of the training of mathematics teachers are clearly identified. Moreover, some of these, now accepted as commonplace, exemplified at that time perceptive thinking. One can only regret that so few of its recommendations have ever been implemented.

9. 'RETIREMENT'

Much of the work that Elizabeth Williams was to do after her retirement lies outside the scope of this book, yet a brief account of some of her activities will serve to indicate both the pattern of events post-1960 and also the remarkable contribution which she has made to mathematics education in England and elsewhere, a contribution recognised by her election as President of the Mathematical Association in 1965 {108} and her appointment as a Commander of the Order of the British Empire in 1958. Earlier in the book we noted how the work of English authors was widely disseminated in the colonies. In the early part of this century, 'Hall and Knight', 'Godfrey and Siddons' and 'Durell' were used throughout the Empire. In some countries, for example Australia and Canada, educational systems existed which could be compared with that in England. Elsewhere, however, opportunities for education were often very limited {109}, and secondary education was provided for only an élite. In the post-1945 era the educational systems of developing countries expanded enormously. Universal primary education was set as a goal and, once that came near to being accomplished, secondary education for all became the next target. Enormous difficulties were experienced in providing sufficient and adequately-trained teachers. Not surprisingly, therefore, developing countries looked to England for advice on how they might tackle a task which it had recently overcome with a fair degree of success. So it was that Elizabeth Williams was asked to spend three months in Kenya in 1956 drawing up a national plan for teacher-training and two years (1963–5) in Ghana as Deputy Director of the School of Education with special responsibility for modernising mathematics teaching in the country's schools and training colleges. The period in Ghana was to coincide with the opening of what was to become a new-style battle for 'mathematical' colonies. The English-speaking countries of Africa, eager that their schools should follow the most up-to-date curriculum, found themselves presented with alternative materials and philosophies. The US-financed African Education (Entebbe) Project offered texts for both primary and secondary levels written in a style which proved anathema to many of the expatriates to be found in the schools of the old British colonies. They looked to Britain for inspiration, and 'battle' commenced. In fact, the Entebbe Primary Project was soon to produce a compromise approach, partly effected by Mrs Williams, who attended the 1964 Entebbe Mathematics Workshop. The secondary battle was not resolved so readily. Unfortunately for the Africans, neither of the 'contestants' was fully aware of the implications of the increased educational provisions in Africa and of the great differences which existed between the conditions and aims of the African educational systems and those of their own country. It was an invigorating contest, but one which was often irrelevant to the needs of the countries concerned.

Making improvements in the widely-attended primary sector was a more important aim than replacing the élite's 'Durell' by Entebbe or an adapted version of SMP, and it was to this goal that Mrs Williams contributed much. During the 1960s she visited many countries in Africa and south-east Asia organising workshops, giving lectures and advising ministries. She was also one of the committee responsible for planning the Commonwealth Conference on Mathematical Education held in Trinidad in 1968, and at the request of CREDO (the Centre for Curriculum Renewal and Educational Development Overseas) produced a series of three books intended for overseas readers which attempted to distil the aims and methods of the Nuffield Mathematics Project (110). These books and, even more, *Primary Mathematics Today* (Longman, 1970) which she wrote with Hilary Shuard {111}, made Mrs Williams' name familiar to teacher-trainers and student-teachers throughout the world, and she made several visits to the USA and to Australia to run courses; journeys which now took less time than it would have done for Charles Hutton to return from London to his native Newcastle.

The greatly improved facilities for international travel, the activities sponsored by the Organisation for European Economic Co-operation (OEEC) (cf. p. 199 above) and UNESCO, and the confrontations (which were always friendly even if occasionally heated) in Africa and elsewhere between advocates of different methods and philosophies now combined to recreate a great interest in comparative, international mathematics education. The education section of the International Congresses promoted by the International Mathematical Union was revivified, and the levels of activity reached in the pre-1914 days were once again attained. In 1968 came the first international journal devoted to mathematics education, *Educational Studies in Mathematics*, and the following year, in Lyons, France, the International Commission on Mathematical Instruction (ICMI) mounted its own international congress on mathematics education. The success of this first meeting was such that it was agreed to hold such a congress every fourth year. The Second International Congress on Mathematical Education (ICME) was held in 1972 in Exeter and Elizabeth Williams was appointed chairman of the committee responsible for organising its programme. It was an appointment which not only recognised her international standing within mathematics education but also the remarkable vitality of one by then in her late seventies.

This increase in international cooperation may well prove to have been the most significant feature of mathematics education during the last two decades. Frequently in this book, we have seen how individuals have drawn inspiration from, or have modelled innovations on, work and workers in other countries. A knowledge of what was happening elsewhere has proved of enormous value; yet now we are offered a more exciting and valuable opportunity still, to work with those overseas to attempt to find solutions for the many pressing problems of mathematics education and together to define the place that mathematics should hold in a person's education. Here perhaps one might well begin by considering Elizabeth Williams' closing words in her Presidential Address to the Mathematical Association:

> You will recall that Newsom recommends that girls should give up mathematics in favour of humane studies. In my view mathematics is a humane study since it shows the development of man's systematized thinking out of which his scientific mastery continues to grow. Abstract the subject may be in its essence but to show its full stature it must be effective in use when a use is found. The fascination of mathematics lies partly in this passage to and fro, from action to thought and from thought to action, new thought to new action. I believe that, impressed as people are by what mathematics can do, much of the admiration given to it today is due to the awareness that it is a supreme body of human thinking that gives men a higher than mountain-top view of the world literally extending man's power to probe and forecast, intellectually producing a confident sense of coherence. For our pupils we would wish some glimpse of this human achievement and, as far as may be, some involvement in it. In this way mathematics can fulfil its true role in education: to enlarge the range and power of each individual's insight into the world of men's horizons.

Postlude

It is possibly still too early to attempt to chronicle and disentangle the various movements that have taken place in mathematics education since 1960. Even if the time were ripe, then it would be impossible to present a detailed, critical account without effectively embarking on the writing of a separate book. Yet to end our story in 1960 would be unfortunate in two respects. First it would leave many important changes unreported and unreferenced. Secondly, and perhaps more importantly, it would not illuminate the way in which so many of the changes of the last two decades grew out of, and were consequences of, beliefs, actions and decisions which have been alluded to in the story of mathematics education prior to 1960. For that reason, I have chosen to end this book with a brief overview of trends in the post-1960 era.

What were the most significant events of these times? If the question had been asked in 1965, then the reply would almost certainly have been the formation of the large 'modern mathematics' projects such as the School Mathematics Project (SMP) and that of the Nuffield Foundation concerned with the age range five to thirteen. These projects brought a new, more open approach to curriculum development and the stress they laid on corporate involvement was an important factor leading to the establishment of a more widely-based 'mathematics teaching community'. However, from the mid 1960s onwards the school system has been restructured and the 'tripartite' system largely replaced by a 'comprehensive' one, a switch which was to have enormous repercussions upon mathematics educators. Immediate responses were demanded from them to this and other changes, such as the raising of the school-leaving age and increased opportunities for further and higher education. Even more problems were posed as a result of the economic crisis of the mid and late 1970s and new employment patterns which restricted opportunities for those leaving school without proper qualifications. Once again the whole purpose of mathematics education in school was called into question; and no obvious consensus of opinion was to be discerned. In retrospect, then, it is the recent changes which have taken place outside the narrow confines of mathematics education which may well prove to have been the most significant of the two decades. One can only regret that to most of these new problems only old solutions have been offered.

There was, of course, no fixed date on which the 'modern mathematics' revolution began. Both the SMP and the Midlands Mathematical Experiment (MME) were, for example, formally established in 1961 (1). Yet the former can be seen as attempting to solve the problem posed by the Jeffery Committee (p. 193 above) of designing an up-to-date mathematics course suitable for the brighter pupils in the grammar schools, and it was influenced to a considerable extent by the three conferences of industrial mathematicians and teachers from schools and universities held in Oxford (1957), Liverpool (1959) and Southampton (1961) (2). The MME had a somewhat different genesis; it displayed more the influence of those who in the mid-1950s formed what was to become the ATM, and also the influence of the OEEC seminars (p. 199 above). The two projects were, however, formed at a time when there was no public money available for curriculum development, although the growing interest of the Ministry of Education in the school curriculum was shortly (1962) to be declared through its formation of a Curriculum Study Group. For the first time since 'payment by results' (p. 121 above) and the 1904 Regulations for Secondary Schools (p. 151 above), central involvement in the curriculum appeared possible. Attempts to ward off this threat to the existing power-structure resulted in the establishment of the Schools Council for the Curriculum and Examinations (1964) (3), a consortium of interested parties in which teachers were dominant. For ten years the Schools Council mounted various curriculum projects within the field of mathematics education (4). It then became both the victim of financial stringency and the object of criticisms which eventually led to its reconstitution. In 1981 its future (and past) became the subject of a governmental inquiry (see p. 280 below).

The history of the Schools Council, then, illustrates the growing awareness that the school curriculum must be the subject of continued scrutiny, appraisal and up-dating and also that it is a matter of considerable social and political importance who should exercise control over that curriculum. In general, when considering the developments in England in the period 1960–80 one must be aware of the way in which actions were closely connected to, and greatly influenced by, politically-initiated educational reforms. Within the classroom, teachers still exercised their traditional control over the curriculum; outside, teachers' unions and other bodies felt increasingly called upon to defend what they saw as their rights.

The absence of public money for curriculum development prior to 1964 meant that the early projects had to seek finance from foundations, trusts and private sources. When doing this they were able to look to the nationally-funded curriculum developments launched in the USA in the late 1950s not only as examples of initiatives to be emulated, but also as providing models for the contemporary management of curriculum development. The complexity of such work was being demonstrated in the

USA, for example the need for trials and for associated in-service training. However, the difficulties of successfully effecting change had not by then become apparent. In retrospect, it can be seen that there were, in fact, more lessons to be learned from English experiences in the period 1900–14, than from visits to the United States to see 'how it should be done'. SMP with its clear links with the university world and with some of the most esteemed schools in the country, succeeded in attracting funds from industry. Other projects were less fortunate and, indeed, less farsighted: the shortage of funds, manpower and (often) a grasp of the problems to be overcome were seriously to affect their work.

Although accepting the US model for the management of curriculum development, the early English projects were not greatly influenced by the mathematical tendencies of the US and Continental innovators. The English authors were well-trained mathematically and, in the tradition of Godfrey, Siddons, Durell, *et al.*, had no doubts about their ability to develop their own courses. They regarded the school mathematics curriculum as something to be determined by schoolteachers. Indeed, there was no significant university contribution to curriculum development in English schools in marked contrast to what occurred in, say, the USA, France and Belgium. As a result, 'modern mathematics' in England assumed a different form from that found in most other countries. Although much new material entered the syllabuses, for example transformation geometry, matrices, statistics, probability, linear programming, set language and elementary graph theory, the underlying emphasis was not on providing an axiomatic approach attuned to the latest thinking among pure mathematicians. Rather, the principles which underpinned the construction of curricula were those set forth half a century earlier by Godfrey, namely interest and utility. The work of Dienes exhibited the influence of Bruner and his 'structures of the discipline' approach (5), but in general curricula were constructed without recourse to theoretical arguments. Teacher domination in the writing teams meant that the materials produced were usually teachable to students of the type to be found in the authors' classrooms, i.e. grammar school pupils.

Perhaps the major misjudgement concerned the rôle of exercises in the acquisition of understanding *and* techniques. The elimination of pointless manipulation has, as we have seen, long been the aim of mathematics educators. Indeed, this continuing battle to achieve a balance between a proper practice of skills and the elimination of drudgery was remarked upon by Ballard in the 1920s:

[It] is going on at the present day, and it will go on in the future . . . (it) is quite inevitable. It is inevitable because the dreamer and the practical man are always with us . . . And the dreamer, the idealist, will always call for elimination, in the interests of intelligence and a broad outlook; and the practical man will always call for restoration, in the interests of efficiency and economy of learning (6).

The swing back to the 'practical man's' position can be detected in the materials produced in the 1970s.

The SMP, the most influential project, had considerable impact not only in the grammar schools, its original target sector, but also in the new comprehensive schools. The Certificate of Secondary Education (CSE) was conceived with the tripartite system in mind, being intended to meet the needs of the secondary modern schools. However, the creation of comprehensive schools posed new problems. Pupils could no longer be divided into GCE and CSE candidates so early or so easily, and attempts had to be made to deal with classes containing pupils who differed widely in ability, knowledge and aspiration. Was setting socially divisive? Was it also mathematically undesirable in that all too early it labelled many children as failures and, in so doing, became self-fulfilling? Attempts to answer such questions led to an upsurge in 'mixed-ability' teaching (7) and to the publication of materials which claimed to be suited for a wide range of ability. The wish to keep options open for as long as possible and to exploit the possibilities of 'double' (CSE and GCE) entry also meant that CSE, far from being designed with the student of average ability in mind, became a pale shadow of O-level in which much of what had been newly introduced became trivialised. Such reforms also effectively brought an end to the tradition successively handed down by the private tutors, the DSA, Perry and the central and technical schools, of teaching technically and vocationally inspired mathematics to selected young adolescents. When the various sectors of education merged it was the academically trained and inspired teachers who had the prestige and who wielded the influence, and it was their kind of mathematics, indeed, their view of education, which was most esteemed.

The problems of differentiation of students and their curricula remain. The solutions offered by the Taunton Commission (p. 131 above) {8} and in the Norwood Report (p. 193 above) have been rejected, but an alternative, compatible with today's social ethos and which can be effectively translated into classroom practice, has still to be found.

The difficulties of effecting substantial changes in attitude and practice have also been demonstrated within higher and further education. I have already referred (p. 201 above) to the problems encountered with respect to the new B.Ed. In the established and new universities the great expansion in student numbers has not always been accompanied by serious consideration of whether degree structures and curricula are still appropriate. Hardy, if he were to return, would find much new and unfamiliar mathematics in university courses, but few changes in regulations and in teaching and examining procedures. A reincarnated Tate would probably regard the Open University as the most interesting innovation of the period. This surely is the contemporary analogue to the Mechanics' Institutes and the 'Lit. and Phils' of the nineteenth century. Regrettably the analogy is perhaps too apt, for again an institution aiming to give a second, or even a first, chance to the working class has tended to become a middle-class preserve. Like so many innovatory institutions, too, the Open University in order to win respect and gain acceptance adopted many existing standards, patterns and procedures in preference to creating new ones. Thus initially its mathematics foundation course was distinguished more for its purity and abstraction, than for any pronounced attempts to present mathematics to a new audience in a radically different way {9}.

The television networks and the Open University have, however, demonstrated most clearly the powers of the new visual media in mathematics teaching. Indeed, the decades since 1960 have seen technological developments within education on a totally unprecedented scale. One need only compare the classroom equipment and aids available to Recorde, Godfrey, Elizabeth Williams (even in her training-college days) and the classroom teacher today, to see how sudden and immense that advance has been. Computers, electronic calculators, mini computers, overhead-projectors, television, video-recorders, and cheap and rapid copying and printing machines have all entered schools within two decades. Each has a significant contribution to make to mathematics education. It is not surprising, however, that amidst the general turmoil created by curriculum revision and school reorganisation these contributions have failed to be adequately identified and appraised.

Yet how are educational aids and ideas to be satisfactorily evaluated? The answer to be found in history is clear: only through classroom use. Unfortunately, as has been shown on many occasions, it is not sufficient to experiment only in specially selected classrooms. What is possible for the individual, enthusiastic innovator cannot always be emulated by the rank and file. There is, then, a need either for large-scale experiments (together with a recognition of the accompanying risks), or for greatly increased understanding of the effects of the widespread dissemination of innovation which will permit us more readily to foresee and forestall problems. Neither of these methods is likely to be effective unless it is allied to critical evaluative and research studies. Attempts to foster research in mathematics education have recently been made through the establishment of a number of research centres for mathematics and science education and the creation elsewhere of special posts. To date, the rewards, measured in terms of findings which can be readily translated into effective changes in classroom practice, have been few. Nevertheless, what has been done reflects a beginning. It is, indeed, probably necessary for any activity to become 'recognised' and 'accepted' before time can be spared and sufficient experience gained to permit that critical self-examination of goals and procedures which, one hopes, would lead to improved effectiveness.

Many of the problems mentioned in the last few pages were considered by the Committee of Inquiry into the Teaching of Mathematics in Schools, established by the Government in 1978 under the chairmanship of Dr W. H. Cockcroft. This inquiry was prompted by a Parliamentary report of 1977 (10) which referred to many unanswered questions about the mathematical attainments of school-leavers and their apparent lack of basic computational skills, the mathematical demands made on adults in everyday life and employment, the lack of qualified teachers, the multiplicity of syllabuses, the lack of communication between further and higher education, employer and schools, and the responsibility of schools to equip children with the skills of numeracy.

The Cockcroft Committee, then, paid particular attention to, and commissioned research into, the mathematical needs of adult life in general and of employment in particular. The lack of confidence in their ability to use mathematics which adults revealed was indeed found to be striking. Not surprisingly, therefore, it was also found that the vast majority never use mathematics except in very specific contexts, the skills used being mainly arithmetical. The need for all to have sufficient confidence to make effective use of that mathematical skill and understanding which they do possess was, therefore, one of the committee's principal messages when it reported (11). So far as the needs of employment were concerned, the committee found little evidence to substantiate the criticisms made of school-leavers and concluded that such criticisms were not soundly based. Schools should not give additional training to prepare pupils for particular types of employment, but work to establish mathematical foundations which would 'enable competence in particular applications to develop within a reasonably short time once the necessary employment situation is encountered' (p. 24).

How, though, was that confidence and competence to be instilled? What changes in current practice were required? In answering these questions the committee offered a synthesis of what was becoming to be considered 'good practice'. At the primary level its recommendations followed the general trends established by the Hadow Report (pp. 185–6 above) and that of the Mathematical Association (pp. 194–6 above); at the secondary level it abhorred the way in which the education of the 'average' and 'below average' pupil had been distorted by the pressures of external examinations and asked for more clearly differentiated curricula for different levels of attainers (p. 208 above) – without, however, giving very clear indications of how the differentiation of students could be achieved in an educationally and socially acceptable form.

Perhaps the major difference between the report and its less specialised predecessors was in the wider view taken of mathematical activities within the classroom. Thus it asked that:

Mathematics teaching at all levels should include opportunities for
- exposition by the teacher;
- discussion between teacher and pupils and between pupils themselves;
- appropriate practical work;
- consolidation and practice of fundamental skills and routines;
- problem solving, including the application of mathematics to everyday situations;
- investigational work (p. 71).

Such goals are unlikely to be attained readily, yet nevertheless they serve as important criteria by which to judge mathematics teaching. The report had much to say about in-service education and the variety of forms this might take. In particular it emphasised the responsibilities of the head of department. Unfortunately, evidence on which to base recommendations concerning the efficacy of the various methods of training appeared to be lacking.

In general the report is distinguished for its balanced approach and for the support it provides to workers in the field of mathematics education. What it does not do – and perhaps this can be seen as lying outside the committee's remit – is to indicate in detail how the suggestions it makes might be implemented. It demands a response from 'teachers, local education authorities, examining boards, central government, training institutions, . . . those who fund and carry out curriculum development and educational research . . . and the public at large, including especially parents, employers and those engaged in public work' (p. 245). In doing so it acknowledges the fact that the teaching of mathematics is now the concern of society at large, and not merely of those who label themselves 'mathematics teachers'. The challenge is a great one. As this history has shown, ideas are apt to run ahead of practice; significant changes are becoming increasingly difficult to implement. It is essential then that careful consideration should be given to those processes of change and I hope that this book will contribute to a greater understanding of them {12}.

APPENDIX

A selection of examination papers, syllabuses, etc.

1802: CAMBRIDGE TRIPOS PAPERS

[Students took up to four papers. The seventh and eighth classes (containing the ordinary degree students) were set only bookwork questions.]

Monday morning problems—Mr Palmer

First and second classes (i.e. the expectant wranglers)

1. Given the three angles of a plane triangle, and the radius of its inscribed circle, to determine its sides.

2. The specific gravities of two fluids, which will not mix, are to each other as $n:1$, compare the quantities which must be poured into a cylindrical tube, whose length is (a) inches, that the pressures on the concave surfaces of the tube, which are in contact with the fluids, may be equal.

3. Determine that point in the arc of a quadrant from which two lines being drawn, one to the centre and the other bisecting the radius, the included angle shall be the greatest possible.

4. Required the linear aperture of a concave spherical reflector of glass, that the brightness of the sun's image may be the same when viewed in the reflector and in a given glass lens of the same radius.

5. Determine the evolute to the logarithmic spiral.

6. Prove that the periodic times in all ellipses about the same centre are equal.

7. The distance of a small rectilinear object from the eye being given, compare its apparent magnitude when viewed through a cylindrical body of water with that perceived by the naked eye.

8. Find the fluents of the quantities

$$\frac{d\dot{x}}{x(a^2 - x^2)}, \text{ and } \frac{b\dot{y}}{y(a+y)^{\frac{3}{2}}}.$$

9. Through what space must a body fall internally, towards the centre of an ellipse, to acquire the velocity in the curve?

10. Find the principal focus of a globule of water placed in air.

11. Determine, after Newton's manner, the law of the force acting perpendicular to the base, by which a body may describe a common cycloid.
12. Find the area of the curve whose equation is $xy = a^x$.
13. What is the value of q that force × (period)^2 = q × radius of circle?
14. Two places, A and B, are so situated that when the sun is in the northern tropic it rises an hour sooner at A than at B; and when the sun is in the southern tropic it rises an hour later at A than at B. Required the latitudes of the places.
15. From what point in the periphery of an ellipse may an elastic-body be so projected as to return to the same point after three successive reflections at the curve, having in its course described a parallelogram?

Monday afternoon problems—Mr Dealtry

Fifth and sixth classes (i.e. the expectant junior optimes)

1. Prove that an arithmetic mean is greater than a geometric.
2. Every section of a sphere is a circle.—Required a proof.
3. If $\frac{3}{4}$ of an ell of Holland cost $\frac{1}{4}$£, what will $12\frac{2}{3}$ ells cost?
4. Prove the method of completing the square in a quadratic equation.
5. Take away the second term of the equation $x^2 - 12x + 5 = 0$.
6. Inscribe the greatest rectangle in a given circle.
7. Sum the following series:

$$1 + 3 + 5 + 7 + \text{\&c. to } n \text{ terms.}$$

$$3 - 1 + \frac{1}{3} - \frac{1}{9} + \text{\&c. ad inf.}$$

$$\frac{1}{1.2.3} + \frac{1}{2.3.4} + \frac{1}{3.4.5} \text{ \&c. ad inf.}$$

8. Find the value of x in the following equations:

$$\frac{42x}{x-2} = \frac{35x}{x-3}$$

$$\frac{1}{2}(x+1) + \frac{1}{3}(x+2) = 16 - \frac{1}{4}(x+3)$$

$$3x^2 - 14x + 15 = 0.$$

9. In a given circle to inscribe an equilateral triangle.
10. Two equal bodies move at the same instant from the same extremity of the diameter of a circle with equal velocities in opposite semi-circles. Required the path described by the centre of gravity; find the path also when the bodies are unequal.
11. Through what chord of a circle must a body fall to acquire half the velocity gained by falling through the diameter?
12. Given the latitude of the place and the sun's meridian altitude, to find the declination.

13. Given the sun's altitude and azimuth and the latitude of the place, to find the declination and the hour of the day.

14. Prove that the velocity in a parabola: velocity in a circle at the same distance $:: \sqrt{2} : 1$.

15. How far must a body fall internally to acquire the velocity in a circle, the force varying $\frac{1}{D^2}$?

(Ball, *A History of Mathematics at Cambridge*, pp. 200–3)

1838: UNIVERSITY OF LONDON MATRICULATION EXAMINATION

SYLLABUS FOR MATHEMATICS AND NATURAL PHILOSOPHY (PASS EXAMINATION)

Mathematics

Arithmetic and Algebra: The ordinary rules of Arithmetic; Vulgar and Decimal Fractions; Extraction of the Square Root; Addition, Subtraction, Multiplication, and Division of Algebraical Quantities; Proportion; Arithmetical and Geometrical Progression; Simple Equations.

Geometry: The first book of Euclid.

Natural Philosophy*

Mechanics: Explain the Composition and Resolution of Statical Forces; describe the Simple Machines (Mechanical Powers), and state the Ratio of the Power to the Weight in each; define the Centre of Gravity; give the General Laws of Motion, and describe the chief experiments by which they may be illustrated; state the Law of the Motion of Falling Bodies.

Hydrostatics, Hydraulics, and Pneumatics: Explain the Pressure of Liquids and Gases, its equal diffusion, and variation with the depth; define Specific Gravity, and show how the specific gravity of bodies may be ascertained; describe and explain the Barometer, the Siphon, the Common Pump and Forcing Pump, and the Air-Pump.

Acoustics: Describe the nature of Sound.

Optics: State the Laws of Reflection and Refraction; explain the formation of Images by Simple Lenses.

* A popular knowledge only of these subjects . . . will be required, such as may be obtained by attending a course of experimental lectures.

SYLLABUS FOR 'CANDIDATES FOR HONOURS IN MATHEMATICS AND NATURAL PHILOSOPHY'

Arithmetic and Algebra; Geometry; Plane and Spherical Trigonometry; Conic Sections; the Elements of Statics and Dynamics; the Elements of Hydrostatics; the Elements of Optics.

SPECIMEN PAPERS FOR THE MATRICULATION EXAMINATION (NOV. 1838)

Arithmetic and Algebra (Pass)

1. Give the definition of a fraction, and explain the method of reducing fractions to a common denominator.

Of the fractions $\frac{2}{3}$, $\frac{5}{7}$, which is the greater?

2. Add together the fractions $\frac{11}{17}$, $\frac{31}{51}$, $\frac{200}{357}$, $\frac{5}{13}$, $\frac{24}{39}$; and divide $43\frac{3}{7}$ by $7\frac{5}{8}$.

3. Reduce the fraction $\frac{4015}{6305}$ to its lowest terms and transform it to a decimal.

4. Prove the rule for the multiplication of decimals. Multiply .576 by 83.4, and divide 222.027 by .0013.

5. Extract the square root of 9,512,295,961.

6. Find the simple interest on 207*l* 12*s*. 6*d*. for $21\frac{1}{2}$ years, at 3 per cent per annum.

7. Prove that $a\,b = b\,a$. State and demonstrate the truth of the rules for the signs in algebraic multiplication and division.

8. Multiply $x^5 + a^5 - a\,x\,(x^3 + a^3)$ by $(x^3 + a^3) + a\,x\,(x + a)$.

9. Find the greatest common measure of $x^5 - y^5$ and $x^2 - y^2$; and reduce to its lowest terms the fraction

$$\frac{3x^2 + 12x + 9}{x^5 + 5x^3 + 6}.$$

10. Solve the following simple equations:

$$\frac{1 + x}{x + 5} = \frac{3 + x}{7 + x};\ 5x + 3 = 11x - 21$$

$$19x + \frac{1}{2}(7x - 2) = 4x + \frac{35}{2}$$

11. Give some account of the different methods employed for eliminating one of the unknown quantities in equations of the form $ax + by = c \qquad a'x + b'y = c'$.

12. Find the values of x and y in the following systems of equations:

$$(1)\ \begin{cases} x^3 + y^3 = 65 \\ x + y = 5 \end{cases} \qquad (2)\ \begin{cases} \frac{3}{x} + \frac{3}{y} = 6 \\ x + y = 2 \end{cases} \qquad (3)\ \begin{cases} x + \sqrt{(x^2 - y^2)} = 8 \\ x - y = 1 \end{cases}$$

13. Explain when four quantities are said to be in proportion.

Show that if $a : b :: c : d$, then $a + b : a - b :: c + d : c - d$, and $a^n : b^n :: c^n : d^n$.

14. The sum of n terms of the arithmetic progression, 1, 3, 5, 7, &c., is n^2. Prove this, and find the sums to infinity of the two geometric series

$$\frac{1}{2} + \frac{1}{4} + \frac{1}{8} + \frac{1}{16}, \text{ \&c.,} \qquad \frac{1}{2} - \frac{1}{4} + \frac{1}{8} - \frac{1}{16}, \text{ \&c.}$$

15. At what time next before 12 o'clock are the hour and minute hands of a watch together?

Problems in Pure Mathematics (Honours)

1. Given the vertical angle, base, and perpendicular, from either of its extremities on the opposite side to construct the triangle.
2. Reduce the square root of 13 to a continued fraction.
3. Prove that in any plane triangle

$$\sin A + \sin B + \sin C = 4 \cos \frac{A}{2} \cos \frac{B}{2} \cos \frac{C}{2}.$$

4. When any regular polygon of an odd number of sides is inscribed in a circle, the sum of the chords drawn from any point in the circumference to the 1st, 3rd, 5th . . . angles is equal to the sum of those drawn to the 2nd, 4th, 6th, . . .
5. Find all the solutions in positive whole numbers of the equation $17x + 5y = 107$.
6. Prove that the chords common to each pair of three intersecting circles meet in the same point, and determine whether a similar property belongs to the other conic sections.
7. Find the locus of the intersections of tangents to a parabola drawn at right angles to each other.
8. Prove that

$$\text{Nap}^n \log x = (x - x^{-1}) - \frac{1}{2}(x^2 - x^{-2}) + \frac{1}{3}(x^3 - x^{-3}) \ldots$$

9. If a straight line be drawn from the extremity of any diameter of an ellipse to the focus, determine the co-ordinates of its point of intersection with the conjugate diameter, and thence show that the part intercepted by the diameters is equal to half the major axis.
10. If perpendiculars be drawn from the extremities, the centre, and any point of the major axis of an ellipse to meet the tangent drawn at the point where the last perpendicular intersects the curve, the intercepted portions of the four perpendiculars will be in geometric progression.
11. Find the sums to infinity of the series

$$\frac{1}{1^2} + \frac{1}{2^2} + \frac{1}{3^2} + \ldots . \qquad \frac{1}{1.2} + \frac{1}{2.3} + \frac{1}{3.4} + \ldots .$$

12. If the extremities of two intersecting straight lines be joined so as to form two vertically opposite triangles, the figure made by connecting the points of bisection of the bases of these triangles with the points of bisection of the given lines will be a parallelogram equal in area to half the difference of the triangles.

(*Parliamentary Papers*, 40 (1840), 'Papers relating to the University of London')

1847: BATTERSEA TRAINING SCHOOL

QUESTIONS PROPOSED TO THE STUDENTS

Arithmetic

(One Question only is to be answered in each Section.)

The method best suited for explanation to learners is to be preferred, and all the work to appear.

Section 1

1. If 670 articles cost 67*l.* 17*s.* 10*d.*, what is the price of each?

2. If 2 cwt. 3 qrs. of sugar cost 7*l.* 15*s.*, what is the least price per pound, in current coin, at which it must be sold in order to gain 7 per cent.

3. If two equally good workmen and three boys, who also work equally well, by labouring 7 hours a day, perform in 7 days what the two men alone could do in 10 days by working $8\frac{1}{2}$ hours a day, what ratio does the work done by a boy bear to that done in the same time by a man?

Section 2

1. What part of a guinea is equivalent to $\frac{5}{9}$-ths of 6*s.* 8*d.*

2. Reduce $2\frac{3}{4}$*d.* to the decimal of a pound, and divide the result by 7,500.

What is meant by incommensurables?

3. Find, by cross-multiplication (or duodecimals), the area of a board, whose length is 6ft. 11in., and breadth 1ft. 6in.

State clearly what the answer is. Explain the steps of the process, and the interpretation of the result.

Do you see any reason why, in representing quantities by means of numbers, incommensurables should occur?

Section 3

1. Explain the reason of each process employed in multiplying 607 by 404.

2. Shew that a fraction may be regarded as a quantity resulting from the division of the numerator by the denominator; and also that, if the numerator of a fraction be multiplied by any number, the result is equivalent to that obtained by dividing the denominator of the same fraction by the same number.

3. Explain the reason of the rules for extracting the square root of any number.

Mechanics

(One Question only is to be answered in each Section.)

Section 1

1. A labourer working with a wheel and axle 8 hours a day, can yield at the rate of 2,600 units of work per minute. How much must he charge per ton for raising coals from a depth of 25 fathoms, in order that he may earn 2*s.* 6*d.* per day?

2. An engine of 5-horse power raises 30 cwt. of coals per hour from a pit whose depth is 240 fathoms, and at the same time gives motion to a forge hammer which makes 25 lifts per minute, each lift being 3 feet; it is required to determine the weight of the hammer.

3. Two men undertake to dig a drain 500 feet long, and to carry the material in barrows to a heap at the end of it. Into what two parts must they divide the work, so that one-half of the labour may fall to the share of each?

Section 2

1. How many cubic feet of water must descend a river every minute to drive a wheel of 4 effective horse power, by means of a fall of 16 feet, the wheel yielding .68 of the work of the fall.

2. How would you calculate the velocity with which a ball leaves the muzzle of an air-gun, knowing the dimensions of the condensing chamber and barrel, the pressure of the condensed air, and the weight of the ball?

3. A square reservoir is to be dug, capable of containing 62,500 cubic feet of water. The material is to be carried an average distance of 10 feet from the edge of the reservoir, and the work of transferring it horizontally is one-sixth of that necessary to transfer it through the same space *vertically*. What must be its dimensions that the work of raising and transporting the material may be a *minimum*?

Section 3

1. Two cords, one 8 feet and the other 10 feet long, are attached to a weight of 100lbs., and fastened at their other extremities to two points in the ceiling 12 feet apart. What is the strain on each cord?

2. A barge, 40 feet long and 10 feet wide, is to be constructed with sheet iron, each square foot of which weighs 20 lbs. What must be its depth that it may just carry 60 tons lading?

3. A train weighing gross 80 tons, is allowed to descend freely an incline of 100 to 1, 500 yards in length. How far will it run along the horizontal line at the bottom of this incline with its acquired velocity, resistance being assumed throughout to be 8lbs per ton?

4. The wall of a reservoir is 15 feet high, and of the uniform thickness of 3 feet, each cubic foot of the material weighing 120lbs. Shores (props) are placed to support it 10 feet apart; each shore is 15 feet long, and its foot rests at a point 8 feet from the base of the wall. What is the thrust on each shore when the reservoir is full?

Differential and Integral Calculus

Section 1

1. Define a differential co-efficient, and a second differential co-efficient. Find the first and second differential co-efficients of

$$\frac{a^2 - x^2}{x}.$$

2. Write down Taylor's theorem; deduce Maclaurin's from it; and apply the latter to the expansion of a^x in a series proceeding by powers of x.

3. Apply Taylor's theorem to the expansion of $\sqrt{a^2 - (x + h)^2}$ in a series proceeding by powers of h. Explain the result when x is made equal to a; and show that the theorem is not untrue in that case.

Section 2

1. Show that the values of x, which make $f(x)$ a maximum or a minimum, render

$$\frac{df(x)}{dx} = 0.$$

Find the least value of

$$\frac{a^2 + 4x^2}{x}.$$

2. State the conditions for determining the maximum and minimum values of functions of one variable; and determine the greatest rectangular beam that can be cut from a given cylindrical piece of timber.

3. Write down the equation to the tangent at any point of the curve whose equation is $y^2 = ax - mx^2$; and show what it becomes when that point is the origin of co-ordinates; and also when it is at a point of the curve corresponding to

$$x = \frac{a}{2m}.$$

Find the greatest value of y; and trace the curve.

Section 3

1. Integrate the differential co-efficients

$$\frac{x}{\sqrt{a^2 - x^2}}$$

and

$$\frac{x^2}{a - x},$$

taking the latter one between the limits

$$x = 0 \quad \text{and} \quad x = \frac{a}{2}.$$

2. Write down the expression for finding the volume of a solid of revolution; explain its meaning in the theory of infinitesimals; and apply it to show that the volume of a right cone is found by multiplying the area of the base by one-third of the height. Show that the volume of any pyramid is found by the same rule.

3. Prove the expression for the differential coefficient of a surface of revolution; and show that if a cylinder circumscribe a sphere, then two planes, each of which is perpendicular to the axis of the cylinder, include between them equal surfaces of the sphere and cylinder respectively.

(*Minutes of Committee of Council*, 1847–8, pp. 359–62)

1884: EXAMINATION FOR THE CAMBRIDGE MATHEMATICAL TRIPOS

Part I

[The subjects in Part I were 'to be treated without the use of the Differential Calculus and the methods of Analytical Geometry'.]

Euclid. Books I to VI. Book XI. Props. I to XXI. Book XII. Props. I. II.

Arithmetic; and the elementary parts of Algebra; namely, the rules for the fundamental operations upon algebraical symbols with their proofs, the solution of simple and quadratic equations, ratio and proportion, arithmetical, geometrical and harmonical progression, permutations and combinations, the binomial theorem, and logarithms.

The elementary parts of Plane Trigonometry, so far as to include the solution and properties of triangles.

The elementary parts of Conic Sections, treated geometrically, but not excluding the method of orthogonal projections; curvature.

The elementary parts of Statics; namely, the equilibrium of forces acting in one plane and of parallel forces, the centre of gravity, the mechanical powers, friction.

The elementary parts of Dynamics; namely, uniform, uniformly accelerated, and uniform circular motion, falling bodies and projectiles in vacuo, cycloidal oscillations, collisions, work.

The first, second, and third sections of Newton's Principia; the propositions to be proved by Newton's methods.

The elementary parts of Hydrostatics; namely, the pressure of fluids, specific gravities, floating bodies, density of gases as depending on pressure and temperature, the construction and use of the more simple instruments and machines.

The elementary parts of Optics; namely, the reflexion and refraction of light at plane and spherical surfaces, not including aberrations; the eye; construction and use of the more simple instruments.

The elementary parts of Astronomy; so far as they are necessary for the explanation of the more simple phenomena, without the use of spherical trigonometry; astronomical instruments.

Part II

Algebra.
Trigonometry, Plane and Spherical.
Theory of Equations.
Easier parts of Analytical Geometry, Plane and Solid, including Curvature of Curves and Surfaces.
Differential Calculus.
Integral Calculus.
Easier parts of Differential Equations.
Statics, including Elementary Propositions on Attractions and Potentials.
Hydrostatics.
Dynamics of a Particle.
Easier parts of Rigid Dynamics.
Easier parts of Optics.
Spherical Astronomy.

Part III*

Group A

Differential Equations.
Calculus of Variations.
Higher Algebra.
Higher parts of Theory of Equations.
Higher Analytical Geometry, Plane and Solid.
Finite Differences.
Higher Definite Integrals.
Elliptic Functions.
Theory of Chances, including Combination of Observations.

Group B

Laplace's and allied Functions.
Attractions.
Higher Dynamics.
Newton's Principia, Book I. Sect. IX., XI.
Lunar and Planetary Theories.
Figure of the Earth.
Precession and Nutation.

Group C

Hydrodynamics, including Waves and Tides.
Sound.
Physical Optics.
Vibrations of Strings and Bars.
Elastic Solids.

Group D

Expression of Functions by Series or Integrals involving sines and cosines.
Thermodynamics.
Conduction of Heat.
Electricity.
Magnetism.

* The Moderators and Examiners may place in the first division [of three, each arranged, alphabetically] any Candidate who has shewn eminent proficiency in any one group.

1900: PERRY'S PROPOSALS FOR A NEW SCHOOL SYLLABUS

Practical Mathematics

Elementary Stage

Arithmetic.—The use of decimals; the fallacy of retaining more figures than are justifiable in calculations involving numbers which represent observed or measured quantities. Contracted and approximate methods of multiplying and dividing numbers whereby all unnecessary figures may be omitted. Using rough checks in arithmetical work, especially with regard to the position of the decimal point.

The use of 5.204×10^5 for 520,400 and of 5.204×10^{-3} for .005204. The meaning of a common logarithm; the use of logarithms in making calculations involving multiplication, division, involution and evolution. Calculation of numerical values from all sorts of formulae.

The principle underlying the construction and method of using a common slide rule; the use of a slide rule in making calculations. Conversion of common logarithms into Napierian logarithms. The calculation of square roots by the ordinary arithmetical method. Using algebraic formulae in working questions on ratio and variation.

Algebra.—To understand any formula so as to be able to use it if numerical values are given for the various quantities. Rules of Indices.

Being told in words how to deal arithmetically with a quantity, to be able to state the matter algebraically. Problems leading to easy equations in one or two unknowns. Easy transformations and simplifications of formulae. The determination of the numerical values of constants in equations of known form, when particular values of the variables are given. The meaning of the expression 'A varies as B.'

Factors of such expressions as $x^2 - a^2$, $x^2 + 11x + 30$, $x^2 - 5x - 66$.

Mensuration.—The rule for the length of the circumference of a circle. The rules for the areas of a triangle, rectangle, parallelogram, circle; areas of the surfaces of a right circular cylinder, right circular cone, sphere, circular anchor ring. The determination of the area of an irregular plane figure (1) by using a planimeter; (2) by using Simpson's or other well-known rules for the case where a number of equidistant ordinates or widths are given; (3) by the use of squared paper whether the given ordinates or widths are equidistant or not, the 'mid-ordinate rule' being used. Determination of volumes of a prism or cylinder, cone, sphere, circular anchor ring.

The determination of the volume of an irregular solid by each of the *three* methods for an irregular area, the process being first to obtain an irregular plane figure in which the varying *ordinates* or widths represent the varying *cross sections* of the solid.

Some practical methods of finding areas and volumes. Determination of weights from volumes when densities are given.

Stating a mensuration rule as an algebraic formula. In such a formula any one of the quantities may be the unknown one, the others being known.

Use of Squared Paper.—The use of squared paper by merchants and others to show at a glance the rise and fall of prices, of temperature, of the tide, &c. The use of squared paper should be illustrated by the working of many kinds of exercises, but it should be pointed out that there is a general idea underlying them all. The following may be mentioned:—

Plotting of statistics of any kind whatsoever, of general or special interest. What such curves teach. Rates of increase.

Interpolation, or the finding of probable intermediate values. Probable errors of observation. Forming complete price lists by shopkeepers. The calculation of a table of logarithms. Finding an average value. Areas and volumes, as explained above. The method of fixing the position of a point in a plane; the x and y and also the r and θ, co-ordinates of a point. Plotting of functions, such as $y = ax^n$, $y = ae^{bx}$, where a, b, n, may have all sorts of values. The straight line. Determination of maximum and minimum values. The solution of equations. Very clear notions of what we mean by the roots of equations may be obtained by the use of squared paper. Rates of increase. Speed of a body. Determination of laws which exist between observed quantities, especially of linear laws. Corrections for errors of observation when the plotted quantities are the results of experiment.

In all the work on squared paper a student should be made to understand that an exercise is not completed until the scales and the names of the plotted quantities are clearly indicated on the paper. Also that those scales should be avoided which are obviously inconvenient. Finally, the scales should be chosen so that the plotted figure shall occupy the greater part of the sheet of paper; at any rate, the figure should not be crowded in one corner of the paper.

Geometry.—Dividing lines into parts in given proportions, and other illustrations of the 6th Book of Euclid. Measurement of angles in degrees and radians. The definitions of the sine, cosine and tangent of an angle; determination of their values by drawing and measurement; setting out of angles by means of a protractor when they are given in degrees or radians, also when the value of the sine, cosine or tangent is given. Use of tables of sines, cosines and tangents. The solution of a right angled triangle by calculation and by drawing to scale. The construction of a triangle from given data; determination of the area of a triangle. The more important propositions of Euclid may be illustrated by actual drawing; if the proposition is about angles, these may be measured by means of a protractor; or if it refers to the equality of lines, areas or ratios, lengths may be measured by a scale and the necessary calculations made arithmetically. This combination of drawing and arithmetical calculation may be freely used to *illustrate* the truth of a proposition.

The method of representing the position of a point in space by its distances from three co-ordinate planes. How the angles are measured between (1) a line and plane; (2) two planes. The angle between two lines has a meaning whether they do or do not meet. What is meant by the projection of a line or a plane figure on a plane. Plan and elevation of a line which is inclined at given angles to the co-ordinate planes. The meaning of the terms 'trace of a line,' 'trace of a plane.'

The difference between a *scalar* quantity and a *vector* quantity. Addition and subtraction of vectors.

Slope of a line; slope of a curve at any point in it. Rate of increase of one quantity y relatively to the increase of another quantity x; the symbol for this rate of increase, namely,

$$\frac{dy}{dx};$$

how to determine

$$\frac{dy}{dx}$$

when the law connecting x and y is of the form $y = ax^n$. Easy exercises on this rule.

In setting out the above syllabus the items have been arranged under the various branches of the subject.

It will be obvious that it is not intended that these should be studied in the order in which they appear; the teacher will arrange a mixed course such as seems to him best for the class of students with whom he has to deal.

(*Nature*, 2 Aug. 1900, 319–20)

1905: STANDARDS OF EXAMINATION IN ARITHMETIC (BOARD OF EDUCATION)

Standard I

The four simple rules. Divisors and multipliers not exceeding 6. No number higher than 99 to be employed in the questions or required in the answers.

Standard II

Compound rules (money). Divisors and multipliers not exceeding 12. Sums of money in the questions and answers not to exceed 10*l*.

Standard III

Simple rules and compound rules (money). Divisors and multipliers not exceeding 99. No number higher than 99,999 to be employed in the question or required in the answer. Sums of money in the questions and answers not to exceed 99*l*.

Standard IV

Compound rules applied to the following weights and measures (length, weight, capacity, time). In length, yards, feet, and inches; in weight, tons, cwts., qrs., lbs., ozs.; in capacity, gallons, quarts, pints; in time, days, hours, minutes, seconds—are the only terms that will be required in this, and in the fifth standard. Divisors and multipliers not to exceed 99.

Standard V

Vulgar fractions (simple fractions only). Practice. Bills of parcels. Common weights and measures.

Standard VI

Decimal fractions (excluding recurring decimals). Simple proportion or single rule of three by the method of unity. Calculation of simple interest upon a given principal. Common weights and measures. Mensuration of rectangles and rectangular solids; the extraction of square and cube roots is not required. (Boys only.)

Standard VII

Vulgar and decimal fractions. Averages and percentages. Investments of savings. Consols.

1910: CAMBRIDGE SENIOR LOCAL EXAMINATION PAPER

[Corresponding to A-level]

Analytical Geometry and Differential Calculus

(*Two hours*)

Analytical Geometry

1. Find the angle between the straight lines

$$ax + by + c = 0, \qquad a'x + b'y + c' = 0.$$

Find the equations of the two lines through the intersection of the lines $x - 2y + 7 = 0$ and $2x - 3y - 3 = 0$ which make an angle of 45° with the line $y = 2x$.

2. Express the area of a triangle in terms of the coordinates of its vertices.

Show that the area of the triangle formed by the lines

$$ax^2 + 2bxy + cy^2 = 0, \qquad y = mx + n$$

is

$$n^2(b^2 - ac)^{\frac{1}{2}}/(a + 2bm + cm^2).$$

3. Find the coordinates of the pole of the line $lx + my = 1$ with respect to the circle $x^2 + y^2 = a^2$.

Find the locus of the poles of chords of this circle which subtend a right angle at the point (h, k).

4. Find the equation of the normal at the point $(am^2, 2am)$ of a parabola, and show that, in general, three normals can be drawn from a point to the curve.

Show that, if Q, R are the feet of the normals drawn from a point P of the curve itself, the line through P and the vertex meets QR on the directrix.

5. Show that the feet of the perpendiculars from a focus to the tangents of an ellipse lie on a circle.

Find, in terms of the eccentric angle of the point of contact, the length of the chord of this circle made by the tangent.

Differential Calculus

6. Define the term *differential coefficient* and give a geometrical interpretation.

Differentiate the following functions with respect to x:

$$\log \sin \sqrt{x}, \quad \tan^{-1}\frac{2x}{1+x^2}, \quad x^{1+x}.$$

7. Employ Maclaurin's Theorem to find the general term of the expansion of $\tan^{-1} x$ in powers of x.

8. Explain a method of determining the maxima of a function of one variable.

O is a fixed point on a circle, P any point on the circle, and OY the perpendicular from O on the tangent at P. Show that the greatest area of the triangle OPY is $3\sqrt{3}/8\pi$ of the area of the circle.

9. Find the equation of the tangent at (x, y) to the curve $x^3 + y^3 = a^3$.

Show that the portion of the tangent intercepted between the coordinate axes is divided at the point of contact in the ratio $x^3 : y^3$.

1911: EXAMINATION FOR THE CAMBRIDGE MATHEMATICAL TRIPOS

Part I

Pure Geometry.—Lines, planes, circles and spheres: including questions on reciprocation, cross-ratios and inversion. The methods of orthogonal projection and of perspective.

Algebra and Trigonometry.—Exercises involving simple algebraic computation, the use of the slide rule and logarithmic tables being allowed. Elementary properties of equations. Use of the binomial, exponential and logarithmic expansions: applications to trigonometric functions. Use and properties of trigonometric functions. Trigonometry of simple figures.

Analytical Geometry.—Lines, planes, circles and spheres, excluding questions requiring the use of oblique co-ordinates. Elementary properties of conic sections, and of an ellipsoid referred to its principal axes, including purely geometrical methods in suitable cases.

Differential and Integral Calculus.—Differentiation and simple integration, partial differentiation. Simple applications to plane curves. Volumes of solids of revolution. Use of Taylor's and Maclaurin's theorems: Maxima and minima for one independent variable. Curvature: plotting of curves from their equations. Simple linear differential equations.

Dynamics.—Mass, momentum, force, energy, moment of momentum, and their fundamental relations. Equilibrium without and with friction. Easy exercises on conservation of energy, of momentum, and of moment of momentum. Motion under gravity, and under simple central forces. Simple and compound pendulums. Elementary uniplanar applications to equilibrium and stability. Simple graphical statics: funiculars, frameworks. Statics of liquids and gases.

Elementary Electricity.—The elementary parts of electricity: namely, properties of the potential: lines of forces; air-condensers; Ohm's law with simple application galvanometers.

Optics.—Reflection and refraction of light: applications to mirrors, prisms, lenses and simple combinations, excluding spherical and chromatic aberrations.

Part II

Schedule A: (Compulsory)

Plane and Solid Geometry, including methods of Pure Geometry. Curvature of curves and surfaces. Plane and Spherical Trigonometry.

Algebra, including Theory of Equations.

Differential and Integral Calculus, including the first variation of integrals.

Elementary parts of the Theory of Functions, including the properties of simple functions of the complex variable, circular functions and their inverses: simple applications to doubly periodic functions. Simple properties of Legendre's functions and Bessel's functions. Use of Fourier's series.

Differential Equations.—Integration of ordinary differential equations of the first order, and of ordinary linear differential equations having constant coefficients. Elementary theory of linear differential equations of the second order, having only regular integrals; solution of simple types of such equations of the second order, having integrals that are not regular. The methods of solving partial differential equations of the first order, and the use of methods of solving simple partial differential equations of the second order.

Dynamics, including Lagrange's equations and simple problems of motion in three dimensions. Elementary analytical statics. Elementary gravitational attractions. Vibrations of a stretched string.

Hydromechanics, including hydrostatics, the general principles of hydrodynamics with simple applications, elementary questions on the vibrations of elastic fluids, plane waves of sound.

Astronomy.—The elementary parts so far as they are necessary for the explanation of simple phenomena.

Electricity.—Fundamental principles of electrostatics, distribution of currents, magnetism, electro-magnetism.

Elementary optics, including the geometrical treatment of interference of waves.

Schedule B: (Voluntary)

Theory of Numbers. Invariants and Covariants. Synthetic Geometry. Algebraic Geometry. Differential Geometry. Groups. Theory of Functions. Elliptic Functions. Differential Equations. Dynamics. Hydrodynamics. Sound and Vibrations. Statics and Elasticity. Electricity and Magnetism. Geometrical and Physical Optics. Thermodynamics. Spherical Astronomy and Combination of Observations. Celestial Mechanics.

1919: REPORT BY THE MATHEMATICAL ASSOCIATION ON 'THE TEACHING OF MATHEMATICS IN PUBLIC AND SECONDARY SCHOOLS'

SUMMARY OF RECOMMENDATIONS*

(1) That a boy's educational course at school should fit him for citizenship in the broadest sense of the word: that, to this end, the moral, literary, scientific (including mathematical), physical and aesthetic sides of his nature must be developed. That in so far as mathematics is concerned, his education should enable him not only to apply his mathematics to practical affairs, but also to have some appreciation of those greater problems of the world, the solution of which depends on mathematics and science.

(2) That the utilitarian aspect and application of mathematics should receive a due share of attention in the mathematical course.

(3) That the mathematical course in the earlier stages should not be concerned exclusively with arithmetic, algebra, and geometry; that such subjects as trigonometry, mechanics, and the calculus should be begun sooner than has been customary and developed through the greater part of the boy's school career, so as to give him time to assimilate them thoroughly, and enable him to cover more rapidly, at a later stage, the higher parts of arithmetic, algebra and geometry.

(4) That no boy should leave school entirely ignorant of applied mathematics (e.g. mathematics relating to machinery, structures, motion of bodies, astronomy, electrical plant, surveying, statistics).

(5) That, while the average boy should receive careful and adequate instruction, the boy of high talent should receive special attention, as the value to the race of carefully trained superior talents is incalculable. A well-equipped secondary school should be so staffed as to be able to educate *pro viribus* both the average boy and the boy of genius.

(6) That a boy who takes mathematics as his main subject in the later part of his school life should also, in general, study science, as well as carry on some form of literary study; and that the general educational purpose underlying the choice of these various subjects should be made manifest to the boy.

(7) That the teaching and organisation of mathematics and physics should be in the closest possible coordination.

* Throughout this Report the word boy is to be taken as referring to pupils of either sex.

(8) That the mathematical teacher should also receive training of a less intensive kind in some subject in which his mathematics can be definitely applied, e.g. geography, physics, chemistry, engineering, manual training, astronomy, etc.

(9) That teachers of insufficient ability or knowledge should not be promoted to be head of departments simply on the ground of long and faithful service. Long and faithful service deserves recognition, but in some other way. The head of a mathematical department has to teach the advanced work–which requires knowledge–and to draw up the general school syllabus–which requires outlook.

(10) That heads of mathematical departments and specialist teachers of the higher branches of mathematics in the advanced departments of the secondary schools should have tolerably short routines in order that they may be able to read more widely in their subject and to study its modern development (cf. 13). Only thus can the knowledge imparted to the boys be kept up to date.

(11) That an external examination syllabus should be frequently revised by a joint body consisting of representatives of the teachers themselves and of the external examining body. Otherwise the syllabus, being stereotyped, tends to become obsolete, and teachers have to teach what they have ceased to value.

(12) That it is more important to teach boys than to examine them; that the number of examinations at present conducted in the majority of schools be reduced, inasmuch as the setting of examination papers consumes the master's time and energy, thereby lowering his teaching capacity, while the continuous pressure of working at examination papers for days on end is an unproductive strain on the boys.

(13) That every secondary school should be provided with a mathematical library, containing books of a more general character than the ordinary textbooks, in order that the pupils and masters may be enabled to widen their horizon and catch a glimpse of the regions beyond.

(14) That portraits of the great mathematicians should be hung in the mathematical classrooms, and that reference to their lives and investigations should frequently be made by the teacher in his lessons, some explanation being given of the effect of mathematical discoveries on the progress of civilisation.

1920: COUNTY MINOR SCHOLARSHIPS AND LOCAL SCHOLARSHIPS EXAMINATION

(West Riding of Yorkshire)

Arithmetic

(*One hour and twenty minutes allowed*)

INSTRUCTIONS TO CANDIDATES

Rulers may not be used for drawing ink lines.

In marking the paper special attention will be given to the methods used in answering the questions.

Appendix

Part I

(*You should answer only* **four** *questions in this part*)

1. The daily wages of each of 1,025 miners were raised from 13*s*. $0\frac{1}{2}$*d*. to 13*s*. 11*d*. How much per week (6 days) was the total increase in wages?

2. (*a*) Express $\dfrac{\text{2 cwt. 28 lb.}}{\text{4 cwt. 14 lb.}}$ as a simple vulgar fraction.

(*b*) Find the value of $13\frac{2}{3} - \frac{3}{4}$ of $5\frac{1}{6}$.

3. (*a*) Which of the following numbers differs most from 23.762?

24.091, 23.617, 23.234, 24.303, 23.404.

Write down the difference.

(*b*) Calculate the value of .63 of £1 in shillings and pence to the nearest penny.

4. A train travelling at 45 miles an hour covers the distance between two telegraph poles in 8 seconds. How far apart are the poles?

5. In a certain district the total length of the roads is $13\frac{3}{4}$ miles. If each road is 12 yards wide what is their total area in acres?

Part II

(*You should answer only* **four** *questions in this part. One, at least, of these should be No.* 7 *or No.* 10.)

6. A piece of iron one foot long becomes $\frac{3}{16}$ of an inch longer when made red hot. How long would a similar piece 5 feet 4 inches long become if made red hot?

7. Draw, to scale, a plan of a kite made with two sticks 25 inches and 17 inches long respectively, crossing at right angles. The sticks are fastened together in the middle of the short one and 8 inches from one end of the long one.

Find from your plan how much string is required to join up the ends of the sticks.

8. A basin with a quantity of water in it weighed 487.36 grammes. With only half the quantity of water it weighed 429.87 grammes. What did the basin weigh?

9. An iron article weighing 4 cwt. 7 lb. is 16 times as heavy as a wooden model of it. The model consists of five parts, four of which weigh 8 lb. 7 oz., 5 lb. 3 oz., 6 lb. 11 oz. and 4 lb. 8 oz. respectively. Find the weight of the fifth part.

10. The plan of a house, drawn to a scale of one inch to four yards, is given. Find the number of square yards covered by the house.

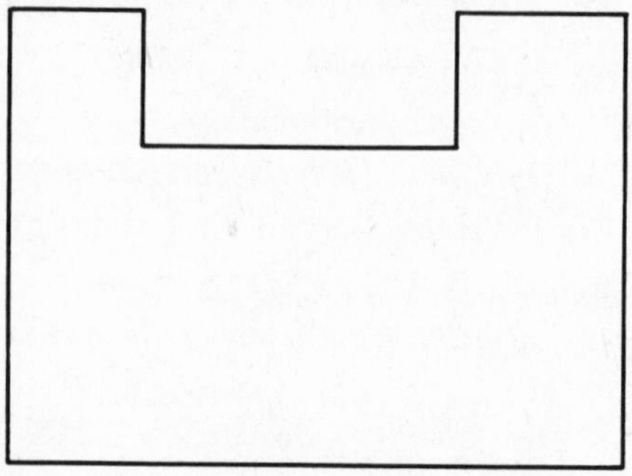

1923: STANDARDISED ARITHMETICAL TEST (LONDON COUNTY COUNCIL)

[Reprinted in case readers are over-impressed by the standards of the County Minor paper reproduced above.]

Fifty Minutes

Mechanical Arithmetic

Work in your head if possible. Scrap paper allowed.

1. 8 + 5.
2. 9 + 6 + 8.
3. 8 + 7 + 13.
4. 34 − 19.
5. 8 + 9 − 3 + 2.
6. 17 + 18 − 14.
7. 3,000 − 1.
8. 156 − 156.
9. 200 − 13.
10. 54 + 368 + 7.
11. 100 − 4 − 96.
12. 201 − 102.
13. 47 × 8.
14. 63 × 5 ÷ 9.
15. 587 × 7.
16. 1 ÷ 1.
17. 5 ÷ 1.
18. 0 ÷ 4.
19. 2,982 ÷ 6.
20. 1 × 1 × 1 × 1 × 1.
21. 4 × 7 × 9 × 0.
22. 18 × 18.
23. 231 × 201.
24. £3 12*s*. + £1 4*s*. 6*d*. + 13*s*. $6\frac{1}{2}$*d*.
25. £5 − £2 12*s*. 6*d*.
26. 2 hr. 35 min. 15 sec. + 46 min. 50 sec.
27. 3 lb. 4 oz. − 1 lb. 12 oz.
28. 3 yards 2 ft. 8 in. − 2 ft. 10 in.
29. 16*s*. $7\frac{1}{2}$*d*. × 2.
30. £11 18*s*. ÷ 7.
31. £2 17*s*. 3*d*. × 5.
32. 1*s*. $3\frac{1}{2}$*d*. × 15.
33. 12 lb. 3 oz. ÷ 5.
34. 7 weeks 4 days 8 hours × 3.
35. 2 ft. 3 in. × 8.
36. 5 yards 0 ft. 6 in. ÷ 6.
37. How much are 50 half-crowns?
38. Reduce £1 5*s*. 2*d*. to pence.
39. $\frac{1}{2} + \frac{1}{3}$.
40. $\frac{1}{3} - \frac{1}{4}$.
41. $\frac{2}{3} + \frac{3}{5}$.
42. $3\frac{5}{36} - 1$.
43. $6\frac{3}{5} - 4\frac{2}{5}$.
44. $1\frac{3}{4} - \frac{1}{2} + \frac{1}{4}$.
45. $2\frac{2}{3} + 1\frac{1}{2} - 2\frac{2}{3}$.
46. $\frac{3}{4} \times \frac{4}{5}$.
47. $\frac{2}{3} \div \frac{5}{6}$.
48. $\frac{3}{8} \times \frac{4}{5} \div \frac{3}{4}$.
49. $8 \times \frac{1}{2}$.
50. $8 \div \frac{1}{2}$.
51. £20 × $\frac{1}{3}$.
52. .5 + .75 + .25.
53. 1.2 + 3.6 − 2.4.
54. 18.2 − 6.7 + 11.8.
55. 10 − 7.35.
56. 3.1 − 3.01.
57. .1 × .1.
58. .6 × .07.
59. 1 ÷ .1.
60. 3 ÷ .03.
61. 7.5 × 8 ÷ .6.
62. .011 × .26.
63. 6 ÷ .5.
64. $\sqrt{81}$.
65. $\sqrt{169}$.
66. $\sqrt{400}$.
67. Square root of 441.
68. $\sqrt{49} \times \sqrt{121}$.

69. 5 per cent of 10*s.*
70. $3\frac{1}{2}$ per cent of £300.
71. 1 per cent of 100.
72. 100 per cent of 1.
73. 100 per cent of 100.
74. 200 per cent of 100.
75. 3 lb. 4 oz. at 1*s.* a lb.
76 4 yd. 1 ft. at 2*s.* 6*d.* a yd.
77. 32 oranges at 8 for 5*d.*
78. 100 oranges at 5 for 1*s.*
79. £3 12*s.* 6*d.* × $\frac{2}{3}$.
80. Reduce $\frac{21}{35}$ to lowest terms.
81. Find the value of £.375.
82. What decimal of £1 is 13*s.* 4*d.*?
83. G.C.M. or H.C.F. of 32, 16, 56.
84. G.C.M. or H.C.F. of 12, 20, 28, 32.
85. L.C.M. of 1, 2, 3, 4, 5.
86. L.C.M. of 7, 8, and 112.
87. L.C.M. of 3, 5, 9, 15.
88. Average of 5, 17, 17, 19.
89. Average of $\frac{1}{4}$, $\frac{1}{2}$, $\frac{3}{4}$.
90. Average of .5, .01, 1.2.
91. 4 : 7 : : 6 : x.
92. 3 : x : : 5 : 10.
93. x : 8 : : 3 : 12.
94. 8 : 7 : : x : 21.
95. Area of plot $2\frac{1}{2}$ yd. sq.
96. Circum. of circle 4 ft. in diam. ($\pi = \frac{22}{7}$
97. Area of circle 4 ft. diam.
98. Area of all the faces of a cube of 2 inch edge.
99. Volume of water in a tank 2 ft. long, 2 ft. wide, $1\frac{1}{2}$ ft. deep.
100. Find radius of circle with area of 154 sq. ft.

Norms giving the average number of sums right by age-group.

Age	9	10	11	12	13	14
boys	11	20	29	38	46	53
girls	11	17	23	29	36	44

(Ballard, *The New Examiner*, 1923)

1934: OXFORD AND CAMBRIDGE BOARD SCHOOL CERTIFICATE PAPER*

ELEMENTARY MATHEMATICS II

[Equivalent to O-level. Candidates took three papers]

2 Hours

Candidates may attempt **seven** *questions only, not more than* **four** *of which are to be selected from the questions numbered* 4–10.

[*Logarithms, slide-rule, or algebra may be used in any question.*]

1. Two square sheets of metal, one of side 37.5 cm., the other of side 27.2 cm., of the same thickness, were melted down together and rolled out into a square sheet of the same thickness as the original sheets. What was the length of the side of the final sheet?

2. If $a = 1 - \dfrac{b}{\dfrac{c}{t} - d}$, find t in terms of a, b, c and d.

If $b{:}c{:}d = 1{:}2{:}3$ and $a = 5$, find the value of t.

* Reproduced by kind permission of the Board.

3. If both pairs of the opposite sides of a quadrilateral are parallel, prove that they are equal.

State and prove the converse proposition.

4. On one particular day £1 was worth 4 dollars 77 cents in New York, one dollar was worth 16.07 francs in Paris, and 78.82 francs were worth £1 in London. What would have been the gain or loss to a merchant if he could have changed £1000 into dollars, the resulting dollars into francs, and the francs back into pounds on that day? [Give your answer correct to the nearest ten shillings.]

5. Use logarithms to evaluate

(i) $3.187 \times \sqrt[3]{0.5276}$;

(ii) $(1.516)^2 - \dfrac{1}{(1.516)^2}$.

6. Prove that in any triangle the sum of the squares on any two sides is equal to twice the square on half the third side together with twice the square on the median which bisects the third side.

ABC is a triangle in which $AB = 10$ cm., $AC = 4$ cm. and $BC = 8$ cm. A point *X* is taken on *BC* such that $CX = 2$ cm. By two successive applications of the first part of this question, *calculate* the length of *AX*.

7. Solve the equations

$$x + y + 1 = 0,$$
$$ax + by + c = 0,$$

for x and y, when a is not equal to b.

If x and y also satisfy

$$a^2x + b^2y + c^2 = 0,$$

shew that c is either equal to a or equal to b.

8. In unloading 5 tons of wood from a barge, it is found that, if x men work for $\frac{1}{4}$ hr and then $x - 4$ men complete the job, it takes $\frac{1}{4}$ hr longer than if $x - 4$ men work for $\frac{1}{4}$ hr and then x men complete the job. If one man takes one hour to unload one ton, find x.

9. (i) Through a point *P* on the side *AB* of a triangle *ABC* is drawn a straight line *PQ*, parallel to *BC*, to meet *AC* at *Q*. Find the ratio of *AP* to *PB*, (*a*) when $PQ = \frac{1}{9}BC$, (*b*) when the area of the triangle $APQ = \frac{1}{9}$ of the area of the triangle *ABC*.

(ii) *ABC* is a triangle and *D* is the middle point of *BC*. *P* is any point on *AB*, and a straight line drawn through *P* parallel to *BC* cuts *AD* at *R* and *AC* at *S*. Prove that $PR = RS$.

[N.B. The triangle *ABC* is **not** isosceles.]

10. The vertical angle of a cone, of height 4.2 in., is 48°. Find the area of the curved surface.

[The area of the curved surface is πrl, where r is the base radius and l the length of the slant height. Take $\pi = 3.142$.]

Appendix

1944: THE JEFFERY PROPOSALS

SUGGESTED ALTERNATIVE SYLLABUS FOR ELEMENTARY MATHEMATICS

There will be three papers of $2\frac{1}{4}$ hours. Each paper may contain questions on any part of the syllabus, and the solution of any question may require knowledge of more than one branch of the syllabus.

Each paper will consist of four questions to be attempted by all candidates and of a second section in which a choice will be allowed. The first four questions will be elementary in type and may consist of two or three short parts.

It is not expected that all candidates will have covered the whole syllabus.

Unless the terms of the questions impose specific limitations,

(*a*) A candidate may use any appropriate method.

(*b*) Tables of logarithms, trigonometrical functions, squares and square roots, compound interest, etc., may be used wherever they give the required degree of accuracy.

(*c*) T-squares, set squares, graduated rulers, diagonal scales, protractors and compasses may be used.

1. Numbers

The ordinary processes of Arithmetic.

The commoner systems of weights, measures, and money, including metric units.

Fractions, decimals, ratio, percentage.

Use of common logarithm and square-root tables.

Significant figures.

(Questions may be set on the applications of these processes to problems of everyday life in the home and community, but such questions will not involve complicated operations or the knowledge of uncommon technical terms. It is not intended that the 'long rules' for the extraction of square roots, the determination of H.C.F., etc., should be included.)

2. Mensuration

The rectangle, triangle and figures derived from them, including easy extensions to three dimensions.

The circle, cylinder, cone and sphere (but formulae for the last two need not be memorised).

(Questions may involve knowledge of the altitude and centre of an equilateral triangle and the ratio of the sides of the right-angled isosceles triangle and the 30°, 60°, 90° triangle.)

3. Formulae and Equations

Construction of a formula through symbolical expression of a functional relation (e.g. V is proportional to x^2), or through generalization of an arithmetical result.

Interpretation, evaluation, and very easy manipulation of a formula.

The use of indices.

(Only simple examples of fractional and negative indices will be set, and questions involving manipulation of such indices or of surds will not be set; but candidates may be expected to understand the use of indices to express such numbers as 3.74×10^8 or 1.35×10^{-6}.)

The use of suffixes.

Common factors; factors of $(a^2 - b^2)$ and of such extensions as occur in mensuration.

Easy trinomial factors.

Fractions whose denominators are single terms or linear expressions.

(Simple manipulations should not include more than two such fractions.)

Simple equations, quadratic equations, and linear simultaneous equations in two variables.

4. Graphs, Variation, Functionality

Graphs from statistical data.

The idea of a function of a variable.

Translation into symbols of relations such as 'y is inversely proportional to x', 'V varies as X^3', and their illustration by sketch-graphs.

Simple cases of the function

$$y = Ax^3 + Bx^2 + Cx + D + \frac{E}{x} + \frac{F}{x^2},$$

where the constants are numerical and at least three of them are zero.

Graphical treatment of these functions.

The gradients of these graphs, by drawing or by calculation.

(Questions will not involve theoretical treatment of limits.)

Application of gradients to (*a*) rates of increase, (*b*) easy linear kinematics including the distance-time and speed-time curves, (*c*) maxima and minima.

The determination of a function from its gradient. The area 'under' a graph.

Application to areas, volumes of revolution, linear kinematics.

(Questions on Calculus will not be set in the compulsory part of the paper. The use of the formal notation of the calculus will not be required, but candidates may be expected to understand the use of the terms 'differentiate' and 'integrate'. Questions on rates of change will not involve the use of the formula

$$\frac{dF(x)}{dt} = \frac{dF(x)}{dx} \cdot \frac{dx}{dt}.)$$

5. Two-dimensional Figures

A sound appreciation of the properties set out below will be expected. Proofs of the items in *italics* may be required. Proofs of the other properties will not be required. Riders may be set which can be solved by the use of these properties, but they themselves will not be set as riders. In solving riders candidates may use any knowledge they possess. Candidates will be expected to understand the relation between a theorem and its converse.

Properties of angles at a point and angles made with parallel lines.

The exterior angle property and angle-sum of a triangle.

Angle-sum properties of polygons.

Congruency of triangles: similarity of triangles.

Symmetry about a point or line.

The isosceles triangle: the parallelogram, rectangle, square.

Parallelograms on the same base and between the same parallels are equal in area.

The straight line joining the mid-points of two sides of a triangle is parallel to the third side and equal to half the length of the third side. (By congruence or areas or similarity.)

Other area properties of rectangles, parallelograms, triangles and trapezia, including the formulae

$$\tfrac{1}{2}bc \sin A \quad \text{and} \quad \sqrt{\{s(s-a)(s-b)(s-c)\}}.$$

Connections between Algebra and Geometry, e.g. $(a+b)^2$, use of co-ordinates, including negative co-ordinates.

The sine, cosine, and tangent of an angle, acute or obtuse.

(Trigonometry of obtuse angles will not occur in the first section of the paper.)

The sine-rule for a triangle.

(With numerical applications involving the use of four-figure tables but not including the ambiguous case.)

The Theorem of Pythagoras. (By proving that the perpendicular drawn to the hypotenuse divides the triangle into similar triangles; or otherwise.)

$$\sin^2 A + \cos^2 A = 1, \qquad a^2 = b^2 + c^2 - 2bc \cos A.$$

(This replaces the 'extensions of Pythagoras'; easy numerical applications using four-figure tables will be set.)

The symmetrical properties of chords of a circle.

An angle at the centre of a circle is twice any angle at the circumference standing on the same arc.

The other 'angle properties' of the circle.

The perpendicularity of tangent and radius; the distance between centres of circles in contact; the equality of tangents from an external point.

The 'alternate segment' theorem.

The 'intersecting chord' theorem for an external point.

$$(OP \cdot OQ = OR \cdot OS = OT^2.)$$

The analogous property for an internal point.

The relationship between the areas of similar triangles.

Corresponding results for similar figures and extension to volumes of similar solids.

The bisector of any angle of a triangle divides the opposite side in the ratio of the sides containing the angle.

The analogous property for an exterior angle.

Knowledge of simple loci, with easy extensions to three dimensions. The method of intersecting loci. Plotting of simple loci other than the circle, e.g. parabola or ellipse.

Knowledge of the following 'ruler and compasses' constructions will be assumed; set squares will be allowed for the construction of parallel lines.

Bisection of angles and straight lines.

Construction of perpendiculars to a given line and of angles equal to a given angle.

Construction of angles of 30°, 45°, 60°.

Construction of triangles, quadrilaterals, and circles from simple data, including the inscribed and circumscribed circles of a triangle.

Division of a straight line into a given number of parts or in a given ratio.

Construction of a triangle equal in area to a quadrilateral or pentagon and of a square equal in area to a given rectangle.

Construction of tangents from an external point.

Construction of a segment containing a given angle.

6. Three-dimensional Figures

The properties of parallel planes: the normal to a plane, the angle between two planes, and the angle between a plane and a straight line.

The forms of the cube, rectangular block, pyramid, tetrahedron, prism, circular cylinder, circular cone, and sphere (including calculations of lengths and angles).

7. Practical Applications

In addition to ordinary riders, questions involving drawing, trigonometry or geometrical reasoning may be set on the following topics:

(*a*) Simple map problems, scales, contour lines, slopes.

(*b*) Determination of positions by two bearings; the nautical mile and knot; the triangle of velocities.

(*c*) Heights and distances.

(*d*) Simple plan and elevation problems.

(*e*) Latitude and longitude; great and small circles on a sphere.

(*f*) The length of an arc in terms of the radius and the measure (in degrees) of the angle at the centre.

(*g*) Three-dimensional problems which can be solved by analysis into plane figures.

Notes

Prelude

1. A description of the curriculum at the school in York in the eighth century is to be found in Sylvester, D. W. *Educational Documents 800–1816*, Methuen, 1970, pp. 2–4. This account, written by Alcuin who was schoolmaster at York and later at Charlemagne's Court, has been frequently reproduced. In it Alcuin mentions how his predecessor, Aethelbert, travelled abroad in search of new ideas. There is also evidence that York attracted students from the Continent, but doubt remains as to whether Alcuin's curriculum was implemented in practice.
2. Book 4, chapter 2. Theodore of Tarsus became Archbishop of Canterbury in 669 and provided a new stimulus to both evangelism and religious education.
3. A brief description of Bede's work and a translation of chapter 2 of *De Temporum Ratione*, 'Calculation or reckoning on the fingers', can be found in Yeldham, F. A. *The Story of Reckoning in the Middle Ages*, Harrap, 1926. Wilson (see note 4) casts doubt on Bede's authorship of this chapter.
4. Boethius produced an *Arithmetic*, based on Nicomachus, a *Geometry* based on Euclid (cf. my p. 21), an *Astronomy* derived from Ptolemy's *Almagest* and a *Music*. The *Arithmetic* influenced many later writers and was itself reprinted on many occasions during the fifteenth and sixteenth centuries. (See Smith, D. E. *Rara Arithmetica*, Chelsea Publishing Co., New York, 1970 (reprint).) The *Euclid* was frequently bound with the *Arithmetic* and D. K. Wilson (*The History of Mathematical Teaching in Scotland to the end of the Eighteenth Century*, University of London Press, 1935) gives references to their use in Scotland at King's College, Aberdeen as late as 1549.
5. Here, again, Boethius utilised Greek ideas for Plato (*Republic*, Book 7) recognised five divisions of mathematics (arithmetic, geometry, solid geometry, astronomy and harmony), whilst Archytas, who lived in the first half of the fourth century BC, listed the four studies, 'geometry, arithmetic and sphaeric, and, not least, . . . music' (see Thomas, I. *Greek Mathematics*, 2 vols., Harvard University Press, 1939, vol. 1, p. 4).
6. The Third Lateran Council of 1179 ordered that in every cathedral there should be a schoolmaster to teach 'the clerke of that church and poor scholars freely'. The Fourth Council of 1215 extended this obligation to other churches of adequate means. See, for example, Lawson, J. and Silver, H. *A Social History of Education in England*, Methuen, 1973, pp. 20–1.
7. See, for example, Lawson and Silver, *op. cit.* (n. 6), pp. 12–14.
8. Witness Fibonacci's *Liber abaci* (1202). It was in this century too that the translation by Adelhard of Bath of Euclid's *Elements* from the Arabic to the Latin began to be truly influential. See, for example, chapter 14 of Boyer, C. B. *A History of Mathematics*, Wiley, 1968.
9. W. W. R. Ball (*A History of Mathematics at Cambridge*, Cambridge University Press, 1889) tells how in 1426 the University of Paris refused a degree to one Paul Nicholas. He took legal proceedings against the university but their right to withhold a degree was upheld. Nicholas can, therefore, claim to be the first known student to have failed his degree course.
10. According to Guy Beaujouan ('Science in the medieval universities' in Crombie, A. C. (ed.) *Scientific Change*, Heinemann Educational Books, 1963, pp. 219–36), fourteenth-century candidates for the Paris bachelor's degree had to swear that they had attended at least a hundred classes in mathematics. After 1452 (Ball, *op. cit.* (n. 9), p. 9) candidates for

the master's degree at Paris had to take an oath that they had attended lectures on the first six books of Euclid.

11. Sacrobosco, John of Holywood, (*c.* 1200–56[44?]) is commonly believed to have been born near Halifax, Yorkshire, and to have studied at Oxford. In 1220 he went to Paris where he taught mathematics at the university, died and was buried. His textbooks were widely used in universities until the seventeenth century (see J. F. Daly's entry in *Dictionary of Scientific Biography*, vol. 12, Charles Scribner's Sons, 1975). A translation of *De Arte Numerandi* can be found in R. Steele (ed.) *The Earliest Arithmetics in English*, Early English Text Society, 1922.
12. Robert Grosseteste (*c.* 1168–1253) was probably educated at Oxford. He later taught there for many years and was celebrated as a scholar and teacher as well as a leading churchman. He made significant contributions to optics, astronomy and calendar reform.
13. Roger Bacon (*c.* 1219–*c.* 1292) studied at Oxford and lectured first at Paris and then at Oxford where he was greatly influenced by Grosseteste. He later returned to France, entered the Franciscan order in 1257, and became embroiled in religious controversy. He was imprisoned sometime between 1277 and 1279. Bacon's interests reflected those of Grosseteste and his mathematical work was mainly concerned with astronomy and optics. Although Bacon 'argued the usefulness of mathematics in almost every realm of academic activity' (A. C. Crombie and J. D. North in *Dictionary of Scientific Biography*, vol. 1, Charles Scribner's Sons, 1970), he does not himself appear to have been a particularly able or creative mathematician.
14. *Opus Tertium* cap vi, quoted in Ball, *op. cit.* (n. 9), p. 3. Some authors, for example J. Høyrup ('Influences of institutionalized mathematics teaching on the development and organization of mathematical thought in the pre-modern period' in *Studien zum Zusammenhang von Wissenschaft und Bildung*, IDM, Bielefeld, 1980, pp. 7–137), challenge the validity of Bacon's statement. (Høyrup's monograph gives a useful summary of mathematics teaching up to medieval times.)
15. Oresme (*c.* 1320–82), Bishop of Lisieux, developed Bradwardine's idea on proportion (see below) and introduced the ideas of irrational powers. He also carried out important work on the concept of velocity, the geometrical representation of functions, and on infinite series. Buridan (*c.* 1295–1385) rejected existing Aristotelian theories of physics and anticipated Newton's first law of motion.
16. Thomas Bradwardine (*c.* 1290–1349) was educated at Oxford and subsequently became a fellow at Merton College. He produced a generalised theory of proportions in his *Tractatus de proportionibus* (1328, printed in 1495) and also wrote other texts including an *Arithmetic* and a *Geometry*, both of which were first printed in 1495 and were reprinted on many occasions. He later became Archbishop of Canterbury.
17. See, for example, Ball, *op. cit.* (n. 9); Rashdall, H. *The Universities of Europe in the Middle Ages,* Oxford University Press, 1936; or Thorndike, L. *University Records and Life in the Middle Ages,* Columbia University Press, 1944.
18. See, for example, Beaujouan, *op. cit.* (n. 10), p. 222.
19. See, for example, Curtis, S. J. *History of Education in Great Britain*, 7th edn, University Tutorial Press, 1967, pp. 51–5; Lawson and Silver, *op. cit.* (n. 6), p. 65; Orme, N. *English Schools in the Middle Ages*, Methuen, 1973, pp. 52–5; Sylvester, *op cit.* (n. 1), pp. 43–7.
20. See, for example, Lawson and Silver, *op. cit.* (n. 6), pp. 72–5, 83.
21. William Byngham writing in 1439: see Leach, A. F. *Educational Charters and Documents 598–1909*, Cambridge University Press, 1911, pp. 402–3.
22. Watson, F. *Vives on Education: A Translation of De Tradendis Disciplinis*, Cambridge University Press, 1913, Book 4, chapter 5.

1. Robert Recorde

1. Or Record, for the two spellings can be found even in the same paragraph of contemporary documents.
2. In June 1548 Recorde was present at a sermon preached by Gardiner, Bishop of

Winchester, who although a supporter of Henry's divorce from Catherine was now a violent critic of the religious policies being pursued under Somerset (see note 3). Gardiner's criticisms led to his being brought for trial in December 1550; Recorde was one of the principal prosecution witnesses. Recorde's evidence begins with a statement of his age, '38 or thereabouts'. If one interprets thirty-eight as being his age at the time of the trial, rather than when the sermon was preached, this places Recorde's birth around 1512 rather than the traditionally ascribed 1510.

3. When Henry VIII died in 1547 he left three children: Edward, son of Jane Seymour, Mary, daughter of Catherine of Aragon (Henry's first wife), and Elizabeth, daughter of Anne Boleyn. Edward, a sickly child, succeeded to the throne at the age of nine and ruled for only six years. Early in his reign, under the Protectorship of Edward Seymour, Duke of Somerset and the young king's uncle, an advanced party of reformers influenced by the Continental Protestants made important doctrinal and other reforms which confirmed the break with Rome. In 1549, however, Somerset fell from favour, and lost first his position to the Duke of Northumberland, and later, in 1552, his head to the executioner. Northumberland continued the conversion to Protestantism whilst increasing his own political power. In an attempt to retain that power and prevent the Catholic Mary from succeeding to the throne, Northumberland attempted to manipulate the succession in favour of his daughter-in-law, Lady Jane Grey. On Edward's death in 1553 Lady Jane Grey was proclaimed Queen. Her 'reign' was brief, a mere nine days. The people resented the manipulation of the succession and Mary was welcomed when she entered London to claim the throne. Northumberland and Lady Jane were executed. Mary forcibly reimposed the rule of Rome on the English Church and became increasingly unpopular. On her childless death in 1558, she was succeeded by Elizabeth who compromised and adopted a less extreme form of Protestantism.
4. Fuller details of Recorde's life are to be found in: Clarke, F. M. 'New light on Robert Recorde', *Isis*, **8** (1926), 50–70; Easton, J. B. 'Robert Recorde', *Dictionary of Scientific Biography*, vol. 11, Charles Scribner's Sons, 1975, 338–40; Kaplan, E. *Robert Recorde: Studies in the Life and Works of a Tudor Scientist*, Ph.D. dissertation, New York University, 1960. The reader is warned that these accounts do not tally exactly. Thus, for example, Kaplan, who provides the most detailed account, nowhere refers to Recorde and Pembroke – information revealed by W. G. Thomas (see Easton *op. cit.*).
5. Until 1751 it was customary for the new year to commence on 25 March. Thus in the old style January 1549 came *after* October 1549. Unless otherwise stated dates in this book have been converted to 'new style'.
6. See Easton, *op. cit.* (n. 4). This incident is not mentioned by Kaplan.
7. See Easton, J. B. 'The early editions of Robert Recorde's *Ground of Artes*', *Isis*, **58** (1967), 515–32 for conjectured date of change of dedication.
8. See Easton, J. B. 'A Tudor Euclid', *Scripta Mathematica*, **27** (1966), 339–55 for conjectured year of publication.
9. See Clarke, *op. cit.* (n. 4); Kaplan, *op. cit.* (n. 4).
10. Henry VIII had corresponded with Gundelfinger (or Gudenfilger) in 1544 (see Kaplan, *op. cit.* (n. 4), p. 16) in one of his many attempts to accelerate England's technological advancement.
11. Easton (1975), *op. cit.* (n. 4).
12. There is no evidence of Recorde ever having married.
13. Kaplan, *op. cit.* (n. 4), p. 4, indicates that Recorde acquired a doctorate in theology.
14. See, for example, Charlton, K. *Education in Renaissance England*, Routledge and Kegan Paul, 1965; Curtis, S. J. *History of Education in Great Britain,* 7th edn, University Tutorial Press, 1967; Lawson, J. *Medieval Education and the Reformation*, Routledge and Kegan Paul, 1967; Leach, A. F. *English Schools and the Reformation*, 1546–8, Westminster, 1896; Watson, F. *The English Grammar Schools to 1660*, Cambridge University Press, 1908; and, for an outstanding survey of Tudor education, Simon, J. *Education and Society in Tudor England*, Cambridge University Press, 1966.
15. Leach, A. F. *The Schools of Medieval England*, Barnes and Noble, 1915, pp. 248–9.
16. See Simon, *op. cit* (n. 14), for details.

17. Taken from *Schools Inquiry Commission* (1868) vol. 7, pp. 262–3.
18. See Watson, *op. cit.* (n. 14), p. 158.
19. Hill, C. *The Intellectual Origins of the English Revolution,* Oxford University Press, 1965, p. 301.
20. Taylor, E. G. R. *The Mathematical Practitioners of Tudor and Stuart England*, Cambridge University Press, 1954, p. 18.
21. See chapter 10 of Simon, *op. cit.* (n. 14). It is probable that the inclusion of mathematics in the curriculum was largely due to Cheke, one of the seven 'visitors'.
22. Ball, W. W. R. *History of Mathematics at Cambridge*, Cambridge University Press, 1889, p. 13. The 'technical' pressures are described in, for example, Taylor, *op. cit.* (n. 20), and Charlton, *op. cit.* (n. 14), chapter 9.
23. William Harrison's *Description of England* (1577), in Sylvester, D. W. *Educational Documents, 800–1816*, Methuen, 1970, pp. 144–9.
24. The edition in the Bodleian Library, Oxford, which is published in facsimile in the 'English Experience Series' and is there dated 1542, is now thought to date from about 1551 (see *Short Title Catalogue*, 2nd edn, Bibliographical Society, 1976). It is this Bodleian edition which is quoted in this chapter and to which page references are given. Recorde's book, then, first appeared some ninety years after Gutenberg's Bible. It is interesting to note that its printer, Wolfe, had recently arrived in London from the Continent and was presumably one of the few London printers experienced in setting mathematical work.
25. Cuthbert Tonstall (Tunstall) (1474–1559) was a Yorkshireman who studied at Oxford, Cambridge and Padua. He was primarily a churchman, holding a succession of important offices before becoming, in 1522, Bishop of London. In 1530 he succeeded Wolsey as Bishop of Durham, and was later imprisoned by Northumberland and deprived of his bishopric. On Mary's coming to the throne in 1553 he was freed and the following year was reinstated as Bishop of Durham. His fortunes were to decline again, however, following Elizabeth's accession and once more he was deprived of his bishopric.
26. Smith, D. E. *Rara Arithmetica*, Chelsea Publishing Co., New York, 1970 (reprint), p. 134.
27. See Richeson, A. W. 'The first Arithmetic printed in English', *Isis,* **37** (1947), 47–56 and Bockstaele P. 'Notes on the first Arithmetics printed in Dutch and English', *Isis*, **51** (1960), 315–21. No copy of the 1537 'edition' is extant; the earliest copy to survive dates from 1539.
28. See Easton (1966), *op. cit.* (n. 8) for a detailed analysis of this work.
29. Such constructions remained part of the artisan's craft for many years. Thus the 'practical' geometries of the nineteenth century (some written with the examinations of the Department of Science and Art (see my pp. 144–7) in mind) showed, *pace* Gauss, how to inscribe a regular *n*-gon in a circle by means of ruler and compass constructions.
30. Easton (1966), *op. cit.* (n. 8), p. 347.
31. Lilley, S. 'Robert Recorde and the idea of progress – a hypothesis and verification', *Renaissance and Mod. Stud.*, **2** (1958), 3–37.
32. Zilsel, E. 'The genesis of the concept of scientific progress', *J. Hist. Ideas*, **6** (1945), 325–49.
33. See pp. 62–3 of Johnson, F. R. and Larkey, S. V. 'Robert Recorde's mathematical teaching and the Anti-Aristotelian Movement', *Hunt. Lib. Bull.*, 7 (1935), 59–87.
34. See, for example, Lilley, *op. cit.* (n. 31), p. 23.
35. A summary of the arguments advanced for and against Recorde being a Copernican are to be found in Kaplan, *op. cit.* (n. 4), pp. 151–60. Arthur Koestler (*The Sleepwalkers*, Hutchinson, 1959) suggests that in Recorde's time the Roman Catholic Church would not have objected to the heliocentric hypothesis; however, in Mary's England religious persecution was rife and brutal.
36. The sign was, however, used independently about the same time by a mathematician at Bologna in his manuscripts. See Cajori, F. *A History of Mathematical Notations*, Open Court, 1928, vol. 1, p. 298.

37. The equals symbol was not immediately adopted by mathematicians. Indeed, it was over sixty years before it next appeared in a published book. Then, in 1618, it is to be found in the English translation of Napier's *Descriptio* and later Harriot and Oughtred both adopted it in their influential books. It then rapidly gained more general favour, and alternative uses to denote, for example, arithmetical difference or parallelism declined. That Harriot and Oughtred should use the sign suggests that they were familiar with Recorde's book. It is noteworthy, however, that Wallis in his 1685 *Treatise of Algebra, both Historical and Practical* appears never to have seen the *Whetstone*. He refers merely to the book written 'about the year 1552 (if I be not misinformed)' (p. 63). This, however, may only serve to strengthen the assertion that it was especially the non-university men who were to use and benefit from Recorde's texts.
38. Kaplan, *op. cit.* (n. 4), contains a section (pp. 52–61) on the use of the vernacular.
39. The form was frequently used up to the eighteenth century. It was revived by Kirkman (see my p. 255) in 1852 and in recent times by, among others, Lucienne Félix and Imre Lakatos who both introduced several scholars each with his own characteristics.
40. See Hill, *op. cit.* (n. 19), p. 21.
41. This was not true, for example, of Wingate's *Arithmetic* which first appeared in 1630 and went through numerous editions, the last editor being Dodson (my p. 256).
42. Sets of questions or exercises to be done by the reader were an eighteenth-century innovation. See, for example, the arithmetic texts by Dilworth (my p. 67) and Walkingame (my p. 246).
43. Is this a harbinger of the 'stages' approach to geometry (see my p. 153).
44. Examples of present-day use of this technique can be found in Swetz, F. (ed.) *Socialist Mathematics Education*, Burgundy Press, 1978.
45. A more detailed account of the political implications of the changes made in the 1552 edition can be found in Easton (1967), *op. cit.* (n. 7).
46. The choice is still being made in developing countries where the move is now away from English (French, Portuguese, etc.) to the vernacular.
47. As had Chaucer in his *Astrolabe*. Other examples of Recorde's creation of terms are given in Easton (1966), *op. cit.* (n. 8) and (from the *Whetstone*) in Baron, M. E. 'A note on Robert Recorde and the Dienes Blocks', *Math. Gazette*, 50 (1966), 363–9.
48. Today's developing countries can, therefore, obtain little guidance from history, only the comfort of knowing that their problems are not new.
49. We note that Dee was a contemporary of Recorde's at Cambridge and there is some evidence (see Kaplan, *op. cit.* (n. 4)) that they were acquainted then.
50. Dee lectured in English according to Ball, *op. cit.* (n. 22), p. 19. Further details are to be found in Taylor, *op. cit.* (n. 20), p. 18.
51. See Simon, *op. cit.* (n. 14), p. 349.
52. Dudley's personal copy of Billingsley's *Euclid* was advertised for sale by Quaritch in 1974 for £5,750.
53. See, for example, Taylor, *op. cit.* (n. 20), p. 320. Taylor reprints an extract from the Preface; the whole is to be found in Dee, J. *The Mathematical Praeface,* Watson, 1975.
54. See Easton (1967), *op. cit.* (n. 7), for further details of Dee's changes and of the reissues.
55. The changes were not always for the better; for example, 'Division is a partition of a greater summe by a lesser' became 'Division is a distributing of a greater summe by the unities of a lesser'.
56. See 'Dee' in Hutton, C. *A Mathematical and Philosophical Dictionary,* 1815.
57. A racy account is contained in an appendix to Edith Sitwell's *The Queens and the Hive,* Macmillan, 1962.
58. See, for example, the booklists to be found in Ball, *op. cit.* (n. 22), pp. 94–6.
59. Typical contents included how to discover the number your companion has in mind, and how to plan river crossings for jealous husbands and their wives or farmers having a fox, a goose and a peck of corn to transport.
60. Nicholas Fiske (1575–1659) was born in Suffolk and practised at Colchester and London.
61. A copy of the advertisement is to be found in Wallis, P. J. 'Arithmetical art', *Maths in School*, **3** (1) (1974), 2–4.

62. This edition is described in Yeldham, F. A. *The Teaching of Arithmetic through Four Hundred Years (1535–1935)*, Harrap, 1936.
63. De Morgan, A. *Arithmetical Books*, 1847, p. 25 (reprinted as an appendix to Smith, note 26).
64. See, for example, De Morgan, *op. cit.* (n. 63). Kersey's edition of Wingate's *Arithmetic* was particularly popular.
65. A book which De Morgan (*op. cit.* (n. 63), p. 67) found to be 'sound' and 'elaborate' but, alas, 'unreadable'.

2. Samuel Pepys

1. Latham, R. and Matthews, W. (eds.) *The Diary of Samuel Pepys,* Bell, 1970–6.
2. See, for example, Leach, A. F. *The Schools of Medieval England*, Barnes and Noble, 1915.
3. It is perhaps of interest to note that early in the nineteenth century the grammar school at Huntingdon offered a free education in 'the *Latin and Greek* Languages' to all 'town-born boys'. The master of the school employed a writing master who taught writing and arithmetic at a charge of 15s (75p) a quarter. (See Carlisle, N. *The Endowed Grammar Schools in England and Wales*, 2 vols., 1818 (republished Richmond Publishing Co., 1972), vol. 1, pp. 557–9.)
4. See Curtis, M. *Oxford and Cambridge in Transition, 1558–1642,* Oxford University Press, 1959.
5. Quoted in Wallis, P. J. Grammar school teachers of mathematics (stencilled notes), Newcastle, 1971. In this brief note, Wallis gives some examples of masters who taught mathematics in their schools in the late sixteenth and early seventeenth centuries. This supplements the information to be found in Watson, F. *The Beginnings of the Teaching of Modern Subjects in England*, Pitman, 1909 (reprinted S. R. Publishers, 1971). The most notable of these masters was William Kempe who was Headmaster of Plymouth Grammar School from 1581–1604. In 1592 Kempe published a translation from the Latin of Peter Ramus' *The Art of Arithmeticke in Whole Numbers and Fractions*, dedicating it to Sir Francis Drake. It seems certain that Kempe taught both arithmetic and geometry to his pupils at Plymouth. William Petty was later (1647) to urge the teaching of arithmetic and geometry in schools not only because of their utility but as 'sure guides and helps to reason, and especial remedies for a volatile and unsteady mind'.
6. Quoted in Picciotto, C. *St Paul's School,* Blackie, 1939, p. 9.
7. The quotations by Mulcaster are taken from Oliphant, J. (ed.) *The Educational Writings of Richard Mulcaster*, Maclehose, 1903. The Ascham quotation is the famous 'Mark all Mathematical heads which be wholly and only bent on these sciences, how solitary they be themselves, how unfit to live with others, how unapt to serve the world'.
8. Quoted in McDonnell, Sir Michael, *A History of St Paul's School*, Chapman and Hall, 1909, p. 205.
9. Brinsley, J. *Ludus Literarius*, 1612.
10. The Mercers' Company, a guild of London tradesmen, was responsible for governing the school. (Although Colet was a churchman, the Dean of St Paul's Cathedral, his father was a mercer.)
11. A sizar was an undergraduate who received an allowance from the college and in return performed certain duties such as bed making, room cleaning, and water fetching. Newton, like Pepys, was a sizar.
12. Quoted in Lawson, J. and Silver, H. *A Social History of Education in England,* Methuen, 1973, p. 141. Further descriptions of Pepys' Cambridge, including an interesting contemporary criticism of the curriculum can be found in Ollard, R. *Pepys,* Hodder and Stoughton, 1974.
13. The early days and early professors of Gresham College are chronicled in Ward, J. *Lives of the Professors of Gresham College*, 1740. The Gresham chairs have been filled, with occasional breaks, down to the present day and the 'professors' (normally people holding chairs elsewhere) have continued to give occasional series of lectures to the general public.

14. For a more detailed account see Taylor, E. G. R. *The Mathematical Practitioners of Tudor and Stuart England*, Cambridge University Press, 1954.
15. Seth Ward (1617–89) was a Cambridge student and, later, lecturer who was best known for his work on astronomy. He was appointed to the Savilian Chair of Astronomy at Oxford in 1649 and, with Wallis, played a part in the establishment of the Royal Society. He left mathematics for divinity and became Bishop of Exeter and, later, Salisbury.
16. John Wallis (1616–1703) was the most distinguished English mathematician before Newton. He wrote on conic sections, on the quadrature of curves, and on algebra, statics and dynamics. His book on algebra included the first historical account of the development of the subject to be written in English. His pupils included Wren, Hooke and Halley.
17. For a description of Cambridge mathematics in the seventeenth century see Ball, W. W. R. *A History of Mathematics at Cambridge*, Cambridge University Press, 1889.
18. Wallis, John, *An Account of Some Passages in His Own Life,* 1697. The importance of London in the early seventeenth century as a centre for scientific studies is well argued in Hill, C. *The Intellectual Origins of the English Revolution*, Oxford University Press, 1965.
19. John Pell (1610–85) entered Cambridge at the early age of thirteen and graduated BA in 1629. He taught at Collyer's School, Horsham and at Samuel Hartlib's school in Chichester, and later was professor at Amsterdam and Breda. Mathematically he is remembered for Pell's equation (although Fermat has a stronger claim to the credit), and he was a prolific (but financially unsuccessful) author. His main contribution to mathematics education was 'An idea of mathematics', published as a broadsheet about 1650 and reprinted in *Durham Research Review*, **18** (1967), 139–48.
20. Elsewhere, Samuel Foster lost his position as Professor of Astronomy at Gresham College for refusing to kneel at the communion table, and Peter Turner, a Royalist, was expelled on political grounds from the Savilian Chair at Oxford. Foster later returned to Gresham College and with Wallis, Turner's successor, helped to found the Royal Society.
21. Pepys was worried about the sharp practices of some of the timber suppliers and he determined to find out more about measuring timber, estimating volumes, etc. It was for this purpose that he required a knowledge of multiplication.
22. For a history of Christ's Hospital and its Mathematical School see Pearce, E. H. *Annals of Christ's Hospital*, Methuen, 1901. (Unattributed quotations on Christ's Hospital in this chapter are from Pearce.)
23. Moore (1627–79) was a north countryman whose interest in mathematics was kindled by a local clergyman and amateur of mathematics. He became mathematical tutor to the Duke of York shortly before the Civil War broke out. In 1647 he published a textbook on arithmetic and in 1649 set up in London as a mathematics teacher. Moore was elected a Fellow of the Royal Society in 1674 and his later pupils and protégés included the astronomers Flamsteed and Halley. His comprehensive textbook, written for Christ's Hospital, was published posthumously. Like Pepys, Hutton and De Morgan, he was a great collector of books and had a mathematics library of some 1,400 volumes.
24. Quoted in Bryant, Sir Arthur, *Samuel Pepys: The Years of Peril*, Cambridge University Press, 1935.
25. Oughtred's *Clavis Mathematica* (1631) was truly a key book for that period and was for many years the basis for lecture courses at Cambridge. Oughtred (1575–1660) was the son of a writing master at Eton College and was educated at that school and at Cambridge. After teaching for some years at Cambridge he became, in 1603, a country parson. Oughtred's interest in mathematics did not cease, however, and he exerted personally, or through his many books, a great influence on English mathematics. He harboured Seth Ward after his expulsion from Cambridge and, according to Aubrey, his pupils also included Wallis and Wren. Both Moore and Newton claim to have had their interest in mathematics aroused by reading Oughtred's *Clavis*.
26. The curriculum is reprinted in Pearce, *op. cit.* (n. 22).
27. See Bryant, *op. cit.* (n. 24), and Pearce, *op. cit.* (n. 22).

28. See Bryant, Sir Arthur, *Samuel Pepys: The Saviour of the Navy*, Cambridge University Press, 1938.
29. Sir Joseph Williamson's Free Mathematical School was founded at Rochester in 1701 to teach 'Mathematics and other things fitting scholars for the sea Service or arts and callings relating thereto'. Neale's Mathematical School, London followed in 1705 and Saunders' School at Rye in 1708. In 1722 Churcher's College, Petersfield was founded to train boys in navigation specifically in connection with the East India Company's work. The schools were not over-successful. Boys from Petersfield did not wish to go to sea, and in 1744 Parliament was petitioned to alter the conditions of Churcher's will. By 1826 the Rochester school gave little instruction in navigation and a similar lack of demand for the subject was reported from Rye where Saunders' school had amalgamated with the local grammar school. (See also Allen, B. The English Mathematical Schools 1670–1720, unpublished Ph.D. thesis, Reading University, 1970.)
30. See Hans, N. *New Trends in Education in the Eighteenth Century*, Routledge and Kegan Paul, 1951, pp. 213–19.
31. See Turnbull, H. W. and Scott, J. F. (eds) *The Correspondence of Isaac Newton*, Cambridge University Press, 1959–67 vol. 3. Newton's views on mathematics education at university can be found in 'Of educating youth in the universities', in Hall, A. R. and Hall, M. B. *Unpublished Scientific Papers of Isaac Newton*, Cambridge University Press, 1978.
32. The revised syllabus is reprinted in Pearce, *op. cit.* (n. 22).
33. See the diary for 21st Oct., 1663. For a more detailed account of Pepys' interests in the field of science, see Nicolson, M. H. *Pepys' Diary and the New Science* Virginia University, 1965.
34. See *Diary,* 11th Aug., 1664.
35. An account of the history of Cocker's book is given in Wallis, P. J. '"According to Cocker" – ghosts and all', *Maths in School,* 3 (2) (1974), 12–14, and its mathematical content is described in Yeldham, F. A. *The Teaching of Arithmetic through Four Hundred Years (1535–1935)*, Harrap, 1936. De Morgan (*Arithmetical Books*, 1847, reprinted as an appendix to Smith, D. E. *Rara Arithmetica*, Chelsea Publishing Co., New York, 1970 (reprint)) claimed that he was 'perfectly satisfied that Cocker's Arithmetic is a forgery of Hawkins'. This somewhat surprising allegation was later refuted by Salmon, in an article which appeared in *The Schoolmaster* (22 Nov. 1919). However, irrespective of whether Cocker did or did not write the book, De Morgan was 'of opinion that a very great deterioration in elementary works on arithmetic is to be traced from the time at which the book called after Cocker began to prevail'.
36. For details of Walkingame's book and its influence, see Wallis, P. J. 'Popularity and the pupil', *Maths in School*, **3** (3) (1974), 7–9 and, for more detail, Wallis, P. J. 'An early best-seller: Francis Walkingame's *The Tutor's Assistant*', *Math. Gazette*, **47** (1963), 199–208. De Morgan, *op cit.* (n. 35), p. 80, said of Walkingame that 'this book is by far the most used of all school books, and deserves to stand high among them'. In 1843 an edition of Walkingame with comic woodcuts was published. Thus 'Subtraction of time' was illustrated by a cartoon representing a watch being stolen. (See De Morgan, *op cit.* (n. 35), p. 96.)
37. Boswell, J. *Journal of a Tour to the Hebrides with Samuel Johnson, Ll.D.* (ed. R. W. Chapman), Oxford University Press, 1930. Despite his appreciation of Cocker it will be noted that mathematics held no place in the curriculum of the school run by Johnson (see Boswell's *Life*).
38. See *Diary*, 23rd Jan., 1661 (1662 new style).
39. See *Diary*, 22nd April, 1662.
40. Wallis' account is reprinted in an unedited version in Lyons, Sir Henry, *The Royal Society 1660–1940*, Cambridge University Press, 1944. See also Hartley, Sir Harold, *The Royal Society – its Origins and its Founders*, Royal Society, 1960, and, for a criticism of the validity of Wallis' account, Purver, M. *The Royal Society: Concept and Creation,* Routledge and Kegan Paul, 1967
41. Brouncker (1620–84) was particularly interested in numerical mathematics and he

contributed work on continued fractions, the solution of the Fermatian equation $a\lambda^2 + 1 = \mu^2$ and infinite series.

42. Lyons, *op. cit.* (n. 40) includes in an appendix data concerning the composition of the fellowship between 1663 and 1880. During that time the percentage of scientific fellows rose slowly from 32.2% to 52.6%. It has since continued to rise. The proportion of mathematicians and astronomers amongst the scientific fellows has since 1663 declined from over a third to about a tenth.
43. See Bryant (1935), *op. cit.* (n. 24).
44. See Lyons, *op. cit.* (n. 40).
45. Halley (1656–1742), the son of a prosperous London soap-boiler, had his interest in astronomy and the construction of instruments encouraged by Jonas Moore. He was elected to the Royal Society in 1678 and was clearly in comfortable financial circumstances when he undertook the printing and publishing of Newton's book, the existence of which owed much to his 'coaxing and goading'.
46. The correspondence with Newton is to be found in Tanner, J. R. (ed.) *The Correspondence of Samuel Pepys, vol. 1, 1679–1700*, Bell, 1926. The background to the correspondence is considered in David, F. N. 'Mr Pepys and Dyse: a historical note' *Annals of Science*, **13** (1957), 137–47.
47. See Lord Braybrooke (ed.) *Memoirs and Correspondence of Samuel Pepys, Esq., FRS*, 2nd edn, 1828, vol. 5. David Gregory was Savilian Professor of Astronomy at Oxford from 1691 to 1708. He was the nephew of James Gregory, remembered for Gregory's series and the Newton–Gregory formula, and brother of James and Charles, professors of mathematics at Edinburgh and St Andrew's respectively.
48. For example, Edmund Stone in his *New Mathematical Dictionary* (1726) had entries under 'counter-tenor' and 'Ionick order'.
49. A similar problem concerning 'honour' and 'international standing' is encountered nowadays by those universities in the developing countries considering a change to teaching in the vernacular.
50. Williams, Reeve *The Elements of Euclid* (after D'Chales), London, 1685.
51. Quoted from the letter from the Vice-Chancellor of Oxford University thanking Pepys for his gift of Wallis' portrait in Braybrooke, *op. cit.* (n. 47).
52. See Taylor, *op. cit.* (n. 14).

3. Philip Doddridge

1. See, for example, Greaves, R. L. *The Puritan Revolution and Educational Thought*, Rutgers University Press, 1969; O'Brien, J. J. 'Commonwealth schemes for the Advancement of learning', *Brit. J. Educ. Studies*, **16** (1968), 30–42; Webster, C. 'Science and the challenge to the scholastic curriculum 1640–1660', in *The Changing Curriculum*, History of Education Society, 1971, 21–35.
2. Petitions were made to Parliament in 1641 for universities at Manchester and York.
3. See Turnbull, G. H. 'Oliver Cromwell's college at Durham', *Durham Research Review*, **3** (1952), 1–6.
4. Longer extracts from the Act of Uniformity (and the Five-Mile Act) can be found in Sylvester, D. W. *Educational Documents 800–1816*, Methuen, 1970.
5. This estimate is based on Calamy, E. *An Account of the Ministers, Ejected or Silenced*, 1713 (reissued in 1934 by the Oxford University Press in a revised version edited by A. G. Matthews). See also McLachlan, H. *English Education under the Test Acts*, Manchester University Press, 1931.
6. See Leach, A. F. *Educational Charters and Documents 598–1909*, Cambridge University Press, 1911, pp. 542–4.
7. See also Ashley Smith, J. W. *The Birth of Modern Education*, London Independent Press, 1954 and McLachlan, *op. cit.* (n. 5).
8. Calamy (*op. cit.* (n. 5)) names Frankland as a tutor at Durham, but there is no evidence of his appointment there (unlike that of Wood, see my p. 37).

9. James Clegg (a student at Frankland's academy), quoted in McLachlan, *op. cit.* (n. 5), pp. 63–4.
10. Other of Frankland's ex-students graduated at Aberdeen, Cambridge, Leyden, Oxford, Padua, and Utrecht. The great majority of his ex-students graduated at universities outside England.
11. See McLachlan, *op. cit.* (n. 5), p. 69. The curriculum did, however, include some physics.
12. See Ashley Smith, *op. cit.* (n. 7), p. 20.
13. Joseph Mottershead, a pupil of Jollie, quoted in McLachlan, *op. cit.* (n. 5), p. 108. Other criticisms of mathematics and mathematicians are to be found in Swift's *Gulliver's Travels*. Gulliver found Laputa to be inhabited by 'absent-minded, impractical monomaniacs on the subject of mathematics' (Eddy, W. A. *Gulliver's Travels: A Critical Study*, Peter Smith, Gloucester, Mass., 1963). The voyage to Laputa and the description of its Royal Academy can be read as an attack on the recently instituted Royal Society and on Baconian scientific thought. In the voyage to Brobdingnag, however, we find Swift describing the people's learning as 'defective': '[mathematics] is wholly applied to what may be useful in life; to the improvement of agriculture and all mechanical arts; so that amongst us it would be little esteemed'. Swift, then, would not appear to be attacking all mathematics, but rather the emphasis given in English educational institutions to academic, abstract and non-utilitarian aspects of the subject.
14. Saunderson (1682–1739) lost his sight through smallpox when only one year old. He attended a local free school at Penistone, near Sheffield before going to Attercliffe Academy. He later moved to Cambridge where his friends included Newton and his successor, Whiston. In 1711 the latter was expelled from the Lucasian Chair (see p. 34) and Saunderson, granted an MA degree by Queen Anne's special patent, was appointed to succeed him. According to Matthew Robinson, writing to William Frend (see my p. 79), Whiston was removed from his post 'because he declared that he did not believe the Trinity' and replaced by Saunderson 'whom everybody perfectly well understood to believe no more in the divinity of the Bible than of the Alcoran'. Saunderson's *Elements of Algebra* was published posthumously in 1740 and contains an interesting account of his life.
15. See, Taylor, E. G. R. *The Mathematical Practitioners of Tudor and Stuart England*, Cambridge University Press, 1954 and Hans, N. *New Trends in Education in the Eighteenth Century*, Routledge and Kegan Paul, 1951, for further details of Martindale, Brancker, Ditton and other dissenting masters of mathematics. An interesting quotation from Martindale's autobiography is to be found in Watson F. *The Beginnings of the Teaching of Modern Subjects in England*, Pitman, 1909 (reprinted S. R. Publishers, 1971), pp. 313–14.
16. Polish-born Samuel Hartlib (1600–70) was one of the leading educators of his age. He pleaded the cause of education for all children and also sought the establishment of a 'universal college' which would act as a scientific research centre and thus do the job which the two universities were failing to do. Hartlib himself ran a school for a time at Chichester at which John Pell (my p. 245) taught.
17. John Milton (1608–74), who like Pepys had attended St Paul's School, although remembered now as one of England's greatest poets, was also an educator of note. He ran a private school for seven years from 1640 and in 1644 wrote, at Hartlib's request, the tract *On Education*. Milton, on coming down from Cambridge, had begun to take a keen interest in mathematics and in his tract he argued the case for the teaching in schools of arithmetic (a subject he himself taught from the Latin textbook by the Swiss mathematician Urstisius (1544–88)) and geometry.
18. Peter Ramus (1515–72) went to Paris University at the age of twelve to act as servant to a rich student. There he attended lectures and at twenty-one was awarded his Master's degree for defending the thesis that 'Everything that Aristotle has said is false'. He attacked not only Aristotelian doctrines but university education as it then was; for he put experience before authority. Besides works on philosophy he wrote books on arithmetic, algebra, geometry, astronomy and physics. Some of these were translated into English during the sixteenth and seventeenth centuries (see my p. 244) Ramus taught at Le Mans and Paris and was a victim of the Massacre of St Bartholomew.

19. The money left by his father had been lost through the speculation of Doddridge's guardian.
20. A letter to the Reverend Thomas Saunders, written in 1728 and quoted in McLachlan, *op. cit.* (n. 5), pp. 135–40.
21. Newton's *Principia* (3rd edn), quoted in Kline, M. *Mathematics: The Loss of Certainty*, Oxford University Press, New York, 1980, p. 59.
22. Isaac Watts (1674–1748) was the son of a man who went to prison for his religious opinions. He was educated at the local grammar school in Southampton and later at Newington Green Academy.
23. See Ashley Smith, *op. cit.* (n. 7), pp. 88–9. Newington Green also included among its old boys Daniel Defoe, the novelist.
24. One recalls that when, in 1619, Savile founded professorships in geometry and astronomy in Oxford 'not a few of our then foolish gentry refused to send their sons thither' to be 'smutted with the black art'; mathematics being 'spells' and the professors of them 'limbs of the Devil' (Anthony à Wood, *History and Antiquities* (ed. Gutch), vol. 2, part 1, 1796, p. 335).
25. E. Wells (1667–1727) was a mathematical tutor at Oxford who in 1712 published his *Young Gentleman's Course of Mathematics*. De Morgan (*Arithmetical Books*, 1847) thought him an 'elegant writer' but was disappointed by his *Gentleman's Course*, for he had imagined that this would demonstrate the place of mathematics within a liberal education and so act as an antidote to those publications which linked the subject too closely with commerce, 'But I was mistaken: it is *gentlemanly* education, as opposed to that of "the meaner part of mankind", that he wants to provide for . . . The gentlemen are those whom God has relieved from the necessity of working.'
26. George Cheyne (1671–1743) was a physician and mathematician whose publications included a treatise on fluxions. This embroiled him in controversy with De Moivre, after which he left mathematics.
27. Watts, I. *Improvement of the Mind*, 1741, vol. 1.
28. Before Doddridge's appointment at Northampton, Watts wrote to him that 'You will have many lads coming from the Grammar Schools, and as many such scholars will not be fit to enter upon your academical course with proper advantage, should not the perfection of the studies of grammar, algebra and geometry be the business of your first half year?' (*Diary and Correspondence of P. Doddridge*, 1803–4, vol. 2, p. 478).
29. At the time of the establishment of Kingswood the great public schools had reached their nadir; in 1751, for example, Winchester had only eighteen non-scholarship pupils. Such pupils as there were rebelled against authority, and riots became common. Vice was widespread (cf. Parson Adams' comments on public schools in Fielding's *Joseph Andrews*) and schools did little to prevent their pupils from supplementing the narrow curriculum of classics with a variety of extra-mural activities: thus at Westminster, 'some [of the boys] have lain out of college whole nights, and come in again early in the morning, having always command of the door'. Wesley planned a school 'fit for the apostolic age' in which children might be educated 'according to the accuracy of the Christian model'. His curriculum was still dominated by the classics but some attention was given to arithmetic, geometry and algebra. (See Wesley's own *Short Account* reprinted in Ives, A. G. *Kingswood School in Wesley's Day and Since*, Epworth Press, 1970.)
30. The Jacobites were supporters of the deposed King James II and his descendants. James and his heir, son of the Catholic Mary of Modena, had seemed likely to impose Roman Catholicism on the English, a threat which the Protestants averted by inviting William of Orange (a Protestant and grandson of Charles I) to take the English throne. William landed at Torbay in 1688 and James fled to France. Support for him was, however, to linger on in England and Scotland for many years.
31. J. Orton's *Memoirs of the Life and Writings of the Reverend Philip Doddridge* (1766), contains an account of Doddridge's curriculum. Extracts from it can be found in Ashley Smith, *op. cit.* (n. 7), pp. 130–2.
32. Quoted in Ashley Smith, *op. cit.* (n. 7), pp. 135–6

33. See the notes on Desaguliers *et al.* (my p. 251) and on the 'Lit. and Phil.' societies (my p. 73). Also see chapter 7, 'Adult education', in Hans, *op cit.* (n. 15).
34. Roebuck and Lucas were two industrialists who made important contributions to the technologies of smelting and chemical manufacture.
35. Quoted in Taylor, E. G. R. *Mathematical Practitioners of Hanoverian England*, Cambridge University Press, 1966, p. 155.
36. Bath, J. L. (ed.) *Autobiography of Joseph Priestley,* Adams and Dart, 1970, p. 72. It is from this work that later unattributed quotations in this section are taken.
37. Gravesande, W. J. *Mathematical Elements of Natural Philosophy Confirmed by Experiment, or an Introduction to Sir Isaac Newton's Philosophy* (trs. J. T. Desaguliers), 1720–21. Watts, I. *Logick: or the Right Use of Reason*, 1725.
38. Priestley, however, never became a very competent mathematician. Thus his *History of Optics*, written to be 'perfectly intelligible to those who have little or no knowledge of mathematics', was found by Thomas Young (*A Course of Lectures in Natural Philosophy*, 1845, vol. 1, p. 381) to be 'deficient in mathematical accuracy . . . the author was not sufficiently a master of the science to distinguish the good from the indifferent'.
39. Priestley's diary for 1755 quoted in Schofield, R. E. *A Scientific Autobiography of Joseph Priestley*, MIT Press, 1966, p. 6.
40. For further details of Warrington Academy see McLachlan, *op. cit.* (n. 5), and Parker, I. *Dissenting Academies in England*, Cambridge University Press, 1914.
41. *Essay on a Course of Liberal Education*, 1768 edn, p. 6.
42. Cyril Burt on 'David Hartley (1705–57)' in *Encyclopaedia Britannica*, 1973 edn.
43. Quoted in Simon, B. *Studies in the History of Education, 1780–1870*, Lawrence and Wishart, 1960, p. 45.
44. Simon, *op. cit.* (n. 43), p. 46.
45. R. L. Edgeworth (1744–1817) was a wealthy Irish inventor who studied at Oxford and later wrote on educational and mechanical subjects. He was an acquaintance and disciple of Rousseau and a member of the Lunar Society.
46. Edgeworth, R. L. and Edgeworth, M. *Practical Education*, 1798, vol. 2, p. 246.
47. 'Joseph Priestley' in Hutton, C. *A Mathematical and Philosophical Dictionary*, 1815.
48. The educational writings of John Locke (1632–1704) greatly influenced Isaac Watts and other dissenters. Tate (see chapter 6) later accused British educators of forgetting Locke and of following not him, but his Continental successors: 'we have unwittingly become the followers of Pestalozzi, when we might have been the disciples of our own immortal Locke' (*Philosophy of Education*). Locke, himself, urged the teaching of mathematics to all students:

    'I have mentioned mathematics as a way to settle in the mind a habit of reasoning closely and in train; not that I think it necessary that all men should be deep mathematicians, but that, having got the way of reasoning, which that study necessarily brings the mind to, they might be able to transfer it to other parts of knowledge as they shall have occasion' (*Conduct of the Human Understanding*).
49. Thus, for example, the curriculum of Mill Hill School, established in 1807 by the Congregationalists, was closer in spirit to that of Northampton Academy than to that of the typical contemporary public school.

4. Charles Hutton

1. Bruce, J. *A Memoir of Charles Hutton, Ll.D., FRS,* Hodgson, Newcastle, 1823. Bruce, too, kept a school in Newcastle where his pupils included Robert Stephenson, the railway engineer.
2. Jurin did not belong to the Mathematical School at Christ's Hospital. After differences with the Governors of the Newcastle Grammar School caused him to resign his post there in 1715, he acquired a considerable reputation as a scientist and mathematician. He was elected an FRS in 1717 and became the Royal Society's Secretary in 1721. In 1734 he published a defence of his former teacher, Newton, against the attacks of Bishop

Berkeley (*Geometry no Friend of infidelity, or a Defence of Sir Isaac Newton and the British Mathematicians*).

3. Jurin's syllabus is reprinted in full in Robinson, F. J. G. 'Trends in education in northern England during the eighteenth century', unpublished Ph.D. thesis, Newcastle University, 1972, vol. 1.
4. After its initial successes, Gresham's College gradually declined as an educational force. Lectures were suspended in 1666 and the Government stopped the salaries of its professors in 1669. Early in the eighteenth century lectures were resumed and soon other lectures, given by private individuals, were offered to the public. The most notable lecturers were J. T. Desaguliers (see Hans, N. *New Trends in Education in the Eighteenth Century*, Routledge and Kegan Paul, 1951, chapter 7) and William Whiston, Newton's successor as Lucasian Professor at Cambridge. Whiston was expelled from his chair in 1711 (see my p. 248) and moved to London where he gave lectures on astronomy and experimental philosophy. Later, other lecturers, such as Benjamin Martin, James Ferguson and Adam Walker helped to disseminate a knowledge of mathematics and, in particular, Newtonian mechanics through their public lectures. Not only did they lecture in London, but they also undertook provincial tours. Ferguson, for example, gave lectures in Newcastle – making use, incidentally, of Hutton's schoolroom – while Walker lectured at Eton and Winchester colleges. The books published by these lecturers give a good indication of the level at which they worked and also indicate that their lectures were attended by both gentlemen and ladies. (See Hans, *op. cit.*)
5. Hutton's advertisement is reprinted in Bruce, *op. cit.* (n. 1).
6. Other typical advertisements can be found in Sylvester, D. W. *Educational Documents 800–1816*, Methuen, 1970.
7. Quoted in Bruce, *op. cit.* (n. 1).
8. Lists of subscribers, particularly when they list professions and addresses, are useful in indicating the presence of mathematical practitioners and teachers. Wallis, for example, has used them to show that mathematics teaching was not so London-based as E. G. R. Taylor (see my p. 250) and other writers might suggest (Wallis, P. J. 'British mathematical biobibliography' *J. Inst. of Navigation*, **20** (2) (1967), 200–5).
9. Gregory (1774–1841) was trained by Hutton and became well known as an applied mathematician. He succeeded Hutton as professor at Woolwich and as editor of the *Ladies' Diary*. Gregory wrote many texts including *Mathematics for Practical Men* (1825) and *Hints to the Teachers of Mathematics* (1840). The latter provided hints on how to make best use of Simson's Euclid and Hutton's *Course*. We note that in 1840 the cadets at Woolwich appeared to study Euclid prior to reading the geometry sections in Hutton (cf. my p. 68). Gregory edited Hutton's *Course* after the latter's death. The 11th edition (1837) is interesting in that it contains both a 'traditional' approach to the calculus using fluxions and fluents and also, as an appendix, a translation of Lubbe's *Lehrbuch des Hohern Kalkuls*. Thus both old and new are represented. Gregory, however, still believed that 'in point of intellectual conviction and certainty, the fluxional calculus is decidedly superior'; to think of calculus 'without motion' was akin to thinking 'of war without bloodshed, gardening without spades'.
10. According to J. Aubrey, the father of William Oughtred (see my p. 245) was a scrivenor (at Eton); and understood common arithmetic, and 'twas no small help and furtherance to his son to be instructed in it when a school boy' (*Brief Lives*, Oxford University Press, 1898 vol. 2). The mathematical knowledge of most scrivenors was, however, extremely limited (cf. my p. 124). (See also Heal, Sir Ambrose, *The English Writing Masters, 1570–1800*, Cambridge University Press, 1931.)
11. This was also the case, for example, at Manchester where, at one time, boys from the grammar school attended the school run by Adam Martindale (my p. 47). Later, in the mid-eighteenth century, mathematics was taught at the grammar school itself, yet even then pupils received further training at a local private school: 'It is doubtful whether many English Grammar Schools possessed equally good mathematical teaching at this period . . . especially when this was reinforced by additional teaching at private mathematical schools' (Mumford A. A. *Manchester Grammar School*, Longman, 1919).

One pupil at Manchester at this time was Henry Clarke (see note 24). It should be emphasised again though that the place of mathematics in a school's curriculum depended upon the wishes and attitudes of the school's master (headmaster). Mathematics was, in fact, to disappear from the curriculum at Manchester and had to be reintroduced into it about 1850 (see Mumford, *op. cit.*).

12. Leeds Grammar School wished to widen its curriculum to introduce modern subjects, in particular, French, because 'the Town of Leeds and its neighbourhoods had of late years increased very much in trade and population . . . and, therefore, the learning of *French* . . . was become a matter of great utility to the Merchants of Leeds'. In 1805 Lord Eldon passed judgement on the case. He ruled that a grammar school should be, in Dr Johnson's terms, 'for teaching grammatically the learned languages'. The founder's wishes were to be obeyed and the classical curriculum retained. As a result, mathematics still was frequently treated as an optional extra and tuition in it was not provided free. However, under an Act of 1812, schools were enabled to apply to revise their curricula and an Act of 1840 set them free to do so. Arithmetic was certainly taught at Leeds Grammar School in the 1820s, for its Master, George Walker, published an *Elements of Arithmetic, Theoretical and Practical, for the Use of the Grammar School, Leeds*; a book which, according to De Morgan (*Arithmetical Books*, 1847) was 'a clear and excellent work, written by a man of real science'. About the same time, mathematics established itself in the curriculum of Eldon's old school at Newcastle. (See Robson, A. 'How they learnt, 1600–1850' *Math. Gazette*, **33** (1949), 81–93.)
13. See White, A. C. T. *The Story of Army Education*, Harrap, 1963. It was, of course, in Madras that Dr Andrew Bell was later to develop the monitorial system (see my p. 102).
14. Standing order of the 25th Foot, 1804.
15. That some training in mathematics was being provided prior to the establishment of the RMA is indicated by Samuel Heynes' description of himself on the title page of his *Treatise of Trigonometry, Plane and Spherical, Theoretical and Practical* (1701) as 'late Reader of the Mathematics to his Majesty's Engineers'.
16. French polytechnics soon acquired a reputation far greater than those of the military academies at Woolwich and Turin (where Lagrange had been a professor of mathematics). Their work was keenly watched in England: 'the establishment is . . . universally admired and considered as a model for imitation. It contains more than 400 young persons, previously educated in the mathematics, and the majority of them intended for engineers . . .; and they labour under the immediate direction of their tutors nine hours every day' (*Monthly Magazine*, Aug. 1798). The polytechnics provided the model for the US Military Academy at West Point (1802), an institution greatly influenced by French mathematical thought. It was following the establishment of West Point that English influence on US mathematics began to wane (see *A History of Mathematical Education in the United States and Canada*, National Council of Teachers of Mathematics (NCTM) 32nd Yearbook, 1970).
17. Simpson (1710–61) was the son of a provincial weaver who could only afford to give the boy the very barest education – to be taught to read. He, too, became a weaver and moved from Leicestershire to London about 1735 (see my p. 93). By then he had begun to teach himself mathematics and, like Hutton, combined learning with teaching: he worked as a weaver by day and taught in the evening. Simpson published a book on fluxions in 1737 (see note 19) and this introduced him to the scientific circles of London. In 1743 he was appointed professor at Woolwich and in 1745 elected an FRS. He published texts on probability, algebra, geometry and trigonometry and is remembered for 'Simpson's rule' in numerical integration.
18. Barlow (1776–1862) was an applied mathematician and instrument maker of great note. He spent most of his working life at the RMA, Woolwich and his publications included: *Theory of Numbers* (1811), *New Mathematical and Philosophical Dictionary* (1813) and *Tables* (1814) which proved to be among the most popular ever printed.
19. See Ball, W. W. R. *A History of Mathematics at Cambridge*, Cambridge University Press, 1889. A description of the Senate House examination in 1772 was given by John Jebb who had been second wrangler in 1757 (see Ball, pp. 191–2). It will be noticed that written

questions were not supplied (although these began to be introduced about that time, the custom of dictating questions survived in some quarters for a further sixty years or so) and that classification generally took place prior to the examination which served only to order students within their class. A candidate was, however, allowed to appeal if he thought he had been assigned to too low a class. The examination lasted three days and the range of subjects for the first or highest class is described by Jebb as follows:

'The moderator generally begins with proposing some questions from the six books of Euclid, plane trigonometry, and the first rules of algebra. If any person fails in an answer, the question goes to the next. From the elements of mathematics, a transition is made to the four branches of philosophy, viz. mechanics, hydrostatics, apparent astronomy, and optics, as explained in the works of Maclaurin, Cotes, Helsham, Hamilton, Rutherforth, Keill, Long, Ferguson, and Smith. If the moderator finds the set of questionists, under examination, capable of answering him, he proceeds to the eleventh and twelfth books of Euclid, conic sections, spherical trigonometry, the higher parts of algebra, and Sir Isaac Newton's *Principia*; more particularly those sections which treat of the motion of bodies in eccentric and revolving orbits; the mutual action of spheres, composed of particles attracting each other according to various laws; the theory of pulses, propagated through elastic mediums; and the stupendous fabric of the world. Having closed the philosophical examination, he sometimes asks a few questions in Locke's *Essay on the human understanding*, Butler's *Analogy*, or Clarke's *Attributes*. But as the highest academical distinctions are invariably given to the best proficients in mathematics and natural philosophy, a very superficial knowledge in morality and metaphysics will suffice.

When the division under examination is one of the higher classes, problems are also proposed, with which the student retires to a distant part of the senate-house, and returns, with his solution upon paper, to the moderator, who, at his leisure, compares it with the solutions of other students, to whom the same problems have been proposed.

The extraction of roots, the arithmetic of surds, the invention of divisors, the resolution of quadratic, cubic, and biquadratic equations; together with the doctrine of fluxions, and its application to the solution of questions 'de maximis et minimis,' to the finding of areas, to the rectification of curves, the investigation of the centers of gravity and oscillation, and to the circumstances of bodies, agitated, according to various laws, by centripetal forces, as unfolded, and exemplified, in the fluxional treatises of Lyons, Saunderson, Simpson, Emerson, Maclaurin, and Newton, generally form the subject-matter of these problems.'

20. See, for example, Montgomery, R. J. *Examinations* Longman, 1965, chapter 2, and Armytage, W. H. G. *Four Hundred Years of English Education*, Cambridge University Press, 1965.

21. From 1853 entry to the India Service was by means of examination; the Civil Service followed suit in 1870. A competitive examination for entrance to the RMA, Woolwich was instituted in 1857. (See Montgomery, *op. cit.* (n. 20), and Roach, J. *Public Examinations in England 1850–1900*, Cambridge University Press, 1971.)

22. *Tables of the Products and Powers of Numbers*, 1781; *Mathematical Tables*, 1785. The latter were reprinted in many editions up to 1894.

23. An account of the controversy is given in Smith, E. *The Life of Sir Joseph Banks*, Bodley Head, 1911.

24. Henry Clarke (1743–1818), a dissenter, was educated at Manchester Grammar School (see note 11) and later taught mathematics at Leeds and at his own school at Salford. He lectured in Manchester and Bristol before being appointed professor at the Royal Military College, Marlow (which was later to move to Sandhurst).

Des Barres was a pupil of Jean Bernoulli, grandson of his famous namesake. He served with Wolfe in Canada and later worked on a survey there with the famous seaman and explorer, James Cook.

Shortly after the Hutton controversy, the Royal Society was again accused of showing prejudice against dissenters. In 1790, Cooper, a chemist from Lancashire and, like

Clarke, a friend of Joseph Priestley's (my pp. 54–7), was 'refused admission into ye orthodox Royal Society on account of his heretical principles and junction with ye dissenters' (letter from R. E. Garnham to William Frend (see my p. 79)). The Royal Society later refused to accept a paper by Cooper at which point Priestley offered his resignation.

25. Raphson, J. *Mathematical Dictionary*, 1702. Ozanam was a French mathematician who published many treatises on mathematics. His *Récréations de Mathématique* appeared in translation in 1708 as *Recreations, Mathematical and Physical.*
26. Harris (1665–1719) was a private teacher of mathematics in London and also gave public lectures on the subject there. His *Lexicon* appeared in 1704 and, although owing much to Ozanam and others, gave a good description of contemporary practical mathematics and the art of instrument making in England. The *Lexicon* boasted a long list of subscribers which testified to the awakening interest in mathematics.
27. Edmund Stone (*c.* 1700–68) was a protégé of the Duke of Argyll on whose estate Edmund's father was a gardener. An early biographer (G. Carey) recounted how the Duke had come across the young, self-taught boy reading Newton's *Principia* and had encouraged him further in his studies. The truth, however, appears to have been more prosaic. Nevertheless, with the Duke's patronage, Stone became an FRS in 1725, the year before he published his *New Mathematical Dictionary*. In this short work, Stone attempted to describe all 'things both in Pure Mathematics and Natural Philosophy so far as it comes under a mathematical consideration'. He still saw mathematics as part of the quadrivium and, as a result, terms such as 'fugue', 'fresco' and 'fuse (of a bomb)' are to be found alongside 'fraction' and 'fluent'.
28. Aug. 1798.
29. See NCTM 32nd Yearbook, *op. cit.* (n. 16).
30. See Howson, A. G. 'The international transfer of learning systems', *Educational Developments International*, 1 (1973), 143–7.
31. Robert Simson (1687–1768) was a Scottish mathematician and professor at Glasgow University. His most notable contributions were in the field of geometry (although the line that bears his name was not his discovery) and, in particular, his translation of Euclid's *Elements*. First published in 1723, this was to become the standard English Euclid for many years: it appeared in several editions (see Wallis, P. J. *A Checklist of British Euclids up to 1850*, Newcastle University, 1967) and formed the basis for later versions by many different authors.

    John Playfair (1748–1819) published his edition of Euclid in 1792, which, like Simson's ran to many editions. He is remembered for the axiom he substituted for Euclid's parallel postulate. Unlike Simson though, whose fame rests on a 'posthumous' discovery, Playfair's axiom had been suggested (in a slightly modified form) by Joseph Fenn in 1769.
32. Carlyle (1795–1881) was a student at Edinburgh University and left there in 1814 without taking a degree. In an attempt to make a living he acted, none too successfully, as a mathematical tutor in Kirkcaldy, Fife. One of his teachers at Edinburgh was the professor of mathematics, John Leslie, who himself produced an *Elements of Geometry* which attempted to break Euclid's stranglehold. Leslie did not go so far as Legendre, however, in encompassing algebraic methods: 'It is a matter of deep regret, that Algebra, or the Modern Analysis . . . has contributed, especially on the Continent, to vitiate the taste and destroy the proper relish for the strictness and purity so conspicuous in the ancient method of demonstration'.
33. See NCTM 32nd Yearbook, *op. cit.* (n. 16).
34. Some of these are reprinted in Ball, *op. cit.* (n. 19). A selection of questions taken from Ball is to be found on my pp. 212–14 and in Hollingdale, S. H. 'The teaching of mathematics in universities: some historical notes' *Bulletin IMA*, 10 (9/10) (1974), 312–18.
35. Preface of *A Course of Mathematics . . . Composed, and More Especially Designed, for the Use of the Gentlemen Cadets in the Royal Military Academy at Woolwich*, 2 vols., 1798.
36. Sir Jonas Moore (see my p. 245) had planned a comprehensive text for use at Christ's Hospital. Moore himself had written only the arithmetic, practical geometry, trigo-

nometry and cosmography sections before his death. The work was completed by Peter Perkins (my p. 36) (who contributed a section on Euclid) and the astronomer John Flamsteed, and appeared as *A New Systeme of the Mathematicks* in 1681 at the high price, for those times, of 35s (£1.75). A detailed description of the book's contents can be found in Robson, *op. cit.* (n. 12).

37. John Ward was a private teacher of mathematics in Chester who in 1707 published his *Young Mathematician's Guide: Being a Plain and Easy Introduction to the Mathematicks.* It contained sections on arithmetic, algebra, geometry (treated in a style more akin to Leslie's (note 32) than Euclid's), conic sections, and 'the arithmetick of infinites' (series, and applications to areas and volumes). The book was very successful: Ward, writing in 1722, indicated that by then it had sold (in three editions) some five thousand copies and had been 'ordered to be publicly read to pupils at universities in England, Scotland and Ireland'. Further editions appeared throughout the eighteenth century, some containing a supplement on the 'History of logarithms'.
38. William Emerson (1701–82) was a highly esteemed mathematical teacher and writer. His many books were widely read and admired by contemporaries, although De Morgan considered him 'as much overrated as Thomas Simpson [note 17] was underrated'. Emerson was an eccentric character who had little time for either the Establishment or for proving rules which 'the practical mechanic does not want'.
39. Bushell, W. F. 'A century of school mathematics', *Math. Gazette*, **31** (1947), 69–89. Bushell here was speaking of universities as well as the upper forms of schools.
40. Thus, for example, A. Nesbit (1778–1859), a private mathematics teacher who taught in Leeds and Manchester in the early nineteenth century, and whose books on mensuration and arithmetic were extremely popular, mentioned Hutton and his works several times in the preface of his *Mensuration*.
41. The key reference on such periodicals remains: Archibald, R. C. 'Notes on some minor English mathematical serials', *Math. Gazette,* **14** (1929), 379–400.
42. Leybourn, T. *The Mathematical Questions Proposed in The Ladies' Diary,* 4 vols., 1817.
43. The contribution of women to the *Ladies' Diary* is described by Perl, T. 'The *Ladies' Diary or Woman's Almanack*, 1707–1841,' *Historia Mathematica,* **6** (1979), 36–53. For a more general discussion of women's interest in mathematics in Britain in the eighteenth century see Wallis, R. and Wallis, P. J. 'Female philomaths', *Historia Mathematica,* 7 (1980), 57–64.
44. Hutton's work began to appear in July 1771 and Clark's followed shortly afterwards.
45. For a less serious question taken from the *Ladies' Diary* for 1752 see Cornelius, M. L. 'How to deal with modern mathematicians', *Math. Gazette*, **58** (1974), 285–7.
46. Leybourn, *op. cit.* (n. 42).
47. '15 young ladies in a school walk out 3 abreast for seven days in succession: it is required to arrange them daily, so that no two shall walk twice [i.e. more than once] abreast'.

    Perhaps the most significant fact about Kirkman (1806–95) is that he was never engaged in teaching mathematics but was rector of a parish in Lancashire. That an amateur mathematician could rank second only to Cayley in England as a group theorist and could communicate on equal terms with Hamilton is indicative of the way in which the gap between professional and amateur mathematicians has grown in the past century. See Biggs, N. L. 'T. P. Kirkman, mathematician', *Bull. LMS*, **13** (1981), 97–120.
48. Quoted in Archibald, *op. cit.* (n. 41).
49. It is, of course, impossible to say how many endowed schools included mathematics in their curriculum in, say, 1810. As we have seen, in some schools (for example, Manchester Grammar School and King Edward VI School, Birmingham) mathematics entered and then was later omitted from the curriculum. None of the leading 'Public Schools' (see my p. 129) possessed mathematics masters at that time (although, since all graduates from Cambridge had studied and been examined in mathematics, the classics teachers would all have had some knowledge of the subject), and mathematics held no fixed place in their curriculum. It would seem that the northern grammar schools were more likely to teach mathematics than those in the south. That this was not entirely due to the demands of the rapidly-growing industrial centres is evidenced by the strength of

mathematics teaching in Westmorland. As a rough quantitative guide, Robinson, *op. cit.* (n. 3), indicates that nine of the twenty-five grammar schools in Cheshire taught some form of mathematics beyond simple arithmetic before 1800.

50. Hans, *op. cit.* (n. 4), gives a list of schools for girls known to have been in existence in the eighteenth century and also includes a chapter on the education of women. Of the 120 women of that period mentioned in the *Dictionary of National Biography* for their 'published work or intellectual and social eminence' between 25% and 30% received some education at a school, the majority being the daughters of professional men. In Hans' words 'even in the best schools for girls neither mathematics nor sciences were taught'.
51. 'I have read over your lectures with great pleasure and the more so, to find that even the learned and most difficult sciences are thus beginning to be successfully cultivated by the extraordinary and elegant talents of the female writers of the present day.' (Hutton to Margaret Bryan, 6 Jan, 1797.)
52. Some evidence of the popularity of Mrs Bryan's school is given by the fact that 157 women (presumably mainly former pupils) are listed amongst the subscribers to her 1806 book on natural philosophy.
53. Kirkman (note 47), for example, published in the journals of the Manchester and Liverpool 'Lit. and Phils'. His 'Theory of groups and many-valued functions' *Mem. Phil. Soc. Manchester* (3rd series), 1 (1862), 274–397 was one of the first expositions of group theory in English.

5. Augustus De Morgan

1. De Morgan was spotted making figures with ruler and compass; in his words he was 'drawing mathematics'. The family friend explained the aims of deductive geometry to the boy who 'was soon intent on [his] first demonstration'. See De Morgan, S. E. *Memoir of Augustus De Morgan*, Longman, 1882. (This book is referred to in the following notes by the abbreviation MAM.)
2. Robert Reece, quoted in MAM p. 7.
3. Dodson (*c.* 1710–57) was a mathematics teacher in Wapping before, in 1755, he was appointed Master of the Royal Mathematical School. At the same time he was elected an FRS. Indeed, about that period the Masters of the Royal Mathematical School at Christ's Hospital appeared almost automatically to have become Fellows of the Royal Society. He published several books and sets of tables, including the *Mathematical Repository* (1748; a collection of problems and worked solutions), and also re-edited Wingate's *Arithmetic*, a book that had first appeared in 1629 and which, in the intervening years, had also been edited by the practitioners John Kersey and George Shelley.
4. MAM p. 233.
5. One notes that De Morgan entered Cambridge midway through the year and that because of the difficulty of getting rooms in Trinity College he was forced to spend his first two years at Cambridge in lodgings.
6. Sylvestre François Lacroix (1765–1843) was, according to De Morgan (*Arithmetical Books,* 1847), 'one of the most systematic and most widely circulated of elementary writers'. His *Traité du calcul différentiel et du calcul integral* contained the idea that the ratio of two quantities, each of which approaches 0, can approach a definite number as a limit (see Kline, M. *Mathematical Thought from Ancient to Modern Times*, Oxford University Press, 1972). However, he still used the symbol 0/0 and did not treat the idea of a limit in the way De Morgan was later to do. Lacroix in this book first introduced the term 'differential coefficient'. In addition to the English version of his *Calculus,* a translation of his *Arithmetic* was also made.
7. Letter to Dr Whewell, 20 Jan. 1861 reprinted in MAM. The rôle of the private tutors at Cambridge is discussed on my pp. 125–7.
8. The unedited correspondence can be found in MAM pp. 386–92.
9. William Whewell (1794–1866) was another who played a prominent part in the 'analytic movement'. Although not remarkable as a mathematician, he exercised considerable

influence in Cambridge (where he held chairs in mineralogy and moral philosophy) and elsewhere. He wrote several textbooks on mathematics including *The Mechanical Euclid* (1837) (see my p. 264). His most notable contribution to mathematics education, the pamphlet *The Study of Mathematics* (1836), launched a famous controversy with the Scot, Sir William Hamilton, in the pages of the *Edinburgh Review*.

10. The law provided employment for several who, for various reasons, could not find suitable appointments as mathematicians. Thus, for example, Sylvester, debarred from graduating at Cambridge because he was a Jew, became a barrister before his appointment in 1855 to the chair at Woolwich held fifty years earlier by Hutton. Cayley, Sylvester's 'twin', was able to graduate and to hold a fellowship. He was, however, unwilling to become a clergyman and, as no mathematical post was available in Cambridge, he too became a barrister. In 1863 Cayley gave up the law, and a considerable income, to return to Cambridge as the first Sadleirian Professor.
11. Bell, E. T. *The Development of Mathematics,* McGraw-Hill, 1940, p. 131.
12. For a full account of Frend's life see Knight, F. *University Rebel: The Life of William Frend*, Gollancz, 1971. De Morgan was a great admirer of this cantankerous, outspoken reformer: 'the noblest man he had ever known'. Frend's reforming zeal, incidentally, extended to the problems of pollution, as shown by his publication, *Is it Impossible to Free the Atmosphere of London in a Very Considerable Degree from Smoke . . .?* (1813).
13. Frend, W. *The Principles of Algebra*, London, 1796.
14. See *The Monthly Magazine*, Nov. 1798. Frend's quixotic gesture also served to show his continued opposition to Isaac Milner, the favourite (and successful) candidate. Milner, whom de Quincey described as a fellow 'opium eater', had been Frend's arch opponent at the time of his banishment from Cambridge.
15. See *The Monthly Magazine*, Aug. 1798. Frend wrote to Lindsey, leader of the Unitarians, for advice on mounting a lecture course. He was told 'No lectures but medical ones are well attended . . . [little] may be expected from so dry a subject as Mathematics in these days' and that Frend, because of his political activities was 'marked out as an infecting person'.
16. Lord Brougham (1778–1868) was one of the most influential and controversial politicians of his age. He drew attention to the importance of popular education (a stance which was satirised by Thomas Love Peacock in his novel *Crotchet Castle*).
17. George Birkbeck (1776–1841) was a pioneer in adult education. When Professor of Natural Philosophy in Glasgow, he instituted lectures on scientific topics for working men. This led in 1823 to the formation of the Glasgow Mechanics' Institution. The following year (with some help from Frend) he founded, and became the first President of, the Birkbeck Mechanics' Institution in London, which eventually developed into Birkbeck College, one of the constituent colleges of the University of London.
18. The wedding took place at the St Pancras Registry Office: 'a new mode of marrying' which 'very much pleased' Frend, although Sophia's relatives 'all Church people' were 'not a little shocked'.
19: MAM p. 28.
20. See Neumann, B. H. 'Byron's daughter', *Math. Gazette*, **47** (1973), 94–7.
21. The obituary notice appeared in the *Cambridge University Reporter*.
22. Quoted in Curtis, S. J. *History of Education in Great Britain*, 7th edn, University Tutorial Press, 1967, p. 221. It must be remembered that Brougham was referring to the provision of elementary education for the general populace. In certain fields of education, as we have seen (p. 71), England could claim some success.
23. The SPCK had among its objectives the founding of schools for the education of poor children in the principles of the Established Church, the establishment of lending libraries and the distribution of Bibles. By 1734 it had established almost 1,500 schools. The curriculum was narrow – religious instruction, reading and writing, with arithmetic for boys and sufficient handicraft to fit girls for domestic service. Children attended the schools for only $1\frac{1}{2}$ to 2 years, merely sufficient time to learn the rudiments. (See Birchenough, C. *A History of Elementary Education in England and Wales*, 2nd edn, University Turorial Press, 1925.) From 1734 the number of such charity schools,

sometimes supported by Roman Catholics or by dissenters (cf. my p. 53), grew rapidly (see my pp. 102–3).

24. See Bamford, T. W. (ed.) *Thomas Arnold on Education*, Cambridge University Press, 1970, pp. 161–2.
25. See Simon, B. *Studies in the History of Education 1780–1870*, Lawrence and Wishart, 1960, pp. 159–63, 227–30.
26. Knight, C. *Results of Machinery, Namely, Cheap Production and Increased Employment Exhibited*, 1830.
27. Francis Place, quoted in Webb, R. K. *The British Working Class Reader*, Allen and Unwin, 1955, p. 144.
28. *Quarterly Journal of Education*, **9** (1835), p. 254.
29. *Quarterly Journal of Education*, **2** (1831), p. 264–5.
30. These were reprinted in *The Schoolmaster* 2 vols., SDUK, 1836.
31. This issue also contained interesting articles on the Ecole Polytechnique which included details of the syllabuses and the books followed, and on the curriculum of a typical German Gymnasium.
32. The interested student will also wish to consult De Morgan's many textbooks, in particular their prefaces.
33. This view of mathematics was currently being shattered by Gauss, Bolyai and Lobachewsky.
34. It is amusing to speculate on how De Morgan would have reacted to workcards, programmed learning, etc.
35. Stories are told of schools in which it was not allowed even to refer, say, to 'the angle BEC' if Euclid called it the angle CEB.
36. I attended a geometry class in India in the 1960s at which the master insisted that boys replying to his questions quoted not only the statements of theorems but also their numbers.
37. Here De Morgan appears to be reaching forward to a Hilbertian view of geometry, but to assume this would be to read too much into his words.
38. This was an argument much heard in the early 1960s. Subsequent research has cast doubt upon its validity.
39. Again this was a viewpoint which gave rise to much argument in the 1960s. Some took the De Morgan view that one should not repel the beginner with a long list of axioms; others argued that it ill serves the axiomatic method if one appears to produce a new axiom everytime one reaches an apparent impasse. The intellectual maturity required to appreciate the benefits of an axiomatic approach were too frequently misjudged.
40. De Morgan's principles would seem based more on sound pedagogy developed at a university level than on experience of what was possible in schools (cf. his remarks on Wilson's *Geometry*, my p. 132).
41. This is an early example of 'Haeckel's Law' applied to education. It is difficult to know who first produced the argument: Herbert Spencer (1854) attributes it to Auguste Comte.
42. This argument is echoed in Branford's *A Study of Mathematical Education* (see my p. 156).
43. Local societies often sprang up as a result of visits from peripatetic lecturers such as Desaguliers. Thus there are references dating from 1719 to the 'Mathematical Society of Manchester'. Other societies were known to exist at, for example, York, Lewes, Wapping and Oldham.(See Cassels, J. W. S. 'The Spitalfields Mathematical Society', *Bull. LMS*, **11** (1979), 241–58; **12** (1980), 343.) Many societies did not restrict their activities to mathematics, but were also concerned with science; indeed, one of the best known, the Lunar Society of Birmingham, gave little attention to mathematics.
44. *A Budget of Paradoxes* (published posthumously in 1872): the extract concerning the Spitalfields club is reprinted in Newman, J. R. *The World of Mathematics*, Allen and Unwin, 1961, vol. 4, pp. 2372–6.
45. See Cassels, *op. cit.* (n. 43).
46. Although De Morgan was not an outstandingly creative mathematician, his contribu-

tions to mathematics were substantial enough to make Sylvester's comment that 'De Morgan did not write mathematics; He wrote *about* mathematics.' (see *Math. Gazette*, **14** (1929), 510) unjust. Sylvester, however, was not one to let the truth stand in the way of a good epigram.

47. De Morgan's major contribution to mathematical bibliography was his *Arithmetical Books* (1847). Other articles by him include 'Notices of English mathematical and astronomical writers between the Norman conquest and the year 1600' (*British Almanac*, 1837), 'Mathematical bibliography' (*Dublin Review*, 1846) and 'On the difficulty of correct descriptions of books' (*British Almanac*, 1853). After De Morgan's death, his library of some 3,000 volumes was bought by Lord Overstone and presented to London University.
48. De Morgan published several articles on decimal coinage in the *British Almanac* and elsewhere.
49. The quotation is taken from Colenso's *Arithmetic* (1872 edition). De Morgan had added an appendix on decimal money to later editions of his own *Arithmetic*.

    Colenso (1814–83) was mathematics master at Harrow School from 1838–42 and later Bishop of Natal. His texts were very popular in Victorian schools, his *Arithmetic* being more widely used than any other in the schools investigated by the 1868 Schools Inquiry Commission. Like Newton, Frend, and (in the twentieth century) Russell and Bishop Barnes of Birmingham, Colenso was a mathematician with unorthodox religious views; indeed, his writings caused him to be tried for heresy.
50. 'If we look at what takes place around us, we shall find that we have no Mathematical Society to look to as our guide. The Royal Society, it is true, *receives* mathematical papers, but it cannot be called a Mathematical Society. The Cambridge Philosophical Society seems to fulfil more nearly the functions of a Mathematical Society, but it is in an exceptional position . . . it is a society in which almost all the members are able to relish its highest discussions.' (De Morgan's inaugural address to the LMS, 16 Jan. 1865.)
51. Hirst (see p. 264) was later to succeed De Morgan as Professor at University College. As we shall see, he was also to play a major rôle in the establishment of the Association for the Improvement of Geometrical Teaching (now the Mathematical Association).
52. See MAM pp. 282–5 and, for the complete text, the first number of the Society's *Proceedings*.
53. The Society still preserves (and uses) the original book signed by members on their admission. It contains a fascinating collection of mathematicians' autographs.
54. The early history of the LMS is described in Glaisher, J. W. L. 'Notes on the early history of the Society', *Journal LMS*, **1** (1926), and Collingwood, Sir Edward, 'A centenary of the London Mathematical Society', *Journal LMS*, **41** (1966). The Society's 'Book' (see note 53) contains the signatures of several schoolteachers (e.g. J. M. Wilson).
55. The De Morgan Medal was instituted in 1884 and is awarded at three-yearly intervals. The first recipient was Cayley, and he was followed by Sylvester, Rayleigh and Klein. Others honoured have included Hardy and Russell and, since 1950, Besicovitch, Titchmarsh, Taylor, Hodge, Newman, Hall, Dame Mary Cartwright (the first woman to receive the Medal), Mahler, Higman, Rogers and Atiyah.

6. Thomas Tate

1. I am indebted to Mr R. Hall, of the Alnwick Public Library, for information concerning Tate's early education.
2. *A Complete Treatise on Practical Arithmetic and Book-keeping*, 1828.
3. See entry by Sarah Wilson in *Dictionary of National Biography*. Edinburgh was, of course, the nearest academic centre to Alnwick and Newcastle. Robert Stephenson (1803–59), Tate's contemporary and fellow Northumbrian, was in 1822 sent to Edinburgh University for six months 'there being then no college in England accessible to persons of moderate means, for purposes of scientific culture' (Smiles, S. *Lives of the Stephensons*, 1874). That six months' study cost George, Robert's father, £80 'but he was amply repaid by the better scientific culture which his son had acquired, and the evidence of ability

and industry which he was enabled to exhibit in a prize for mathematics which he had won'. It is unlikely, however, that Tate benefited so much from his Edinburgh visit, for Kay-Shuttleworth (see my p. 111) refers to him as 'self-taught'.

4. See Wilson, S. 'George Tate (1805–71)' *Dictionary of National Biography*.
5. Tate G. *A History of Alnwick*, 1866–9, p. 414.
6. *Mechanics' Magazine*, 27 Aug. 1825.
7. His topics included: On gravity; Kepler's Laws; Physical astronomy; The Cartesian system of the universe; Central forces; Mechanics and natural philosophy; Mathematical evidence. These lectures were probably given in the 1850s.
8. Letter from the Reverend F. Temple to R. R. W. Lingen, *Minutes of the Committee of Council*, 1856–7, pp. 32–3.
9. Temple, *op. cit.* (n. 8), indicates that Tate first taught in Alnwick in 1830. However, M. Brierley ('Lancashire mathematicians', *Papers Manchester Lit. Club*, **4** (1878), 7–30) suggests that Tate was in York by 1829. Brierley names Tate as the first editor of the *York Courant's* mathematical column as do R. C. Archibald ('Notes on some minor English mathematical serials', *Math. Gazette*, **14** (1929), 379–400) and M. P. Black and A. G. Howson ('A source of much rational entertainment', *Math. Gazette*, **63** (1979), 90–8). Unfortunately, the columns of the *York Courant* were unsigned and there is no reference in the paper as to the identity of the first editor. Certainly, Tate edited the column in the 1830s before his move to London (see Abram, W. A. 'Memorial of the late T. T. Wilkinson, FRAS of Burnley', *Trans. Hist. Soc. Lancs and Cheshire*, ser. 3, vol. 4 (1876), 77–94, p. 86), but the evidence now suggests to me that he did not initiate the column.
10. Wetherill, J. H. 'The York Medical School', *Med. History*, **5** (1961), 253–69.
11. At that time 2s (10p) would have bought, for example, 4 lb of beef.
12. The paper by Black and Howson, *op. cit.* (n. 9), describes the *York Courant* mathematical column, its content and contributors: Wilkinson wrote a series of articles in the *Mechanics' Magazine* **54–59** (1848–53) describing some thirty such periodicals.
13. This was probably about the time when Tate became the column's editor.
14. Brierley, *op. cit.* (n. 9), p. 20.
15. The quotation comes from the *Universal Magazine of Knowledge and Pleasure* and was a first warning to those who set questions 'calculated merely as a trial of skill between the professors of [mathematics]'. No heed was taken of the warning and that mathematical column too was closed. (See Goldsmith, N. A. 'The Englishman's mathematics as seen in general periodicals in the eighteenth century', *Math. Teacher,* **46** (1953), 253–9.) Brierley, *op. cit.* (n. 9), tells of the accusations of plagiarism made against Wilkinson and some of his friends in connection with their correspondence with periodicals such as the *York Courant* and the *Educational Times:* 'a vice . . . practised to a considerable extent among the present degenerate race of Lancashire mathematicians'.
16. *Report of the Proceedings at the Annual Meeting of the Leeds Mechanics' Institution on the 20th September, 1830.*
17. See Harrison, J. F. C. *Learning and Living 1790–1960*, Routledge and Kegan Paul, 1961.
18. Hudson, J. W. *A History of Adult Education,* Longman, 1851 (reissued, Woburn Press, 1969).
19. Quoted in Tylecote, M. *The Mechanics' Institutes of Lancashire and Yorkshire before 1851*, Manchester University Press, 1957.
20. Tate, T. T. *The Philosophy of Education*, 2nd edn, Longman, 1857, pp. 215–16.
21. When at Battersea, Tate gave evening classes, at the request of the directors of one of the railway companies, to its employees, engineers, stokers, etc. This had the two-fold aim of improving their efficiency, it being felt that many of the frequent accidents 'were attributable to the ignorance of the men', and also of giving the work-force something more valuable to do in the evenings than 'stupifying themselves with dram drinking'. (Kay-Shuttleworth, J. P. *Four Periods of Public Education*, Longman, 1862, p. 343.)
22. Thus, for example, John Stuart Mill argued in 1859 that 'a general State education is a mere contrivance for moulding people to be exactly like one another: and as the mould

in which it casts them is that which pleases the predominant power in the government . . . in proportion as it is efficient and successful it establishes a despotism over the mind, leading . . . to one over the body'.

23. In 1816 approximately one in two English children attended a school of some kind. (See, for example, Adamson, J. W. *English Education 1789–1902*, Cambridge University Press, 1930, p. 27.)
24. Thus, for example, in 1834, Carlisle, which had a school population of 2,680 drawn from about 20,000 inhabitants, boasted over fifty day schools and twelve Sunday Schools.
25. Tate, T. T., *op. cit.* (n. 20), pp. 2–3.
26. *Hansard* **20,** 30 July 1833. Roebuck's important speech is reprinted in De Montmorency, J. E. G. *State Intervention in English Education*, Cambridge University Press, 1902, pp. 325–51.
27. Quoted in Sutherland, G. *Elementary Education in the Nineteenth Century*, Historical Association, 1971, p. 10.
28. See, for example, Birchenough, C. *A History of Elementary Education in England and Wales*, 2nd edn, University Tutorial Press, 1925, p. 291. Bell's National Schools proposed six (or eight) arithmetical grades (Birchenough, *op. cit.*, p. 286).
29. Birchenough, *op. cit.* (n. 28), p. 307.
30. The pupil-teacher scheme was instituted in 1846. Bright pupils were apprenticed to headteachers for up to five years (13–18). They were examined yearly and if they acquitted themselves well their heads received a bonus. At the close of their apprenticeships pupil-teachers were eligible to compete for Queen's Scholarships which admitted them to a three-year training-college course.
31. Report by HM Inspector of Schools, Reverend H. Moseley, *Minutes of the Committee of Council*, 1847–8, p. 8. Moseley contrasted King's Somborne School with the usual run of National Schools in which not one child in twenty could write down the number ten thousand and ten and where 'not often have I succeeded in getting a sum of money correctly multiplied by a number of one figure'.
32. Dawes' (and Tate's) pioneering work in science teaching is described in Layton, D. *Science for the People*, Allen and Unwin, 1973.
33. As late as 1852, 53% of pupils in inspected schools had attended for less than one year and 73% for less than two. Less than one-sixth of the pupils were over eleven, the vast majority of these being pupil-teachers. (*Minutes of the Committee of Council*, 1852–3, appendix 1, table 1).
34. This minute is reprinted in Kay-Shuttleworth, *op. cit.* (n. 21).
35. *Report of the 1834 Select Committee on Education*, p. 232.
36. See, for example, Birchenough, *op. cit.* (n. 28), p. 486.
37. See Kay-Shuttleworth, *op. cit.* (n. 21), pp. 293f.
38. It is interesting to note that the classically-trained inspector, the Reverend J. Allen, had to seek Tate's aid 'both in framing my questions [on mechanics] and in ascertaining the merits of the answers sent in to them by the pupils'. (*Minutes of the Committee of Council*, 1843–4, p. 16.)
39. Report by the Reverend J. Allen, *Minutes of the Committee of Council*, 1843–4, pp. 15–16.
40. Report by the Reverend H. Moseley, *Minutes of the Committee of Council*, 1845–6, p. 251.
41. See Birchenough, *op. cit.* (n. 28), p. 436.
42. The classics also featured in the curriculum of other colleges, for example, Cheltenham, and York and Ripon.
43. 'League tables' of examination results obtained in the varying colleges are to be found in the *Minutes of the Committee of Council*. Battersea (in 1851–52, for example) came near the bottom in arithmetic (in the percentage of students classed 'good' or 'fair'), was third behind Chelsea and Cheltenham in geometry, third to Chester, and York and Ripon in mensuration, fourth to Cheltenham, Chelsea, and York and Ripon in higher mathematics and came top in industrial mechanics and algebra. However, when 'attainment' was adjusted to allow for the number of months the students had attended college, Battersea came top in all except higher mathematics. Readers were warned of the difficulty of interpreting such tables!

44. Baden Powell (1796–1860) was Savilian Professor of Geometry at Oxford from 1827 to 1860. He was father of the founder of the Boy Scouts movement.
45. Simon, B. *Studies in the History of Education 1780–1870*, Lawrence and Wishart, 1960, p. 92.
46. *Minutes of Committee of Council* 1847–8, vol. 2, p. 537.
47. Frederick Temple (1821–1902) was a double first in classics and mathematics at Oxford in 1842. He worked for the Committee of Council from 1848. When Principal of Kneller Hall (1849–55) he had Tate on his staff; later, when Headmaster of Rugby (1857–61), J. M. Wilson. In 1861 he became Bishop of Exeter from which see he was translated first to London and then, in 1896, to Canterbury. His son, William, was also first a headmaster and later Archbishop of Canterbury.
48. *Report of the Newcastle Commission*, 1861, p. 151.
49. Kneller Hall was established as a training college to provide masters for what, in effect, were schools for pauper and criminal children. An outstanding staff was recruited: Temple as Principal, Palgrave (of the *Golden Treasury*) as Vice-Principal, and Tate as First Master (with salaries of £800, £500 and £250 p.a. respectively). It had a brief and disastrous history, both the staff and the students being misled. Initially high standards were demanded of entrants and these were reflected in the levels of work attained: 'In no other training school has the annual examination . . . exhibited a higher degree of efficiency' (*Minutes of the Committee of Council*, 1854–5). Yet the number of entrants dwindled, the Government, preoccupied with the Crimean War, neglected the college, and Temple was attacked by various sections of his own church for his ecumenical leanings (for example, for forbidding bonfires on 5 November and so kowtowing to Rome, and for admitting dissenters). He resigned and the college was closed in 1855: 'What a triumph . . . ! Down goes the godless College! Hurrah for the true Church' (Sandford, E. G. *Frederick Temple*, Macmillan, 1907, p. 172).
50. The first edition appeared in 1854 (Longman), the second (from which all the quotations are taken) in 1857 and a third in 1860. Two editions appeared in the United States, in 1884 and 1885, the preliminary note to which claimed that 'No English book on education has been oftener called for during the past five years'.
51. *The Educational Expositor*, which was edited by Tate and his assistant at Kneller Hall, J. Tilleard, appeared from 1853 to 1855 and was 'specially designed for Schoolmasters and Schoolmistresses, mothers of families, and all interested in education'. Vol. 3 contained the first parts of Tate's *Mathematics for Working Men* and vol. 2 some trenchant criticisms of the educational exhibition and lectures arranged by the Society of Arts. After the latter the 'indignation of the London schoolmasters ran high' for they had been bitterly disappointed by the lectures given by the distinguished speakers: the teachers had expected the lecturers to know something about their work and to bear on it!
52. Pole, W. *Sir William Fairbairn, Bart*, Longman, 1877, pp. 211, 421.
53. Temple, *op. cit.* (n. 8).
54. Letter from Canon Moseley to R. R. W. Lingen, *Minutes of the Committee of Council*, 1856–7, p. 32.
55. Arnold, M. 'Report' in *Minutes of the Committee of Council*, 1869–70, p. 291.

7. James Wilson

1. *J. M. Wilson: An Autobiography*, Sidgwick and Jackson, 1932, p. 68. Unattributed quotations in this section are also taken from Wilson's autobiography.
2. The College claims connections with an earlier grammar school endowed by the mathematician Isaac Barrow.
3. Wilson gives an account of this and of his reactions in his autobiography. Another pupil at King William's in the 1840s, F. W. (Dean) Farrar, was later to write a moral Victorian tale *Eric, or Little by Little* (Edinburgh, 1858) describing conditions at the school; Wilson found this book 'no caricature of this school, though . . . a caricature of the human boy'.
4. Wilson, *op. cit.* (n. 1), pp. 22–4. Evans, the Sedbergh headmaster, had been Third Wrangler in 1828.

5. There is no general survey concerning the early nineteenth century to which one can refer, but the report of the Taunton Commission (*Parliamentary Papers*, 1867–8, vol. 28, part 15) indicates, for example, that whereas in the 1860s some Yorkshire schools, such as St Peter's York, and the grammar schools at Barnsley, Bradford, Halifax (Heath) and Leeds offered wide-ranging mathematics courses, others taught little; for instance, King's School, Pontefract and Keighley Grammar aspired only to Euclid Book 1 and Skipton Grammar to some algebra. Sedbergh was in a 'temporary and wholly exceptional state of inefficiency' and had only twenty or so students. It claimed, however, to teach Colenso's *Algebra* and Euclid Books 1–4, 6, 11. The textbooks used were mainly those by Euclid (usually in Potts' or Simson's editions), Colenso, Lund and Todhunter – although other authors including Bridge, Barnard Smith, Wood, Nesbit and Tate were represented.
6. Thus, for example, Thomas James, Head of Rugby School, 1778–94, gave Saturday morning lectures on mathematics and appointed arithmetic masters. However, there is no mention of mathematics teaching at Rugby in Carlisle's survey of 1818 and the first regular mathematics master was not appointed until 1845. Similarly, George Butler, Headmaster of Harrow, read Euclid to his sixth-form before Marillier arrived in 1819. James was to complain of the 'mental strain' of lecturing for an hour on mathematics: 'it not only kept my mind upon the full stretch during the delivery, . . ., but even *wearied* my body to excess, and made it hot, or, at any rate, perspire too much' (quoted in Rouse, W. H. D. *A History of Rugby School*, Duckworth, 1898, p. 139).
7. The teaching of mathematics was not always encouraged: R. A. Ingram (Senior Wrangler in 1784) complained in 1792 that some schools were now studying mathematics to the neglect of the classics, 'an evil of some magnitude'.
8. Patchett Martin, A. *Life of Lord Sherbrooke (Robert Lowe)*, 1893, vol. 1, p. 13.
9. Trollope, T. A. *What I Remember*, 1887, vol. 1, p. 125. T. A.'s more famous brother Anthony wrote (*Autobiography*, 1883) that in his twelve years of schooling (mostly at Harrow and Winchester) 'no attempt had been made to teach me anything but Latin and Greek . . . I do not remember any lessons in arithmetic'.
10. For example, mathematics teachers at Eton could not wear academic dress in chapel or say prayers in their boarding houses. See Howson, A. G. 'Mathematics: the fight for recognition', *Maths in School*, **3** (6) (1974), 7–9.
11. A number which the Taunton Commission felt was 'out of all proportion to [the school's] needs . . . even in the most favourable circumstances'.
12. All unattributed quotations in this section are taken from chapter 3 of Wilson. *op. cit.* (n. 1).
13. Although Wilson's fellowship attracted no duties, he always drew attention to the fact that he was 'late Fellow of St John's College, Cambridge'. The title gave evidence of academic (if not always intellectual) respectability. The harking-back-to-a-glorious-Cambridge-youth syndrome is illustrated in the story told of one of England's finest mathematicians who devoted over thirty years to building up the mathematics department of a provincial university. On his retirement it was agreed to name a building in his honour and he was asked to suggest suitable wording for the dedicatory stone: 'Nothing extravagant', he is said to have replied, 'just my name and "Sometime Fellow of Trinity College, Cambridge".'
14. According to the Report of the Clarendon Commission, Wilson, then aged twenty-six, was, in 1862, receiving £647 p.a. (plus presumably free accommodation and meals). (The stipends for the ten new chairs recommended by the Cambridge University Commission varied from £400 to £800 p.a.) Mayor, the senior mathematics master, who kept a boarding-house, received £1412 p.a., while Temple, the Head, in addition to being provided with a 'handsome residence', received a salary of almost £3000 p.a. The lowest assistant master's salary was under £300 p.a. (modern languages).
15. Quoted in Adamson, J. W. *English Education, 1789–1902*, Cambridge University Press, 1930, p. 176.
16. However, in 1853 the tradition that members of King's College (Eton's sister college) were exempted from taking the Tripos examinations was ended.

17. See Simon, B. *Studies in the History of Education, 1780–1870*, Lawrence and Wishart, 1960, p. 299.
18. The unattributed quotations in this section are taken from chapter 4 of Wilson, *op. cit.* (n. 1). This paragraph underlines a recurring problem in mathematics education: the energetic schoolteacher working in privileged conditions is often able to turn students into mathematical scholarship winners but not into mathematicians. In this case, neither Cope nor Darnell distinguished himself at university.
19. See, for example Montgomery, R. J. *Examinations*, Longman, 1965.
20. The Earl of Clarendon quoted in Simon, *op. cit.* (n. 17), p. 304. In 1861, two old boys (one E. A. Abbott) of the City of London School had between them been Senior Wrangler, First Smith's Prizeman, Senior Classic and First Chancellor's Medallist – a clean sweep never achieved before by one school.
21. This, too, was still the view of many mathematicians, for example, G. B. Airy (Plumian Professor at Cambridge) who wanted freshmen to have a good grounding in algebra (as far as quadratic equations), arithmetic, plane trigonometry and one or two books of Euclid: 'the mind is not prepared [for anything more advanced] at that early age'. Classics should still form the basis of an education.
22. Wilson, *op. cit.* (n. 1), p. 66.
23. An exception was Italy where, on the recommendation of a special government commission to inquire into geometry teaching in schools, it was decreed in 1867 that all classical schools would adopt Euclid '*pure* and *simple*'. See Hirst, T. A. 'Chairman's Address', *AIGT: First Annual Report*, 1871, pp. 8–13.
24. Whewell, W. *Of a Liberal Education*, 1845, p. 65. Whewell's admiration of Euclid was such that he published *The Mechanical Euclid* (1837), a textbook of mechanics 'demonstrated after the manner of the Elements of Geometry'. On the degree of acceptance of Euclid, it is worth noting that the Yorkshire schools described in the Taunton Report as teaching geometry used Euclid without exception. See also note 9 on p. 251.
25. An extract from Williamson's (1781) denunciation of Clairaut's *Eléments* (1741) can be found in A. G. Howson's 'Foreword' to Maxwell, E. A. *Geometry by Transformations*, Cambridge University Press, 1975.
26. Wallis, P. J. *A Checklist of British Euclids up to 1850*, Newcastle University, 1967. The data given here have been updated by Peter Wallis.
27. Potts, R. *Euclid's Elements of Geometry*, Cambridge University Press, 1845.
28. Wilson, J. M. *Elementary Geometry*, Macmillan, 1868. Other editions appeared in 1869, 1873 and 1878.
29. It has been suggested that similar objections could be levelled against most present-day university analysis courses.
30. Other 'alternative' geometries published at that time included Wormell, R. *An Elementary Course of Plane Geometry*, 1868 and Wright, R. P. *The Elements of Plane Geometry*, 1868.
31. Extracts from De Morgan's review of Wilson's *Geometry* (*Athenaeum*, 18 July 1868) are reprinted in Carroll, L. *Euclid and his Modern Rivals*, Dover, 1973 (a reprint of the 1885 second edition). It should be stressed, however, that the faults which De Morgan found in Euclid were not those to be later uncovered by Pasch and others.
32. 'Elementary geometry' in Todhunter, I. *The Conflict of Studies*, Macmillan, 1873. Again, extracts from this essay are reprinted as an appendix to Carroll, *op. cit.* (n. 31).
33. T. A. Hirst contributed an anonymous review of Carroll's book to *Nature*, 20 (1869), 240–1.
34. The address was reprinted in the *Educational Times*, **21** No. 90 (Sept. 1868), 125–8.
35. Wilson spoke in January 1870 in response to an invitation offered by the Lord Provost, selected Members of Parliament and professors from the Scottish universities.
36. T. A. Hirst (1830–92), attended the West Riding Proprietary School in Wakefield and then worked as a surveyor in Halifax before studying mathematics in Germany and France. On his return to Britain he taught at the Quaker school, Queenswood (Hampshire), where a fellow teacher was Richard Wright (note 30), and at University

College School (London). After leaving University College, London, he became Director of Studies at the Royal Naval College, Greenwich.

37. The early history of the AIGT is described in Brock, W. H. 'Geometry and the universities: Euclid and his modern rivals 1860–1901', *History of Education*, **4** (2) (1975), 21–35.
38. Wilson, J. M. 'Early history of the Mathematical Association', *Math. Gazette*, **10** (1921), 239–44.
39. *British Association Reports*, **42** (1873), 459–60.
40. *AIGT Report*, **2** (1872), p. 21.
41. *AIGT Report*, **5** (1875), p. 12.
42. *AIGT Report*, **8** (1878), pp. 18–19.
43. *British Association Reports*, **45** (1876), pp. 8–13.
44. See, for example, his addresses reprinted in the first two *AIGT Reports* (1871–2).
45. See Brock, *op. cit.* (n. 37).
46. See, for example, Bamford, T. W. *The Rise of the Public Schools*, Thomas Nelson, 1967, chapter 5; Bishop, G. D. *Physics Teaching in England from Early Times up to 1850*, RPM Publishers, 1961.
47. John Tyndall (1820–93) was educated at a National School in his native Ireland and at Preston Mechanics' Institute before taking up a post as a mathematics teacher at Queenswood College (see note 36). He later obtained a Ph.D. (under Bunsen) at Marburg, before returning to Britain to become established as a scientist and a populariser of science. He died when his wife inadvertently gave him chloral instead of milk of magnesia.
48. Farrar, F. W. (ed.) *Essays on a Liberal Education*, 1867. An account of Wilson's science teaching at Rugby is to be found in Gosden, P. H. J. H. *How They were Taught*, Basil Blackwell, 1969, pp. 119–22.
49. See Bamford, *op. cit.* (n. 46), pp. 202–3.
50. For example, when at Rugby, Percival decreed that the boys should lengthen their football shorts and tie them below their knees with elastic, lest the sight of bare flesh should prove an incentive to vice.
51. John Perry (1850–1920) left school at fourteen to become an apprentice in a Belfast foundry. When eighteen he began to study at Queen's College, Belfast. In 1871 he was appointed to the staff at Clifton College. At that time public-school heads had often to choose between appointing non-scientists (but gentlemen) to teach science or to appoint men 'of different social standing from [their] regular staff' (Turner, D. M. *History of Science Teaching in Britain*, Chapman and Hall, 1927, p. 94) who often then had 'an uncomfortable time in the common room'. Clifton (see also note 52) appointed science specialists.
52. In addition to Perry other future FRSs included J. G. MacGregor (1877–9), W. A. Shenstone (1880–1908), (Sir) W. A. Tilden (1872–80) and A. M. Worthington (1880–5). Tilden and Shenstone were non-university men; MacGregor and Worthington (a pupil of Helmholtz), like Tyndall, did postgraduate work at German universities.
53. *Journal of Education*, 1 Jan. 1887, pp. 11–12 (Problem: 'No solution may exceed one page' – which rules out Haken and Appel's proof!), 1 June 1889 (Temple's 'proof').
54. A sixth-former whose betting debts had just been revealed, attempted to kill Wilson in November 1882. (Wilson (1932), *op. cit.* (n. 1), pp. 117–21).
55. Bamford, *op. cit.* (n. 46), p. 150. Public-school headmasters were frequently found a bishopric (e.g. Percival), and this, in the cases of Longley (Harrow), Tait (Rugby), Temple (Rugby) and Benson (Wellington), was the first step to the Archbishopric of Canterbury. (Temple's son, William, was also to become a public-school head (Repton) and then Archbishop.) Failing a bishopric, one could hope for a deanery (e.g. Farrar (Marlborough) and Goulbourn (Rugby)).
56. Some of his young parishioners told Wilson of this star pupil at Manchester Grammar School and asked him to supply them with problems with which to catch Whittaker out. The first, that the sums of every third coefficient in the binomial expansion, whether one started with the first, second or third term, were either equal or differed by one, was

solved by Whittaker. The second, to show that every prime of the form $4n + 1$ was the sum of two squares did the trick. Wilson was later to write an article for the *Manchester Guardian* about Whittaker's solution of Laplace's equation (Wilson (1921), *op. cit.* (n. 38)).

8. Charles Godfrey

1. Hutton, T. W. *King Edward's School, Birmingham, 1552–1952*, Basil Blackwell, 1952, p. 104.
2. He was Eleventh Wrangler in 1865 when Lord Rayleigh was Senior Wrangler.
3. Anthony Thacker (see Robson, A. 'How they learnt, 1600–1850'. *Math. Gazette*, **33** (1949), 81–93). The study of mathematics was also encouraged at King Edward's during the headship of Prince Lee (later first Bishop of Manchester, cf. my p. 265); see Hutton, *op. cit.* (n. 1), p. 85.
4. Mayo, C. H. P. 'Rawdon Levett', *Math. Gazette*, **11** (1923), 325–8.
5. Godfrey, C. 'Rawdon Levett', *Math. Gazette*, **11** (1923), 328–9 (reprinted **55** (1971), 129–30).
6. Godfrey, *op. cit.* (n. 5). In a similar vein, W. F. Bushell ('A century of school mathematics', *Math. Gazette*, **31** (1947), 69–89) tells how in his last term at Charterhouse before going up to Cambridge his mathematics teacher, C. O. Tuckey, 'produced a ruler with strange markings on it' – a protractor. Siddons later told D. A. Quadling that he (Siddons) had introduced Tuckey to a protractor the previous week!
7. Siddons, A. W. 'Progress', *Math. Gazette*, **20** (1936), 7–26, (pp. 14–5).
8. Hardy, G. H. *A Mathematician's Apology* (with foreword by C. P. Snow), Cambridge University Press, 1967, p. 19.
9. Forsyth, A. R. 'Old Tripos days at Cambridge', *Math. Gazette*, **19** (1935), 162–79, (p. 167). (Pearson, K. 'Old Tripos days at Cambridge as seen from another viewpoint', *Math. Gazette*, **20** (1936), 27–36, presents a different view of Cambridge mathematics.)
10. A. R. Forsyth, (1858–1942) was Senior Wrangler in 1881. The following year he was appointed the first professor of mathematics at Liverpool University College. After four terms in the post he returned to Trinity College, Cambridge. In 1893 he published his *Theory of Functions* which E. T. Whittaker described as having a greater effect on British mathematics than any other book since Newton's *Principia*. It pointed British mathematics in a new direction. Unfortunately, having pointed out the way Forsyth was soon outdistanced by his competitors and, though appointed Sadleirian Professor in 1895, he gradually sank into mathematical oblivion. He resigned his chair in 1910 and took up a post in India. In 1913 he returned to England as professor of mathematics at Imperial College.
11. Forsyth, *op. cit.* (n. 9), p. 170.
12. Hardy, *op. cit.* (n. 8), p. 23.
13. Hardy, *op. cit.* (n. 8), p. 23.
14. *The Times* 27 April 1906. See also Hassé, H. R. 'My fifty years of mathematics', *Math. Gazette*, **35** (1951), 153–64.
15. *The Times* 20 Oct. 1906.
16. A letter from the Cambridge professors of mathematics, *The Times* 22 Oct. 1906.
17. Hassé (*op. cit.* (n. 14)) tells of the in-fighting including special trains arranged to take voters from London to Cambridge (and one which would allow them to dine in their old colleges before making a leisurely return).
18. Hardy, G. H. 'The case against the Mathematical Tripos', *Math. Gazette*, **13** (1926), 61–71. (reprinted **32** (1948), 134–45.)
19. Among those who attended a 'new' university before going up to Cambridge were Todhunter and Hayward (London) and J. J. Thompson and Eddington (Manchester). The custom persisted until after the Second World War; for example, eight or nine Presidents of the Mathematical Association had graduated elsewhere before taking the Cambridge Tripos course.
20. *The Times* 20 Oct. 1906.
21. Newman, M. H. A. 'G. Hardy', *Math. Gazette*, **32** (1948), 51.

22. Governmental minute of 1859 (see Argles, M. *South Kensington to Robbins*, Longman, 1964, p. 20.)
23. In later years the 'payments by results' scheme was abandoned in favour of a system of inspection. (Argles, *op. cit.* (n. 22), p. 22.)
24. The lecturers included scientists such as T. H. Huxley. (See, for example, Adamson, J. W. *English Education 1789–1902*, Cambridge University Press, 1930, pp. 394–5.)
25. In 1897, 34.1% of pupils in higher-grade schools were children of skilled or unskilled manual workers compared with 6.8% in the grammar schools. (Banks, O. *Parity and Prestige in English Secondary Education*, Routledge and Kegan Paul, 1955, p. 29.) The higher-grade schools saw themselves as serving a new and important purpose: 'As a class we are as much preparatory schools for the university colleges e.g. Birmingham, Leeds, Liverpool, Manchester, etc. especially their technical and scientific sides, as the grammar schools are for the older universities' (President of the Association of Headmasters of Higher Grade Schools, 1895; quoted in Banks, *op. cit.*, p. 24).
26. Scott, R. P. (ed.) *What is Secondary Education?*, Incorporated Assn. of Headmasters, 1899, quoted in Banks, *op. cit.* (n. 25), p. 17.
27. Turner, D. M. *History of Science Teaching in England*, Chapman and Hall, 1927, p. 115.
28. Association of Assistant Masters, 1900; quoted in Banks, *op. cit.* (n. 25), p. 32.
29. See Brock, W. H. and Price, M. H. 'Squared paper in the nineteenth century: instrument of science and engineering and symbol of reform in mathematical education', *Educ. Studies Math.*, **11** (1980), 365–81.
30. Statement of 14 Dec. 1887 quoted in Turner, H. H. 'John Perry', *Proc. Roy. Soc.*, **111** A (1926), i–viii.
31. Perry, J. 'The teaching of mathematics', *Nature*, 2 August 1900, 317–20.
32. Hardy (1967), *op. cit.* (n. 8), p. 17. It is interesting to reflect what 'best mathematical school' meant. Winchester, for example, has had some distinguished mathematics masters, C. V. Durell, Bryan Thwaites, T. A. Jones and J. H. Durran who have contributed greatly to curriculum development, but has not produced a stream of outstanding mathematicians (although Freeman Dyson and Sir James Lighthill were both pupils there) to match those from, say, St Paul's.
33. Extracts from the headmaster's annual confidential reports on Godfrey are to be found in Howson, A. G. 'Charles Godfrey and the reform of mathematical education', *Educ. Studies Math.*, **5** (1973), 157–80.
34. See Godfrey, C. 'The teaching of mathematics – a compromise', *Math. Gazette*, **2** (1901), pp. 106–8.
35. Siddons, A. W. 'A short memoir' in *Godfrey and Siddons*, Cambridge University Press, 1952.
36. Michael Price has drawn my attention to the significance of Forsyth's approaching Godfrey rather than the Mathematical Association, at that time in the doldrums.
37. It had repercussions in the USA, on the Continent where the Germans coined the term Perryismus, and in Japan where Perry had spent the years 1875–82. See, also, Wolff, G. *Der Mathematische Unterricht der Höheren Knabenschulen Englands*, Teubner, Leipzig, 1915, pp. 67–84; Young, J. W. A. *The Teaching of Mathematics in the Elementary and the Secondary School*, 2nd edn, Longman, 1920, pp. 87–121; and *A History of Mathematical Education in the United States and Canada*, NCTM 32nd Yearbook, 1970, pp. 388–92.
38. The talk and the discussion which followed Perry's address are reprinted in Perry, J. *The Teaching of Mathematics*, Macmillan, 1902, as are written comments by a number of mathematicians, scientists, engineers and educators. (The criticisms quoted in the text are taken from this report.) A particularly interesting contribution came from the American, D. E. Smith, who wrote (pp. 89–91): 'As one who has seen . . . [mathematics teaching] in England, France, Germany and America, I should say that the subject is more poorly taught in England than in any of the countries named, and that it is best taught in Germany . . . England seems chained by her examination system'. Smith felt that Perry's tone 'grated' and that 'he went to an extreme', but 'in fighting conservatism one must sometimes be ultra-radical'.
39. Siddons, A. W. 'From a public school point of view', *Math. Gazette*, **2** (1901), 108–11. The

way in which Cambridge's shadow dominated the thoughts of possible reformers is also illustrated by Horace Lamb in his contribution to Perry (1902), (*op. cit.* (n. 38)): 'A good textbook to replace Euclid is much wanted; but it should be issued with some authority, and Cambridge is practically the only place that could confer this'.

40. Langley, E. M. 'The teaching of mathematics', *Math. Gazette*, **2** (1901), 105–6 (reprinted **55** (1971), 125–7).
41. Siddons, A. W. 'The first twenty years of the Teaching Committee', *Math. Gazette*, **36** (1952), 153–7.
42. *Math. Gazette*, **2** (1902), 167–8 (reprinted **55** (1971), 127–9). It was 'not proposed to interfere with the logical order of Euclid's series of theorems'.
43. British Association, *Report of the Seventy-Second Meeting* (Belfast, 1902), reprinted *Math. Gazette*, **2** (1902), 197–200.
44. Accounts of how Forsyth guided the proposals through the Cambridge Senate can be found in Siddons (1952), *op. cit.* (n. 35), and Howson, *op. cit.* (n. 33). Oxford and other bodies had changed their requirements before Cambridge, but it was, of course, the Cambridge decision which was crucial for the leading schools which wished to send pupils there.
45. The book remained in print until 1973 and sold over a million copies.
46. *Math. Gazette*, **2** (1903), 369, reprinted **55** (1971), 239–40.
47. An account of the Cockerton case and the events leading up to it can be found in Eaglesham, E. *From School Board to Local Authority*, Routledge and Kegan Paul, 1956.
48. The debate and movement leading to the framing of the 1904 Regulations are described in Banks, *op. cit.* (n. 25), chapter 3.
49. Armstrong, H. E. 'Prof. John Perry, FRS', *Nature*, **105** (1920), 751–2.
50. Perry was to complain in 1908 ('The correlation of the teaching of mathematics and science', *Math. Gazette*, **5** (1909), 1–40) that the MA syllabuses were too orthodox, that the proposals made by the BA had been 'little' comprehended and that 'some teachers think that squared paper [see note 29] was invented merely to illustrate the solutions of certain simultaneous and quadratic equations'. The demise of the 'higher-grade' school spirit was to be lamented in the Government report on *Secondary Education with Special Reference to Grammar Schools and Technical High Schools* (Spens Report), HMSO, 1938.
51. An attempt to unravel and describe the complex happenings in this period is to be found in Price, M. H. 'The reform of English mathematical education in the late nineteenth and early twentieth centuries', unpublished Ph.D. thesis, Leicester University, 1981, and the serious student should consult this work. Perry's gradual fall from power and the way in which he became discredited within the mathematical community is detailed by Price. Eventually Perry was forcibly retired from his chair in London (see *Nature*, 12 Sept. 1912); Sir William White being convinced (*Proc. 5th ICM (1912)*, Cambridge University Press, 1913, vol. 1, 145–61) that engineers should be taught mathematics by professional mathematicians because 'this method must lead to a broader view of science and a greater capacity for original and independent investigation'. The ultimate irony was when Forsyth replaced Perry at Imperial College. This institution then had engineers taught by mathematicians, whilst at Cambridge the system that Perry had advocated, i.e. the establishment of a separate mathematics group within the engineering department, was adopted.
52. Godfrey, C. 'Geometry teaching: the next step', *Math. Gazette*, **10** (1920), 20–4.
53. Wolff, *op. cit.* (n. 37), pp. 98–9.
54. Fletcher, W. C. *Elementary Geometry*, Edward Arnold, 1902.
55. Siddons (1936), *op. cit.* (n. 7), p. 21.
56. Presidential Address to Section A of the British Association, September 1910. Reprinted in *Nature*, 10 Sept. 1910, 284–91.
57. The notion of the 'three stages' of teaching geometry was a very powerful one and, despite objections (see p. 166), it was gradually refined to yield the Stages A, B and C of the Mathematical Association *Report on the Teaching of Geometry* (Bell, 1923). An account

of how these ideas developed is to be found in Howson, A. G. 'Milestone or millstone', *Math. Gazette*, 57 (1973), 258–66.

58. C. G. Allum in the Board of Education's *Special Reports* series (1900, pp. 249–56) suggested that the majority of boys should be able to cover the first three books of Euclid by the age of thirteen, while the more mathematically inclined would add the fourth and sixth books without much trouble!
59. Contribution to the discussion 'The correlation of the teaching of mathematics and science' (see note 50).
60. See Board of Education, *Special Reports on Educational Subjects*, HMSO, 1912, vol. 1, 393–428 which contains accounts of 'Practical mathematics' at several leading public schools. Such work was stimulated by revised Army requirements.
61. Mayo, C. H. P. *Reminiscences of a Harrow Master*, Rivingtons, 1928, chapter 7: 'I had little sympathy with these so-called reforms to the extent to which we were forced to carry them; they seemed to me not only useless in themselves, but to be contrary to the purpose for which mathematics existed as part of the general curriculum . . . elementary and pottering experimentation gave no mental grip, and their value was almost negligible'. Reactions to other curricular innovations, in particular graphical work and geometrical drawing, are described in Price, *op. cit.* (n. 51), chapter 5.
62. The curriculum devised at the Royal Naval Colleges at Osborne and Dartmouth is described by J. W. Mercer in the Board of Education *Reports*, *op. cit.* (n. 60), vol. 1, 183–223. A slightly edited version is reprinted in Griffiths, H. B. and Howson, A. G. *Mathematics: Society and Curricula*, Cambridge University Press, 1974, pp. 159–78.
63. Godfrey, C. and Siddons, A. W. *Elementary Algebra*, Cambridge University Press (two parts, 1912, 1913).
64. Godfrey, C. and Price, E. A. *Arithmetic*, Cambridge University Press, 1915.
65. Fletcher, W. C. 'The position of mathematics in secondary schools in England' in Board of Education *Reports, op. cit.* (n. 60), vol. 1, 90–103.
66. Barnard, S. 'The teaching of algebra in schools' in Board of Education *Reports, op. cit.* (n. 60), vol. 1, 312–37.
67. The sources of these quotations are given in Howson, A. G. 'Curriculum development and curriculum research', in Bell, A. W., Bishop, A. and Howson, A. G. *A Review of Research in Mathematical Education*, Nelson/NFER (to appear). The last, on in-service training, is by Mrs (Lady) Napier Shaw and comes from Perry (1902), *op. cit.* (n. 38), p. 52. A brief description of 'in-service' facilities in the opening decades of the century is contained in Price, *op. cit.* (n. 51), chapter 2.
68. Broadbent, T. A. A. 'The Mathematical Gazette: our history and our aims', *Math. Gazette*, 30 (1946), 186–94. The *Gazette* was first issued in April 1894 with what in present-day usage strikes us as an odd subtitle, 'A terminal journal for students and teachers'.
69. Nunn, T. P. *The Teaching of Algebra*, 2nd edn, Longman, 1919, p. vii.
70. B. Branford (186(8)–1944) graduated at Edinburgh University and lectured in mathematics at Leeds University before, in 1901, becoming Director of Higher Education and Principal of the Technical College in Sunderland (where he initiated 'sandwich' courses for apprentices). From 1905 to 1929 he was Divisional Inspector for the London County Council. His early interests in mathematics education soon widened and he was later to publish on many subjects.
71. G. St L. Carson (1873–1934) gained a first in Mathematics at Liverpool before entering Trinity College, Cambridge, in 1893. He was Third Wrangler in 1896 and then embarked on a most varied career. From 1898 to 1901 he lectured at University College, Sheffield, before moving into industry. In 1905 he was appointed Head of the Mathematics Department at Battersea Polytechnic which he left in 1908 to join Tonbridge School. He was Lecturer in Education and Reader in Mathematics at Liverpool University (1914–17) and then an HMI.
72. *Journal of the Association of Teachers of Mathematics*, 1 (1911).
73. The women were, of course, all unmarried; it was to be many years before married women were permitted to teach (cf. my p. 183).

74. Thus, Barnard, *op. cit.* (n. 66), wrote: 'Mr Godfrey refers contemptuously to those who regard mental discipline as the chief aim of mathematical education. Yet it is difficult to assign any other reason for teaching mathematics to the ordinary boy'.
75. (Sir) John Adams' (cf. my p. 181) *The Herbartian Psychology* (1897) had proved extremely influential in discrediting faculty psychology and drawing attention to Herbart's work.
76. Carson (see Nunn, *op. cit.* (n. 69), p. 162) had suggested denoting 'negative five' by $\bar{5}$, but then could not suggest an easily printed symbol for 'positive five'.
77. Dieudonné, J. *New thinking in School Mathematics*, Organization for European Economic Cooperation (OEEC), 1961, 31–46; reprinted in Howson, A. G., Keitel, C. and Kilpatrick, J. *Curriculum Development in Mathematics*, Cambridge University Press, 1981, pp. 101–7. According to the publisher, Dobbs' book 'failed abominably', whilst Nunn's sold 'only in trifling numbers'. Boon's book proved really successful only when reissued in 1960. (See Price, *op. cit.* (n. 51), p. 269.)
78. See, for example, Godfrey (1920), *op. cit.* (n. 52).
79. See, for example, Roach, J. *Public Examinations in England, 1850–1900*, Cambridge University Press, 1971, pp. 60–1.
80. Sandford, E. G. (ed.) *Memoirs of Archbishop Temple*, Macmillan, 1906, vol. 2, p. 541.
81. Roach, *op. cit.* (n. 79), chapter 4.
82. It is of interest that the Boards for a time extended their activities to examining schools as well as pupils (see Roach, *op. cit.* (n. 79)).
83. See Hawkins, C. 'Examinations from the school point of view' in Board of Education *Reports, op. cit.* (n. 60), vol. 1, 439–542. This account includes a valuable selection of examination papers.
84. Macaulay, F. S. 'Examinations for mathematical scholarships' in Board of Education *Reports, op. cit.* (n. 60) vol. 2, 210–62, p. 211. The new universities lacked endowments to provide scholarships on a sufficient scale to exert any marked influence on the school curriculum.
85. The pros and cons of external examinations were advanced in the 1911 report *Examinations in Secondary Schools* (HMSO). These are reprinted in Tibble, J. W. 'The educational effects of examinations in England and Wales' in *World Year Book of Education*, Evans Bros., 1969 and in *Secondary School Examinations other than the GCE* (Beloe Report), HMSO, 1960.
86. For example, Hartog, P. J. *Examinations and their Relation to Culture and Efficiency*, Constable, 1918.
87. This he saw as an essential task 'until the time comes when a teacher is trusted to examine his own class' (Godfrey, C. and Siddons, A. W. *The Teaching of Elementary Mathematics*, Cambridge University Press, 1931, p. 36).
88. *School Certificate Mathematics. The Report of a Conference of Representatives of the Examining Bodies and Teachers' Associations,* Cambridge University Press, 1944.
89. See Montgomery, R. J. *Examinations*, Longman, 1965, p. 180.
90. Much has been written of the teacher who uses his autonomy to innovate; rather less about the 'autonomous conservator'.
91. Moore, E. H. 'On the foundations of mathematics', *Bull. Amer. Math. Soc.*, 1903, 424f, reprinted in NTCM *The First Yearbook,* (1926), 32–57.
92. Halsted, G. B. *Rational Geometry*, Wiley, 1904.
93. Stamper, A. W. *A History of the Teaching of Elementary Geometry*, Teachers College Press, Columbia University, 1909.
94. Méray, C. *Nouveaux éléments de géométrie*, Paris, 1874.
95. *Proc. 4th ICM (1908)*, Lincei, Rome, 1909, vol. 3.
96. Fehr, H. 'La Commission Internationale de l'Enseignement Mathématique de 1908 à 1912' in *Proc. 5th ICM (1912)*, *op. cit.* (n. 51), vol. 2.
97. Godfrey, C. 'On the work of the International Commission on Mathematics Teaching', *Math. Gazette*, **6** (1912), 243.
98. This is probably still the case and presumably reflects the difference between a theoretical approach to education requiring the establishment of a technical language and a pragmatic approach which gets by mainly on folklore and the recounting of anecdotes.

99. Thus, for example, the German and French reports each ran to five volumes and that of the USA to eleven. (A list of publications can be found in *Proc. 5th ICM (1912), op. cit.* (n. 51), pp. 642–53.) The British contribution was the Board of Education *Reports*, 2 vols., *op. cit.* (n. 60). The splendid survey by Wolff of British education (note 37) was part of the German contribution.
100. Godfrey, C. 'Methods of intuition and experiment in secondary schools' in *Proc. 5th ICM (1912), op. cit.* (n. 51), vol. 2, pp. 633–41.
101. The questionnaire was sent to 112 independent and 439 state schools and replies were received from 74 of the former and 295 of the latter. It is probable that the schools replying would be the more forward-looking and adventurous ones.
102. The teaching of simple statistics had been advocated by, among others, Whitehead, and indeed at Tonbridge School special classes on statistics were being given to senior boys in the arts streams who had dropped the study of mathematics.
103. By 'average', one here means of 'average ability among the 25% most able pupils'. Among those who opposed the early introduction of the calculus was G. Chrystal who wrote (*Algebra*, A & C Black, 1889) 'A practice has sprung up of late (encouraged by demands for premature knowledge in certain examinations) of hurrying young students into the manipulation of the machinery of the Differential and Integral Calculus before they have grasped the preliminary notions of a *Limit* and of an *Infinite Series* . . . Besides being to a large extent an educational sham, this course is a sin against the spirit of mathematical progress'. This is not, of course, an outlook of which Perry would have approved. The place of calculus in the curriculum in the early 1900s is described in Price, *op. cit.* (n. 51), chapter 7.
104. One notes that CIEM tended to place more emphasis on higher education than does ICMI nowadays, but, like ICMI, tended to avoid discussing how mathematics should be taught at university to would-be mathematicians. Presumably it was thought that no improvements could be, or were likely to be, made!
105. Godfrey, C. 'The teaching of calculus', *Math. Gazette*, 7 (1913), 233–40. The O and C Board had introduced some calculus into their additional mathematics syllabus from 1910.
106. It is interesting to contrast the 'analysis' revolution in the universities and its effects on schools with the 'algebraic structure' movement in the universities in the 1950s.
107. Godfrey (1913), *op. cit.* (n. 105). The last sentence is frequently quoted in England as a guiding principle of mathematics education, although I have never seen Godfrey – or, indeed, any predecessor of his – given the credit for coining it.
108. Greenhill, Sir George, 'The use of mathematics', *Math. Gazette*, 7 (1914), 110. The 1916 ICM was cancelled because of the First World War. After the war a new series of congresses was begun, the first being in Strasbourg in 1920. Little time, however, was devoted there to pedagogical matters and the CIEM did not meet again until the 1928 ICM at Bologna.
109. T. A. A. Broadbent, *op. cit.* (n. 68), was to refer to the *Gazettes* of 1918–30 as being 'less bright'.
110. Godfrey, C. and Siddons, A. W. *Practical Geometry*, Cambridge University Press, 1920, p. vi.
111. Godfrey (1920), *op. cit.* (n. 52).
112. *The Teaching of Geometry in Schools*, Bell (for the Mathematical Association), London, 1923.
113. The report and its effects are described more fully in Howson (1973), *op. cit.* (n. 57).
114. Siddons, A. W. 'Charles Godfrey, MVO', *Math. Gazette*, **12** (1924), 137–8. Godfrey was appointed a Member of the Royal Victorian Order (MVO) in 1910 for his part in educating the future king.
115. See, for example, the criticisms contained in *Math. Gazette*, **12** (1924). Those of H. F. Baker are reprinted in *Math. Gazette*, **55** (1971), 146–52.
116. The fourth edition appeared in 1944 (the only textual change from the second edition being the addition of a footnote) and had been reprinted seven times by 1963.

9. Elizabeth Williams

1. Elizabeth's parents had only elementary education though her mother had taught for a time in the village school. However, it is indicative of the widening educational opportunities that three of her close male relatives were to enjoy a university education: one was a classical scholar at Magdalen College, Oxford, and two were at Cambridge (one later becoming an FRS and a professor of nuclear physics).
2. Children could, however, be partly or wholly excused from attendance from the age of eleven, provided they obtained a certificate stating they had received a certain standard of education. In 1914 there remained over 70,000 half-timers in the country. Not until the 1918 Education Act was full-time attendance required of all children until they had attained the age of fourteen. (In 1900 local authorities were allowed to raise the school leaving age to fourteen.)
3. Elementary education in all state-funded schools became free only after 1918. In 1902 there were still over 600,000 elementary-school children paying fees.
4. Many 'private schools' survived in the suburbs and small towns as socially superior alternatives to the 'free-for-all' board schools.
5. The London School Board had recommended the pattern 5–7 (co-educational infants' school), 7–10(11) (mixed or single-sex junior school), 11–13(14) (single-sex senior school). However, in many areas the junior and senior schools were amalgamated as at Shaftesbury Road.
6. The Home and Colonial Infant School Society (established 1836) had propagated Pestalozzian ideas from its foundation. The playground of its model school (see Lawson, J. and Silver, H. *A Social History of Education in England*, Methuen, 1973, p. 281) with its climbing frames, ropes, parallel bars, etc., demonstrates the society's forward-looking nature. Froebel's ideas first reached England in the 1850s and were taken up by the Home and Colonial and the British Societies in the next two decades. Yet the latter society had not always devoted much attention to infant education, having at one time ruled that entrants to its schools should be aged at least six or, should there be severe pressure on accommodation, eight.
7. The conditions were alleviated in 1890 and removed entirely in 1898.
8. *Instruction of Infants*, quoted in Birchenough, C. *A History of Elementary Education*, 2nd edn, University Tutorial Press, 1925, pp. 375–6.
9. MacCarthy, E. F. M. *Thirty Years of Educational Work in Birmingham*, Birmingham, 1900, p. 26.
10. Unattributed quotations in this chapter are taken from private correspondence with Mrs Williams.
11. Brock, W. H. and Price, M. H. ('Squared paper in the nineteenth century . . .', *Educ. Studies Math.*, **11** (1980), 365–81) make no reference to activities at this level.
12. Mary Everest Boole (1832–1916), the wife of the mathematician George Boole, taught at Queen's College (my p. 172) after her husband's death. From 1883 until her death she produced a series of books on education. (A selection of writings on mathematics education is contained in Tahta, D. G. *A Boolean Anthology*, Association of Teachers of Mathematics, 1972.) She is especially (but unrepresentatively) remembered for her work on curve-stitching, described by her friend Edith Somervell in the books *A Rhythmic Approach to Mathematics* and *Boole Curve-sewing Cards* (both G. Philip & Sons, 1906).
13. According to P. Ballard ('The teaching of elementary mathematics in London public elementary schools' in Board of Education, *Special Reports on Educational Subjects*, HMSO, 1912, vol. 1, pp. 3–30) a random survey of seventy London schools in 1910 found that thirty taught geometry as a separate subject and sixteen algebra. The latter was disappearing as a separate subject as attempts (not altogether successful) were made to graft it on to arithmetic rather than, as previously, to treat it as 'the unintelligent manipulation of symbols'. Geometry meant mensuration and practical geometry: 'Theoretical geometry is only taken in the Central Schools.'
14. 'The only unanimity of practice which the Board . . . desires to see in Public Elementary

Schools is that each teacher shall think for himself, and work out for himself such methods of teaching as may use his powers to the best advantage and be best suited to the particular needs and conditions of the school. Uniformity in details of practice . . . is not desirable even if it were attainable', Board of Education *Handbook of Suggestions for Teachers*, HMSO, 1905.

15. The pupil-teacher system (my p. 261) continued to grow during the late nineteenth century and survived in some authorities until the Second World War. Pupil-teachers centres (akin to secondary schools) were established from 1876 onwards. The pros and cons of the system were set out in the *Final Report* (1888) of the Cross Commission established 'to inquire into the working of the Elementary Education Acts'.
16. Tawney, R. H. (ed.) *Secondary Education for All,* Labour Party/Allen and Unwin, 1922.
17. K. Lindsay (*Social Progress and Educational Waste*, Routledge, 1926) revealed, for instance, that one school in middle-class Lewisham obtained as many scholarships as were awarded in the whole East End borough of Bermondsey.
18. Ballard, *op. cit.* (n. 13), pp. 20–1.
19. The pay, qualifications and social status of those teaching in secondary schools were greater than those of the elementary-school teachers. See, for example, chapter 6, 'The role and status of the teacher' in Wardle, D. *English Popular Education: 1780–1970*, Cambridge University Press, 1970.
20. See, for example, chapter 9, 'Reform of female education' in Archer, R. L. *Secondary Education in the Nineteenth Century*, Cambridge University Press, 1921. Significantly, the terms of reference of the Taunton Commission made no reference to girls' education; it had been forgotten. The Commissioners were persuaded to investigate girls' schools (Miss Beale in her evidence doubted whether girls could proceed as far as boys in higher mathematics) and their report was damning: girls' education was marked by 'want of thoroughness and foundation, want of system; slovenliness and showy superficiality; inattention to rudiments, undue time given to accomplishments' (p. 548). The Commissioners, however, could list only twelve endowed secondary schools for girls and these had on average only fifty-two pupils. (See, for example, Adamson, J. W. *English Education: 1789–1902*, Cambridge University Press, 1930, chapters 10 and 12).
21. In 1871 Miss Buss relinquished her financial interest in the school and with a slight change of name – 'Girls' for 'Ladies' – it became a public institution.
22. Archer, *op. cit.* (n. 20), p. 243.
23. The Endowed Schools Act of 1869 referred to the need to extend educational provision for girls and, as a result, girls' schools also came to be created alongside ancient boys' schools, for example, Haberdashers' Aske's Hatcham Girls' School.
24. Girls were admitted to the Cambridge 'locals' in 1865, to those of Oxford in 1870 and to the examinations of the O and C Board in 1876. Elizabeth Garrett had been refused admission to the London University matriculation examination in 1862 and this was not opened to women until 1869. The early results obtained by the girls in arithmetic were disastrously bad, but improvements soon came.
25. Quoted in Archer, *op. cit.* (n. 20), p. 254.
26. *Introductory Lectures Delivered at Queen's College, London*, 1849. Maurice, F. D. 'Queen's College London: its objects and method', pp. 1–27; Hall, T. G. 'On mathematics', pp. 323–45. The staff of King's College greatly assisted the new Queen's College.
27. Ford, M. M. 'The position of mathematics in the education of girls', *Journal ATM*, 3 (Sept. 1912), 32–40. The same points were (first ?) made in Gwatkin, E. R. 'The value of the study of mathematics in public secondary schools for girls' in Board of Education *Reports*, *op. cit.* (n. 13), vol. 1, 560–74.
28. Quoted in Ford, *op. cit.* (n. 27), p. 35.
29. Burstall, S. 'The place of mathematics in the education of girls and women' in Board of Education *Reports*, *op. cit.* (n. 13), vol. 1, 575–81.
30. 'Girl' was a four-letter word to Perry and he avoided its use throughout his contribution to the 1901 discussion. (His practical mathematics syllabus is described, however, as being suitable for 'boys and girls'.) Headmistresses do get a mention, but only because they would 'wail if we took away from them their golden calf, their external

[examination]' (p. 93). It will also be noticed that Godfrey's letter (my p. 149), was not signed by any school-mistress, and that there were no women on the MA's original Teaching Committee. By 1911 the Committee had eight women members and in 1916 the MA produced a report on *The Teaching of Elementary Mathematics in Girls' Schools.*

31. Thus, for example, the Oxford Higher Local Examination did not include calculus although the corresponding Cambridge and O and C examinations did. However, the 1910 paper did contain the questions 'Prove De Moivre's Theorem. Find all the roots of the equation $x^5 + 1 = 0$' and 'If p_n/q_n denotes the nth convergent of a continued fraction, show that $p_n/q_n - p_{n-1}/q_{n-1} = (-1)^n/q_nq_{n-1}$.

    It is also interesting to note (with reference to recent discussions on examining at 16+) that the Oxford papers contain asterisked simpler questions which were not to be attempted by those who aspired to a first or second class mark.

32. The County Major Scholarship was the second 'rung' in the educational ladder provided by the local education authorities. The County Minor facilitated entry to secondary school, the Major to university.

33. The way in which trigonometry became established in the curriculum is described in Price, M. H. 'The reform of English mathematical education in the late nineteenth and early twentieth centuries', unpublished Ph.D. thesis, Leicester University, 1981, chapter 7.

34. Regrettably, this handicap, reinforced by a variety of prejudices, continues to be manifest. See, for example, Kelly, A. (ed.) *The Missing Half: Girls and Science Education*, Manchester University Press, 1981.

35. A. N. Whitehead (1861–1947) was educated at Sherborne School and Cambridge. Although mainly remembered mathematically for *Principia Mathematica* (1910–13) which he wrote with Russell, he excelled in a variety of mathematical fields before, in 1924, becoming Professor of Philosophy at Harvard.

36. Respectively, *Math. Gazette*, **7** (1913), 87–94; **8** (1916), 191–203; and **9** (1917), 20–33. The second essay is reprinted in *Math. Gazette*, **32** (1948), 110–19. The third essay and Whitehead's talk to the ICM meeting at Cambridge in 1912 are reprinted in his *The Aims of Education*, Williams and Norgate, 1932 (E. Benn Ltd, 1950).

37. Whitehead (1916), *op. cit.* (n. 36), p. 193. The sentence is quoted by Edith Biggs in *Mathematics in Primary Schools*, Curriculum Bulletin No. 1, HMSO, 1965, p. 1.

38. See, for example, Rich, R. W. *The Training of Teachers in England and Wales during the Nineteenth Century*, Cambridge University Press, 1933.

39. The idea had been mooted before. In 1838 the winner of a national competition for the best essay on 'The expediency and means of elevating the profession of the educator in public estimation' had considered appointing a professor of education to each university but rejected this suggestion as unlikely to produce much of practical benefit. The College of Preceptors approached the Privy Council in 1872 seeking the establishment of chairs of education in English universities.

40. The lectures on 'practice' given by Joshua Fitch (*Lectures on Teaching*, Cambridge University Press, 1881) included some interesting observations on the teaching of mathematics.

41. See Rich, *op. cit.* (n. 38). Archer (*op. cit.* (n. 20)) compared the scheme to trying to teach cricket by lecturing on techniques and on the lives of great cricketers.

42. *Royal Commission on Secondary Education* (Bryce), *Minutes of Evidence*, HMSO, 1895, vol. 3, p. 61.

43. *Report of Proceedings of the Training of Teachers Joint Committee*, 1897. (See, for example, Rich, *op. cit.* (n. 38), pp. 271–3.)

44. In addition to running day training colleges, several institutions offered diplomas in education to graduates, for example, London (1883) and Manchester (1895). By 1900 there were twenty-one institutions with secondary-training departments.

45. Sir Percy Nunn (1870–1944) studied at Bristol University College before becoming a schoolteacher, and then joining the staff of the London Day Training College. He became Director of this in 1922 and in 1932 saw it transformed into the London Institute of Education.

46. 'The training of teachers of mathematics' in Board of Education *Reports, op cit.* (n. 13), vol. 2, pp. 291–307. This is an important and interesting paper for it sets out to justify professional training, describes contemporary provisions and presents Nunn's views on the forms it might take. Nunn also considers issues such as concurrent *v.* consecutive training and the mathematical knowledge required of teachers.
47. The Board of Education stipulated that those studying to be elementary-school teachers should have at least eight weeks school practice whilst 'secondary school' students must spend at least sixty school days 'in contact with class work under proper supervision'.
48. Descriptions of the Dalton Plan can be found in Lynch, A. J. *Individual Work and the Dalton Plan*, Philip, 1924; Parkhurst, H. *Education on the Dalton Plan*, Bell, 1922; Kimmins, C. W. and Rennie B. *The Triumph of the Dalton Plan*, Nicholson and Watson, 1932; Yeldham, F. A. 'The Dalton Plan and the teaching of mathematics', *Math. Gazette*, **11** (1922), 45–50 (reprinted **55** (1971), 200–7). An interesting evaluation is given in Spriggs, G. W. 'Problems of individual education with special reference to work in mathematics', *Math. Gazette*, **15** (1930), 38–55.
49. It is interesting to conjecture what beliefs underlay this ban on married women. Mrs Williams sees it largely 'as a safeguard for the unmarried women who had to make a living. (There seemed to be so many maiden aunts in those days . . . [following] the first world war.) It was assumed that the married woman was "kept" by her husband and . . . anything that might detract from her status was "taboo".'
50. An earlier Bill, introduced in 1917, had sought to increase the powers of the Board of Education at the expense of the many local authorities, in the hope of providing more uniform educational opportunities throughout the country. This attempt at greater centralisation was, however, rejected. See Bernbaum, G. *Social Change and the Schools*, Routledge and Kegan Paul, 1967, p. 18.
51. See Bernbaum, *op. cit.* (n. 50), chapter 3. The continuation schools in Rugby owed their existence not so much to an 'enlightened' authority, but to the enthusiasm for the scheme shown by the big employers.
52. Tawney, *op. cit.* (n. 16).
53. Bernbaum, *op. cit.* (n. 50), pp. 15, 92–3.
54. Tawney, *op. cit.* (n. 16), pp. 28, 30.
55. It also recommended that modern-school pupils should have their own leaving examination (a suggestion which eventually bore fruit forty years later in the CSE).
56. *The Education of the Adolescent* (Hadow Report), HMSO, 1926, pp. 214–20.
57. The feasibility of covering the same 'fundamental' work in the first two years of both a selective grammar school and of an unselective modern school does not appear to have been seriously considered. Decreeing that it was possible did, however, serve to overrule some of the objections to selection at 11+.
58. See, for example, Bernbaum, *op. cit.* (n. 50), pp. 44–51.
59. *The Primary School* (Hadow Report), HMSO, 1931, pp. 175–82.
60. According to Mrs Williams: 'one of the skills I learned at our own "little" school was the management of mixed groups with wide ranges of age and ability. So often the practical experience needed as a basis for study . . . could be shared by such a group, [but further] development of ideas and manipulative powers would then depend on such factors as ability and interest. The importance of separating into ability sub-groups is evident to anyone who has witnessed the differences of concentration and consecutive thinking among children.'
61. *Labour's Policy for the School*, Labour Party, 1938.
62. Trades Union Congress, *Annual Report*, 1934, p. 145.
63. Trades Union Congress, *Annual Report*, 1939, p. 189.
64. *Secondary Education with Special Reference to Grammar Schools and Technical High Schools* (Spens Report), HMSO, 1938.
65. *Math. Gazette*, **18** (1934), 112–8.
66. Bryan, G. H. in Board of Education *Reports*, *op. cit.* (n. 13), vol. 2, 308–26.
67. Frobisher, L. J. and Joy, R. R. *Mathematical Education: a Bibliography of Theses and Dissertations*, Mathsed Press, 1978. See also Price, *op. cit.* (n. 33).

68. There was no significant increase in the numbers of postgraduate degrees awarded in mathematics education until the late 1960s when for the first time the English universities gave over ten (in fact, nineteen) such degrees in one year. By 1975 the number had risen to forty-eight (eight being doctorates).
69. Hamilton, D. 'Educational research and the shadows of Francis Galton and Ronald Fisher', in Dockrell, W. B. and Hamilton, D. *Rethinking Educational Research*, Hodder and Stoughton, 1980, pp. 153–68.
70. W. Taylor ((ed.) *Research Perspectives in Education*, Routledge and Kegan Paul, 1973, p. 12) contends that before 1940 'psychologists came close to making claims that their concerns and research interests constituted the sole basis of educational practice'.
71. *Education: its Data and First Principles*, 3rd edn, Edward Arnold, 1945, p. 235.
72. Cyril Burt ('Historical notes on faculty psychology' in the Spens Report, *op. cit.* (n. 64), 429–38, p. 435) also discounts any influence of the Gestalt psychologists on contemporary (1938) educational textbooks.
73. *Language and Thought of the Child*, (1926); *Judgment and Reasoning of the Child*, (1928); *The Child's Conception of the World*, (1929); *The Child's Conception of Physical Causality*, (1930); *The Moral Judgment of the Child*, (1932); (all published by Kegan Paul).
74. Isaacs, S. 'Review of *The Child's Conception of the World*', *Mind*, **38** (1929), 506–13.
75. The paper by H. R. Hamley 'On the cognitive aspects of transfer of training' (Spens Report, *op. cit.* (n. 64), 439–52) is an interesting personal description of aspects of educational research in the 1920s and 30s. The major influences described – works, authors, etc. – are (with the exception of Piaget) those to which Mrs Williams has referred in letters to me. She, Hamley, Burt and Nunn were all much influenced by Spearman (*The Abilities of Man*, Macmillan, 1927) and his attempt to identify specific factors of ability, leading him to such conclusions as 'there appears no real basis for the common opinion which would take arithmetic and geometry to furnish one single ability. Their union as "mathematics" seems rather to be merely one of practical convenience' (*op. cit.*, p. 232). A contemporary description of Spearman's researches so far as they referred to mathematics is contained in Hughes, A. G. 'The psychology of mathematical ability', *Math. Gazette*, **14** (1928), 205–15. Another important field of research lay in the construction of standardised tests in mathematics. For example, P. B. Ballard, ('Norms of performance in the fundamental processes of arithmetic', *J. Exp. Pedagogy*, **2** (1914), 396–405, **3** (1915), 9–20 and *The New Examiner*, University of London Press, 1923) had prepared such tests for use within the London County Council schools and the results obtained served to generate 'an investigation of backwardness in arithmetic among London elementary school children' (title of a 1927 dissertation) – a constantly recurring problem!
76. In 1938 over 14% of male teachers in English elementary schools were graduates. (See, for example, Osborne, G. S. *Scottish and English Schools*, University of Pittsburgh Press, 1966, p. 283.) Most of these would have taught pupils aged 11+. It must also be remembered that within such elementary schools, male teachers were very much in the minority.
77. The scheme was also intended as a means of solving the vast employment problem resulting from demobilisation. A government film, aimed at recruiting teachers, was made at Goldsmiths' 'starring' Elizabeth Williams as the 'tutor'.
78. The effects of this programme have yet to be evaluated. Certainly, it supplied the schools with many efficient teachers who in the war had learnt to exercise authority and responsibility. On the other hand, the weaknesses in the academic training of these and the 'two-years' teachers contributed to the failure in the 1960s to grasp the opportunities offered by the new Certificate of Secondary Education.
79. *Teachers and Youth Leaders* (McNair Report), HMSO, 1944.
80. Significantly, in March 1947 the London County Council announced plans for the establishment of 103 comprehensive schools (although by 1957 it had opened only 11). Other such schools were established in the immediate post-war years in, for example, Westmorland and the West Riding of Yorkshire.
81. Secondary Schools Examinations Council *Curriculum and Examinations in Secondary*

Schools (Norwood Report), HMSO, 1943. This report was very much the work of its chairman, Sir Cyril Norwood. Despite its critics ('Seldom has a more unscientific or more unscholarly attitude disgraced the report of a public committee', S. J. Curtis) it was highly influential in the way that it gave 'theoretical' support for the tripartite system previously formulated by Hadow. '[According to Norwood] the Almighty [had] benevolently created three types of children in just those proportions which would gratify educational administrators' (S.J.C.), and, moreover, which class a child belonged to was clearly to be observed by the time he was eleven. The report had relatively little to say about mathematics, yet its influence on subsequent mathematics education has been considerable. For a critique of Norwood and a review of the subsequent developments within the educational system see, for example, Rubenstein, D. and Simon, B. *The Evolution of the Comprehensive School, 1926–72*, Routledge and Kegan Paul, 1973.

82. University of Cambridge Local Examinations Syndicate, 1944. See also 'Possible changes in the mathematical syllabus for the School Certificate examination', *Math. Gazette*, **28** (1944), 125–43 and 'Syllabuses for examinations taken by sixth-form pupils', *Math. Gazette*, **29** (1945), 161–77.
83. The Jeffery Committee also recommended that the beginnings of calculus should be included in the School Certificate examination (which in 1951 evolved into GCE O-level). See my pp. 234–7.
84. *Social Trends*, HMSO, 1970, p. 124.
85. The Mathematical Association report *Mathematics in the Secondary Technical School*, (Bell, 1949) suggests suitable objectives and syllabuses for such schools.
86. The effects of the war in '[cutting off the modern school] from its mathematical history' is briefly considered in Adams, L. D. 'Full cycle', *Math. Gazette*, **44** (1960), 161–72.
87. I remember when nine being in a class with many twelve-year-olds.
88. Note the new remit resulting from the reorganisation following the 1944 Act. The year 1946 also saw the delayed publication of the 1939 ATCDE report *Arithmetic in Schools* (written by a group chaired by Mrs Williams). This was an influential report which sold well until superseded by the MA report.
89. This, of course, can be seen as illustrating either the dominance of Cambridge and its graduates within mathematics education, or the Association's preoccupation with preparing the next generation of Cambridge scholarship winners and students.
90. From the late 1960s onwards the Association made great attempts to widen its target audience through, for example, its new publications *Mathematics in School* and *Mathematics Round the Country*, and its Diploma in Mathematical Education intended for primary and middle-school teachers.
91. The ATM did, however, share many members, including Elizabeth Williams, with the Mathematical Association. The involved mathematics educator realised that both associations had much to offer even though their target audiences and their views on 'authority' might differ.
92. Our continued failure to help children recognise to which problems their arithmetical techniques are relevant is illustrated in Hart, K. M. (ed.) *Children's Understanding of Mathematics, 11–16*, John Murray, 1981.
93. The report, then, stressed the responsibility of the autonomous teacher to prepare appropriate courses. A similar view of curriculum development was later to be taken by, for example, the Nuffield Mathematics Project and the Mathematics for the Majority Project. It was indeed a 'counsel of perfection' which failed to take into account the pressures on, the abilities of, and the support which was required by and could be given to the average teacher.
94. *Maths Teaching* **11** (Nov. 1959), 37–8. A series of articles on 'Secondary modern school mathematics' appeared in *Maths Teaching* **10–13, 15–19, 21**. The articles, by a variety of authors, present an interesting account of the identified problems and postulated solutions.
95. The Newsom Report (*Half our future*, HMSO, 1963) painted a less rosy picture. It found (p. 240) that in the girls' secondary modern schools sampled, 31% of all forms still learned only arithmetic. The average number of minutes per week devoted to

mathematics was 260 in boys' schools, 215 in co-educational schools and 180 in girls' schools.

96. The difficulties of providing a 'topic-based' course with a satisfactory mathematical and pedagogical structure became increasingly obvious in the following two decades.
97. *New Thinking in School Mathematics*, OEEC (Paris), 1961.
98. *Synopses for Modern Secondary School Mathematics*, OEEC (Paris), 1961.
99. Thus, for example, in the inspectorate's *Teaching Mathematics in Secondary Schools*, HMSO, 1958 we are told the 'girls' schools are more reluctant than boys' schools to adopt systems of grading in sets, and the better girls proceed too slowly' (p. 27). It is also interesting to observe that both this report and that of the Mathematical Association devote a chapter to the history of mathematics and its bearing on teaching – appeals which largely fell on deaf ears.
100. *Secondary School Examinations other than the GCE* (Beloe Report), HMSO, 1960.
101. See also, for example, Thwaites, B. *Mathematics: Divisible or Indivisible*, University of Southampton, 1961, and the leader of *The Times*, 24 May 1961.
102. Thus the number of mathematics graduates from English and Welsh universities increased from 1,754 in 1966 to 2,832 in 1973 (in 1960 there were only 650 *honours* graduates). The percentage entering teaching declined to 20% by 1974 (see *The Training and Professional Life of Teachers of Mathematics*, The Royal Society, 1976, p. 9).
103. In 1975 the shortfall of graduate mathematics teachers was given as 1,582 (Royal Society, *op. cit.* (n. 102), p. 8).
104. Such courses were offered by a number of universities in the 1970s. However, they attracted few students, most preferring to delay opting for school teaching until they were in possession of a more-marketable degree in mathematics.
105. See, for example, The Royal Society (*op. cit.* (n. 102)) and Griffiths, H. B. and Howson, A. G. *Mathematics, Society and Curricula*, Cambridge University Press, 1974, 219–21.
106. See, for example, *Report on In-Service Training for Teachers of Mathematics*, Joint Mathematical Council, 1965 and *Teacher Education and Training* (James Report), HMSO, 1972. The recommendations of the latter were accepted by the government of the day but have still to be implemented.
107. For example, by university schools of education, colleges of education, private bodies (e.g. the School Mathematics Project); the Department of Education and Science and local education authorities (the last in a great variety of forms). The result of this was that when economic circumstances worsened in the late 1970s in-service education proved an obvious victim for cuts.
108. Mrs Williams was the first married woman to be President of the MA (her Presidential Address 'The changing role of mathematics in education' appears in *Math. Gazette*, 50 (1966), 243–55) and, indeed, was probably responsible for other 'married firsts'.
109. For example, even by the early 1960s only 43% of Ugandan children attended primary schools. (See, Coombs, P. H. *The World Educational Crisis*, Oxford University Press, 1968, p. 205).
110. *Mathematics: The First Three Years*, Chambers and Murray, 1970. *Mathematics: The Later Primary Years*, Chambers, Murray and Wiley, 1972. *Mathematics: From Primary to Secondary*, Chambers, Murray and Wiley, 1978. Mrs Williams is also known in many developing countries through her widely-used *Highway Mathematics* textbooks. These first appeared in 1956 and have frequently appeared in revised editions.
111. The third edition of this book appeared in 1982 and included some new material contributed by Mrs Williams when in her mid eighties on how the relationships and operations to be found in ordinary objects and events can be made the basis of mathematics learning.

Postlude

1. The early Director's Reports of the SMP are reprinted in Thwaites, B. *SMP: The First Ten Years*, Cambridge University Press, 1972. Those of the MME were published by Harrap, 1963.

2. Proceedings of all three conferences were published, those of the Southampton meeting appearing as Thwaites, B. (ed.) *On Teaching Mathematics*, Pergamon Press, 1961.
3. See, *Report of the Working Party on the Schools' Curricula and Examinations* (Lockwood Report), HMSO, 1964.
4. Full details of these, which included the Mathematics for the Majority Project, the Mathematics for the Majority Continuation Project, the Sixth Form Project and the Mathematics Curriculum Critical Review Project, are included in the reports issued by the Schools Council. Descriptions of the projects are also to be found in, for example, Griffiths, H. B. and Howson, A. G. *Mathematics, Society and Curricula*, Cambridge University Press, 1974; Howson, A. G. 'Change in Mathematics Education since the late 1950's – United Kingdom', *Educ. Studies Math.*, **9**, 1978, 183–223; Howson, A. G., Keitel, C. and Kilpatrick, J. *Curriculum Development in Mathematics*, Cambridge University Press, 1981; Mathematical Association, *Mathematics Projects in British Secondary Schools*, Bell, 1967 and 1976; Stenhouse, L. *Curriculum Research and Development in Action*, Heinemann Educational, 1980; Watson, F. R. *Developments in Mathematics Teaching*, Open Books, 1976.
5. See Howson *et al.*, *op. cit.* (n. 4), for a discussion of the various approaches exhibited by the different projects.
6. Ballard, P. B. *The Changing School*, Hodder and Stoughton, 1925, pp. 223–36.
7. See, for example, Schools Council *Mixed ability teaching in mathematics*, Methuen/Evans, 1977.
8. The committee recommended that a boy should follow one of three mathematics courses depending on the social and financial status of his father.
9. One of the many contributions which the Open University has made to mathematics education is in promoting an interest and understanding of the history of mathematics and in encouraging its teaching.
10. *The Attainments of the School Leaver* (Tenth Report of the Expenditure Committee), HMSO, 1977 (see, especially, p. 29).
11. *Mathematics Counts* (Cockcroft Report), HMSO, 1982.
12. In this brief 'postlude' it has not been possible to refer in detail to many significant events occurring in the two decades 1960–80. Accordingly, some of these are listed below in order to direct reader's attention to them and to a consideration of their effects. The books by Griffiths and Howson, Howson *et al.* and Watson, and the paper by Howson (all *op. cit.* (n. 4)) provide appropriate bibliographical references. Fletcher, T. J. and Howson, A. G. 'Developments in the teaching of secondary mathematics in schools in England and Wales: 1958–79' (Klett, to appear) is a collection of readings designed to illustrate changes during this period and their effects, and *Comparative Studies of Mathematics Curricula – Change and Stability: 1960–80*, IDM, Bielefeld, 1980, contains articles describing changes in the teaching of geometry, algebra, statistics and applications of mathematics in English schools from 1960–80. More detailed descriptions of mathematical changes are given in the publications of the Schools Council Mathematics Curriculum Critical Review Project (Blackie).
    (a) The establishment of the Institute of Mathematics and its Applications, the Joint Mathematical Council of the United Kingdom, the Royal Society/IMA Committee on Mathematical Education, the Technician and Business Education Councils and the (National) Council for Educational Technology. The growth of polytechnic education and the formation of the Council for National Academic Awards.
    (b) The creation of teachers' centres and the evolution of new 'models' for curriculum development. The appointment of LEA 'mathematics advisers'.
    (c) The publication of various official reports including: *15–18* (Crowther Report), HMSO, 1959; *Half our Future* (Newsom Report), HMSO, 1963; *Children and their Primary Schools* (Plowden Report), HMSO, 1967; *Enquiry into the Flow of Candidates in Science and Technology into Higher Education* (Dainton Report), HMSO, 1968; *Teacher Education and Training* (James Report), HMSO, 1972; *Mathematics 5–11* (DES), HMSO, 1979; *Aspects of Secondary Education in England* (DES), HMSO, 1979.

(d) The attempts to reform the sixth-form examination system (Qs and Fs, Ns and Fs), to introduce a common system of examinations at 16+ and to establish a Certificate of Extended Education. The development of new examination procedures (Mode 3 CSE, Applicable Mathematics AO-level, etc.) and the establishment of the Assessment of Performance Unit (APU).

(e) The development of a national olympiad and participation in the International Mathematical Olympiad. Regional and other schemes to identify and foster precocious mathematical talent.

(f) The (significant but limited) attempts to increase school/industry cooperation and understanding through the establishment of joint committees, short-term secondments of teachers to industry, periods in industry and commerce as part of teacher-training and specific attempts to recruit as mathematics teachers those having some years experience of working in industry and business.

A change of considerable importance occurred in April 1982 when the Secretary of State for Education and Science announced that the Schools Council was to be replaced by an Examinations Council and a School Curriculum Development Council. (This amounted to a rejection of the recommendations of the inquiry into the Schools Council (p. 206 above).) Any attempt to deal separately with the curriculum and examinations would, of course, seem to run counter to all the lessons to be revealed by a study of this book. More importantly, however, the arrangements proposed by Sir Keith Joseph would appear to presage further moves towards 'central' control of the curriculum. In the first draft of my 'Postlude', the opening paragraphs were somewhat different from those printed on p. 205, for I saw as 'the most significant event' of the last two decades, the way in which the Department of Education and Science was seeking to exercise greater control over the school curriculum. Friends, whose views I much respect, persuaded me to rewrite the passage on the grounds that I read too much into what were chance decisions. Recent moves have served to reinforce my earlier apprehensions. This is not to say that greater central control of the curriculum is necessarily wrong. What it does mean, however, is that educators should be aware of any drift and of the probable consequences of that movement and should frame appropriate responses. A move towards a centralised curriculum would indeed be the most significant of contemporary trends in English mathematics education.

Name index

Abbott, E. A., 264n
Abram, W. A., 260n
Adams, Sir J., 181, 270n
Adams, L. D., 277n
Adamson, J. W., 261n, 263n, 267n, 273n
Addison, J., 49
Adelhard of Bath, 239n
Airy, Sir G. B., 72, 79, 85, 94, 264n
Alcuin, 239n
Allan, B, 246n
Allen, J., 261n
Allum, C. G., 269n
Archer, R. L., 273n
Archibald, R. C. 255n, 260n
Archytas, 239n
Argles, M., 267n
Armstrong, H. E. 147, 151, 268n
Armytage, W. H. G., 253n
Arnold, M., 121, 262n
Arnold, T., 86
Ascham, R., 24–5, 31, 244n
Ashley Smith, J. W. 247–9n
Ashworth, C., 54–5
Aubrey, J., 251n
Augustine, St, 1

Babbage, C., 71, 77, 82, 94
Bacon, F., 113–15
Bacon, R., 3, 240n
Baker, H., 28
Baker, H. F., 271n
Ball, Sir W. W. R., 143, 214, 239–40n, 242–3n, 245n, 252n, 254n
Ballard, P. B., 192, 207–8, 232n, 272–3n, 276n, 279n
Bamforth, T. W., 258n, 265n
Banks, Sir J., 66, 73
Banks, O., 267–8n
Barlow, P., 65, 71, 98, 252n
Barnard, S., 155, 159, 167, 269–70n
Barnes, Bishop, 259n
Baron, M. E., 243n
Barrow, I., 34, 49, 262n
Bath, J. L., 250n
Beale, D., 135, 161, 172–3, 179, 273n
Beaujouan, G., 239–40n
Bede, Venerable, 1, 239n
Bedford, Duchess of, 48
Bedford, Earl of, 7
Bell, A., 102–3, 105, 252n, 261n
Bell, A. W., 269n
Bell, E. T., 79, 257n
Bell, G. M., 153
Berkeley, Bishop, 250–1n
Bernbaum, G., 275n
Bewick, T., 63
Biggs, E., 178, 274n
Biggs, N. L., 255n
Billingsley, H., 16, 25, 243n
Birchenough, C. A., 257n, 261n, 272n
Birkbeck, G., 81, 257n
Bishop, A., 269n
Bishop, G. D., 265n
Black, M. P., 260n
Bland, M., 75–6
Bockstaele, P., 242n
Boethius, A., 1, 18, 21, 27, 239n
Bonnycastle, J., 78, 89–91, 100
Boole, M. E., 170, 272n
Boon, F. C., 159, 270n
Borel, E., 165
Boswell, J., 39, 246n
Bourdon, P. L. M., 82
Bowen, E. E., 180
Bower, T. W., 124
Boyer, C. B., 239n
Boyle, R., 40, 49
Bradwardine, T., 3, 240n
Brancker, T., 47, 248n
Branford, B., 141, 156, 159, 258n, 269n
Braybrooke, Lord, 247n
Brewster, G. W., 152
Bridge, B., 75–6, 98, 263n
Brierley, M., 260n
Briggs H., 33
Brinsley, J., 244n
Britten, B., 101
Broadbent, T. A. A., 155, 269n, 271n
Brock, W. H., 265n, 267n, 272n
Bromwich, T. J. I'A., 143
Brougham, Lord, 72, 81, 85–6, 257n
Brouncker, Viscount, 40, 246–7n
Bruce, J., 250–1n
Brunelleschi, F., 16
Bruner, J. S., 207n
Bryan, G. H., 275n
Bryan, M., 72–3, 256n
Bryant, Sir A., 245–7n
Buridan, J., 3, 240n
Burstall, S., 174–5, 273n
Burt, Sir C., 190, 250n, 276n
Bushell, W. F., 255n, 266n
Buss, F. M., 172–3, 176, 273n
Butler, G., 263n

Butler, S., 25
Byngham, W., 240n
Byron, Lady, 80, 82

Cajori, F., 242n
Calamy, E., 247n
Cardan, G., 12
Carlisle, N., 244n, 263n
Carlyle, T., 68, 254n
Carroll, L., 132, 264n
Carson, G. St L., 141, 156–7, 159, 192, 195, 269–70n
Cartwright, Dame, M., 259n
Cassels, J. W. S., 258n
Cavalieri, B., 34
Cayley, A., 71, 126–7, 134, 136, 142–3, 148, 177, 255n, 257n, 259n,
Cecil, Sir W., 24
Chapman, S., 145
Charles I, 30
Charles II, 40, 45
Charlett, A., 42
Charlton, K., 241n
Chaucer, G., 243n
Cheke, Sir J., 12, 24, 242n
Cheyne, G., 51, 249n
Chrystal, G., 271n
Clairaut, A. C., 130, 132
Clarendon, Earl of, 264n
Clark, S., 70, 255n
Clarke, F. M., 241n
Clarke, H., 66, 252–4n
Clayton, Sir R., 35
Cleomedes, 17
Clifford, W. K., 134
Cockcroft, W. H., 210
Cocker, E., 28, 39, 246n
Colenso, Bishop, 124, 259n, 263n
Colet, Dean, 30, 244n
Collingwood, Sir E., 259n
Comenius, J. A., 183
Comte, A., 258n
Coombs, P. H., 278n
Cornelius, M. L., 255n
Crabbe, G., 101
Crombie, A. C., 239–40n
Cromwell, O., 30
Curtis, M., 244n
Curtis, S. J., 240–1n, 257n, 277n

Daboll, N., 67
Daltry, C. T., 157
Daly, J. F., 240n
David, F. N., 247n
Dawes, R., 104, 261n
Dee, J., 24ff, 243n
Defoe, D., 249n
De Moivre, A., 76, 249n
De Montmorency, J. E. G., 261n
De Morgan, A., 28, 46, 75ff, 125–6, 132, 143, 244–6n, 249n, 252n, 255ffn, 264n
De Morgan, G., 95–6
De Morgan, S. E., 81, 256ffn
de Quincey, T., 257n
Des Barres, J. F. W., 66, 253n
Desaguliers, J. T., 250–1n, 258n
Descartes, R., 34
Dienes, Z. P., 207, 243n
Dieudonné, J., 159, 270n
Dilworth, T., 67, 243n
Ditton, H., 47, 248n
Dobbs, W. J., 159, 181, 270n
Dockrell, W. B., 276n
Doddridge, P., 45ff, 247ffn
Dodgson, C., 132, 264n
Dodson, J., 76, 243n, 256n
Dollond, J., 93
Durcell, C. V., 67, 202–3, 207, 267n
Durran, J. H., 267n

Eaglesham, E., 268n
Easton, J. B., 16, 241ffn
Eddy, W. A., 248n
Edgeworth, F. Y., 162, 189
Edgeworth, M., 57, 250n
Edgeworth, R. L., 57, 250n
Edward VI, 7, 241n
Eldon, Lord, 59, 64, 124, 252n
Elizabeth I, 12, 24, 241–2n
Emerson, W., 68, 253n, 255n
Erasmus, 9
Euclid, 3, 12, 15–16, 18ff, 24–5, 34, 42, 49, 53, 55, 68, 72, 75, 78, 91–2, 97ff, 104, 112, 120, 124, 130ff, 134, 141, 147, 149ff, 160, 163–4, 167, 173, 176, 179, 239ffn, 251n, 253–4n, 258n, 263–4n, 269n
Evans, J. H., 123–4, 262n
Evelyn, J., 40–1

Fairbairn, Sir W., 118
Farrar, Dean, 137, 262n, 265n
Fawcett, P., 177
Fehr, H., 163–4, 270n
Félix, L., 243n
Fenn, J., 254n
Ferguson, J., (Alnwick) 97
Ferguson, J., (London) 251n, 253n
Fielding, H., 249n
Fine, O., 17
Fisher, Sir R., 189
Fiske, N., 26–7, 243n
Fitch, Sir J., 274n
Flamsteed, J., 245n, 255n
Fletcher, T. J., 279n
Fletcher, W. C., 152, 154, 268–9n
Ford, M. M., 174, 273n
Forsyth, A. R., 143, 148ff, 266–8n
Foster, S., 40, 245n

Frankland, R., 47ff, 247–8n
Frend, W., 79ff, 92, 248n, 254n, 257n, 259n
Frisius, G., 24
Frobisher, L. J., 188, 275n
Frobisher, M., 18
Froebel, F., 170, 181, 272n

Galton, F., 189
Garnham, R. E., 254n
Garrett, E., 273n
Gattegno, C., 195–6
George III, 67
George VI, 154
Glaisher, J. W. L., 259n
Godfrey, C., 141ff, 175, 182, 202, 207, 209, 266ffn, 274n
Goldsmith, N. A., 260n
Gosden, P. H. J. H., 265n
Grace, J. H., 143
Gravesande, W. J., 54, 250n
Greaves, R. L., 247n
Greenhill, Sir G., 164, 166, 271n
Gregory, D., 38, 42–3, 247n
Gregory, J., 247n
Gregory, O., 63, 65, 71, 86, 251n
Gresham, Sir T., 33
Grey, Lady Jane, 7–8, 241n
Grey, M., 173
Griffiths, H. B., 269n, 278–9n
Grosseteste, R., 3, 240n
Gundelfinger, J., 7, 241n
Gunter, E., 33
Gwatkin, E. R., 273n

Hadow, Sir H., 184–6, 277n
Haeckel, E., 258n
Hall, A. R., 246n
Hall, H. S., 67, 138, 152, 202n
Hall, M. B., 246n
Hall, R., 259n
Hall, T. G., 173–4, 273n
Halley, E., 41, 245n, 247n
Halsted, G. B., 163, 270n
Hamilton, D., 189, 270n
Hamilton, Sir W., 257n
Hamilton, Sir W. R., 255n
Hamley, H. R., 190, 276n
Hans, N., 246n, 248n, 251n, 256n
Hardy, G. H., 79, 143–4, 148, 165, 208, 259n, 266–7n
Harriot, T., 243n
Harris, J., 67, 254n
Harrison, J. F. C., 260n
Harrison, W., 242n
Hart, K. M., 277n
Hartley, D., 56–7, 250n
Hartley, Sir H., 246n
Hartlib, S., 25, 48, 245n, 248n
Hartog, P. J., 270n
Hartwell, R., 24, 27–8
Hassé, H. R., 144, 266n
Hatton, E., 24, 27–8
Hawkins, C., 270n
Hawkins, J., 39, 246n
Hawtrey, S., 124
Hayman, H., 138
Hayward, R. B., 134, 136, 266n
Heal, Sir A., 251n
Heath, R., 69
Heaviside, O., 149
Henry VIII, 5, 8, 241n
Herbart, J. F., 158, 181, 270n
Herschel, J., 71–2, 77, 85, 93–4
Heynes, S., 252n
Hilbert, D., 163–4
Hill, C. 242–3n, 245n
Hilton, H., 176, 178
Hirst, T. A., 95, 134–6, 259n, 264n
Hobson, E. W., 152
Hollingdale, S. H., 254n
Hooke, R., 245n
Hope, C., 198–9
Hopkins, W., 125
Horner, W. G., 71
Horsley, Bishop, 65–6
Howson, A. G., 254n, 260n, 263–4n, 267–71n, 278–9n
Howson, J. S., 174
Høyrup, J., 240n
Hudson, J. W., 260n
Hudson, W. H. H., 149
Hughes, A. G., 276n
Hutton, C., 18, 26–7, 59f, 97–8, 203, 243n, 245n, 250ffn, 257n
Hutton, T. W., 266n
Huxley, T. H., 137, 267n

Ingram, R. A., 263n
Isaacs, S., 190, 276n
Iselin, J. F., 134–5
Ives, A. G., 249n

James II, 249n
James, T., 263n
Jebb, J., 252–3n
Jeffery, G. B., 163
Jennings, J., 48ff
Jex-Blake, T. W., 138
John of Holywood, 3, 17, 240n
Johnson, F. R., 242n
Johnson, S., 39, 246n, 252n
Jollie, T., 47, 49, 248n
Jones, J., 134
Jones, T. A., 267n
Jordan, C., 143, 165–6
Joseph, Sir K., 280n
Joy, R. R., 188, 275n
Jurin, J., 60–1, 250–1n

Kaplan, E., 241ffn
Kay-Shuttleworth, J. P., 106ff, 126, 260–1n
Keitel, C., 270n, 279n
Kelly, A., 274n
Kelvin, Lord, 149
Kempe, A. B., 139
Kempe, W., 244n
Kersey, J., 244n, 256n
Kidder, R., 32
Kilpatrick, J., 270n, 279n
Kimmins, C. W., 275n
Kirkman, T. P., 71, 243n, 255–6n
Klein, F., 163–5, 259n
Kline, M., 249n, 256n
Kneller, Sir G., 44
Knight, C., 258n
Knight, F., 257n
Knight, S. R., 67, 202
Koestler, A., 242n
Köhler, W., 189

Lacroix, S. F., 77, 131, 256n
Lakatos, I., 243n
Lamb, H., 149, 268n
Lancaster, J., 102–3, 105
Langley, E. M., 149–50, 268n
Langley, J., 32
Laplace, P. S., 70, 72, 77, 127
Larby, E. M., *see* Williams, E. M.
Larkey, S. V., 242n
Larmor, Sir J., 143–4, 149
Latham, R., 244n
Laverty, W. H., 135
Lawson, J., 239–41n, 244n, 272n
Layton, D., 261n
Leach, A. F., 240–1n, 244n, 247n
Leake, J., 36
Lee, P., 266n
Leeke, J., 25
Legendre, A.–M., 68, 130–2, 254n
Leibniz, G. W., 49, 57
Leicester, Earl of, 24–5, 243n
Leslie, J., 254n
Levett, R., 133–4, 136, 141–2
Leybourn, T., 71, 255n
Lilley, S., 16–17, 242n
Lily, W., 31
Lindsay, K., 273n
Locke, J., 58, 113, 250n, 253n
Lodge, A., 149
Louis XIV, 37
Lovelace, Lady, 82
Lowe, R., 95, 121
Lucas, Sir H., 34
Lucas, S., 53, 250n
Lynch, A. J., 275n
Lyons, Sir H., 246–7n

Macaulay, F. S., 270n
MacCarthy, E. F. M., 134, 272n
Maclaurin, C., 54–5, 252n
McClelland, W., 188
McDonnell, Sir M., 244n
McLachlan, H., 247–50n
McMillan, M., 183
Mair, D. B., 159, 163
Marillier, J. F., 263n
Martin, B., 251n
Martindale, A., 47, 248n, 251n
Mary, Queen, 5, 7–8, 241–2n
Masères, Baron, 80
Maskelyne, N., 65–6
Matthews, A. G., 247n
Matthews, W., 244n
Maurice, F. D., 173, 273n
Maxwell, E. A., 264n
Maxwell, J. C., 125
Mayo, C. H. P., 266n, 269n
Mellis, J., 24, 26
Méray, C., 163, 270n
Mercator, G., 24
Mercer, J. W., 269n
Mill, J. S., 260n
Milner, I., 257n
Milton, J., 48, 58, 248n
Moberly, G., 137
Montessori, M., 181, 183
Montgomery, R. J., 253n, 264n, 270n
Moore, E. H., 163, 270n
Moore, Sir J., 35, 39, 68, 245n, 247n, 254–5n
More, Sir T., 5
Moseley, Canon, 119, 261n
Mottershead, J., 248n
Mulcaster, R., 31–2, 244n
Mumford, A. A., 251–2n

Napier, J., 27, 39, 243n
Napier Shaw, Lady, 269n
Nesbit, A., 100, 255n, 263n
Neumann, B. H., 257n
Newcastle, Duke of, 120
Newman, J. R., 258n
Newman, M. H. A., 259n, 266n
Newton, Sir I., 16, 34, 38, 41, 49, 53, 57, 65, 67, 72, 77–8, 98, 174, 240n, 244ffn, 248–51n, 253–4n, 259n, 266n
Newton, J., 58
Newton, S., 38
Nicolson, M. H., 246n
North, J. D., 240n
Northumberland, Duke of, 8, 24, 241n
Norton, R., 24, 26
Norwood, Sir C., 277n
Nunn, Sir P., 141, 155–7, 159, 167, 180–1, 183, 189–90, 195, 201, 269–70n, 274–6n

O'Brien, J. J., 247n
Oliphant, J., 244n

Ollard, R., 244n
Oresme, N., 3, 240n
Orme, N., 240n
Orton, J., 249n
Osborne, G. S., 276n
Oughtred, W., 37, 243n, 245n, 251n
Ozanam, J., 67, 254n

Paget, E., 37–8
Parker, I., 250n
Parkhurst, H., 275n
Parkinson, S., 125–6
Patchett Martin, A., 263n
Payne, J., 137
Peacock, G., 71, 77, 79
Peacock, T. L., 257n
Peano, G., 164
Pearce, E. H., 245–6n
Pearson, K., 266n
Peirce, B., 71, 132n
Pell, J., 34, 36, 245n
Pembroke, Earl of, 7–8
Pendry, F., 180–1
Pepys, J., 32
Pepys, S., 29ff
Percival, J., 138, 265n
Perkins, P., 36, 255n
Perl, T., 255n
Perry, J., 138, 147ff, 153, 155, 158, 168, 175, 181, 194, 208, 222–4, 265n, 267–9n, 271n, 273–4n
Pestalozzi, J., 103, 107, 110, 113, 170, 250n
Peter the Great, 37
Petty, Sir W., 30, 40, 244n
Piaget, J., 190, 195, 276n
Picciotto, C. 244n
Place, F., 258n
Plato, 239n
Playfair, J., 68, 70–1, 254n
Pole, W., 262n
Pollock, Sir F., 77–8
Pope, A., 49
Potts, R., 131, 263–4n
Powell, B., 112, 127, 137, 262n
Price E. A., 154, 269n
Price, M. H., 267–8n, 271–2n, 274–5n
Priestley, J., 54ff, 250n, 254n
Proclus, 17
Ptolemy, 12, 17, 239n
Purver, M., 246n

Quadling, D. A., 266n

Ramus, P., 24, 48, 244n, 248n
Ranyard, A. C., 95
Raphson, J., 254n
Rashdall, H., 240n
Rayleigh, Lord, 259n, 266n
Recorde, R., 6f, 130, 209, 240ffn
Reece, R., 256n
Reid, E. J., 172
Rennie, B., 275n
Rheticus, 21
Rich, R. W., 274n
Richard of Wallingford, 17
Richeson, A. W., 242n
Rignano, E., 190
Roach, J., 253n, 270n
Robinson, F. J. G., 251n, 256n
Robinson, M., 248n
Robson, A., 252n, 255n, 266n
Roebuck, J. A., 102, 104, 120, 261n
Roebuck, J., 53, 250n
Rouse, W. H. D., 263n
Routh, E. J., 125, 143
Rubenstein, D., 277n
Rudd, T., 25
Russell, B., 143, 259n, 274n

Sacrobosco, *see* John of Holywood
Sadler, J., 32
Salmon, G., 125, 134
Salmon, W. H., 246n
Sanderson, F. W., 152
Sandford, E. G., 262n, 270n
Sandwich, Lord, 35
Saunderson, N., 47, 248n, 253n
Savile, Sir H., 33, 249n
Sawyer, W. W., 192
Scheubel, J., 18
Schofield, R. E., 250n
Scott, C. A., 177
Scott, J. F., 246n
Scott, R. P., 267n
Serle, G., 25
Shafto, R., 62, 64–5
Shakespeare, W., 14, 20
Shandarkar, D., 188
Shelley, G., 256n
Shuard, H., 203
Siddons, A. W., 142, 149ff, 166–8, 202, 207, 266ffn
Silver, H., 239–40n, 244n, 272n
Simon, B., 250n, 258n, 262n, 264n, 277n
Simon, J., 241–2n
Simpson, T., 65–6, 69–70, 78, 93, 252–3n, 255n
Simson, R., 68, 251n, 254n, 263n
Sitwell, E., 243n
Smiles, S., 259–60n
Smith, A., 65
Smith, D. E., 156, 163–4, 192, 239n, 242n, 246n, 267n
Smith, E., 253n
Smith, H. J. S., 134
Somerset, Duke of, 7–8, 22, 241n
Somervell, E., 272n
Somerville, M., 71–2

Spearman, C., 190, 276n
Spencer, H., 258n
Spenser, E., 31
Spriggs, G. W., 275n
Stamper, A. W., 270n
Steele, R., 240n
Stenhouse, L., 279n
Stennett, S., 58
Stephenson, R., 250n, 259–60n
Stevens, F. H., 138, 152
Stevin, S., 16, 26
Stifel, M., 18
Stokes, G. G., 95, 125–6
Stone, E., 67, 247n, 254n
Stowe, D., 106
Sutherland, G., 261n
Swetz, F., 243n
Swift, D., 248n
Sylvester, D. W., 239–40n, 242n, 247n, 251n
Sylvester, J. J., 46, 65, 68, 71, 126, 134, 136, 257n, 259n

Tahta, D. G., 272n
Tait, P. G., 125
Tanner, J. R., 247n
Tartaglia, N., 16
Tate, G., 97, 260n
Tate, T., 97ff, 157, 209, 250n, 259ffn, 263n
Tawney, R. H., 273n, 275n
Taylor, E. G. R., 242–3n, 245n, 247–8n, 250–1n
Taylor, S., 82–3
Taylor, W., 276n
Temple, F. (Archbishop), 112–13, 118–19, 126, 129, 131, 138–9, 160–1, 260n, 262–3n, 265n
Thacker, A., 266n
Theodore, Archbishop, 1, 239n
Thomas, I., 239n
Thomas, W. G., 241n
Thompson, S. P., 149
Thomson, J. J., 125
Thorndike, E. L., 188–9
Thorndike, I., 240n
Thwaites, B., 267n, 278–9n
Tibble, J. W., 270n
Tilleard, J., 262n
Todhunter, I., 125, 132–3, 263–4n, 266n
Tonstall, C., 12–13, 242n
Trollope, A., 263n
Trollope, T. A., 263n
Tucker, R., 133
Tuckey, C. O., 266n
Tufnell, C., 106
Turnbull, G. H., 247n
Turnbull, H. W., 246n
Turner, D. M., 265n, 267n
Turner, H. H., 267n
Turner, P., 245n
Turton, T., 77, 79
Tylecote, M., 260n
Tyndall, J., 137, 265n

Vince, S., 78
Vives, J. L., 5, 168

Wade, J., 24, 26
Wagner, R., 20
Walford, J. D., 124
Walker, A., 251n
Walker, G., 252n
Walkingame, F., 39, 243n, 246n
Wallis, J., 33–4, 38ff, 43–4, 243n, 245n, 247n
Wallis, P. J., 243–4n, 246n, 251n, 254–5n, 264n
Wallis, R., 255n
Ward, J. (Chester), 68, 255n
Ward, J. (London), 244n
Ward, S., 33–4, 245n
Wardle, D., 273n
Watson, F., 240–2n, 244n, 248n
Watson, F. R., 279n
Watson, H. C., 138
Watts I., 49ff, 249–50n
Webb, R. K., 258n
Webster, C., 247n
Wells, E., 51, 249n
Wesley, J., 52, 249n
Wetherill, J. H., 260n
Whalley, R., 6ff
Whewell, W., 72, 79, 130, 256–7n, 264n
Whiston, W., 248n, 251n
White, A. C. T., 252n
White, Sir W., 268n
Whitehead, A. N., 143, 178, 271n, 274n
Whittaker, E. T., 139, 143, 265–6n
Wilderspin, S., 103
Wilkinson, T. T., 260n
William III, 249n
Williams, E. M., 157, 169ff, 209, 272ffn
Williams, R., 247n
Williamson, J., 130, 264n
Willoughby, F., 41
Willsford, T., 24, 27
Wilson, D., 189
Wilson, D. K., 239n
Wilson, J. M., 123ff, 141, 168, 258–9n, 262ffn
Wilson, S., 259–60n
Wingate, E., 243–4n, 256n
Wolfe, R., 242n
Wolff, G, 152, 267n, 271n
Wolsey, Cardinal, 5, 242n
Wood, A. à, 249n
Wood, J., 78, 98, 263n
Wood, R., 36–7, 247n
Woodhouse, R., 77
Woodworth, R. S., 188
Wormell, R., 133, 264n
Wren, Sir C., 40, 245n

Wright, E., 27
Wright, R. P., 132, 264n

Yeldham, F. A., 239n, 244n, 246n, 275n

Young, J. W. A., 267n
Young, T., 250n

Zilsel, E., 16, 242n

Subject index

abacal arithmetic, 14
Académie des Sciences, 41
Acts
  Chantries (1547), 9
  Education: Endowed Schools (1869), 273n; Forster (1870), 146, 169; (1880), 169; (1902), 151, 171, 173; (1918), 173, 183, 272n; Butler (1944), 191, 193–4
  Five-Mile (1665), 46
  to Prevent the Growth of Schism (1713), 46;
  Toleration (1689), 48
  of Uniformity (1662), 46
Admiralty, 29, 35
adult education, 55, 60, 69ff, 73, 80, 86, 93, 97ff, 209, 251n, 257–8n, 260n, 279n
African education, 202–3
aims
  of education, 9, 92, 107
  of elementary education, 103, 113–14
  of mathematics education, 5, 12, 31, 50–3, 88, 110, 114, 120, 122, 158–9, 250; for girls, 174–5; in primary schools, 186, 196; in secondary schools, 187, 197–8
algebra, 2, 18, 42, 48–51, 53–4, 62, 68, 75–6, 78, 89, 90, 92, 98, 104, 112, 117, 120, 123, 125, 141, 147, 155, 158, 160, 175–6, 182, 185, 198, 222, 234–5
algebraic v. geometric methods, 78, 131, 254n
algorism, 3
analysis, 143, 166, 264n, 271n
Analytical Society, 77
Anti-Euclid Association, 133
apparatus, 91, 103, 183, 266n
applications of mathematics, 5, 15, 17, 26, 31, 39, 44, 50, 51, 61–2, 64, 88, 98, 110, 114, 120, 122, 135, 153, 158–9, 185, 207, 228
applied mathematics, 175, 178, 183, 228
arithmetic, 1–3, 12, 13ff, 18, 26–30, 35, 39, 62, 68, 75, 89–91, 103, 110, 112, 116, 119–20, 122, 158, 160, 174–6, 184, 197, 222, 224–5, 229–32
Army education, 64ff, 252n
  see also Military Academies
Assessment of Performance Unit (APU), 280n
Association for the Improvement of Geometrical Teaching (AIGT), 134ff, 141
Association of Teachers in Colleges and Departments of Education (ATCDE), 193, 277n
Association of Teachers of Mathematics (ATM), 196, 206, 277n
    for the SE part of England, 156, 178
Association for Teaching Aids in Mathematics (ATAM), 196
astrology, 3, 27
astronomy, 1, 3, 12–13, 17ff, 27, 40, 42, 49–51, 72, 137, 142, 242n
attitudes to mathematics, 34, 47, 51–3, 76, 244n
axiomatic approach, 200, 258n
axioms in geometry, 16, 92

Beloe Report (1960), 199
Board of Education circular on geometry (1909), 153
book-keeping, 26, 62
British Association, 118, 133–4, 136–7, 148ff, 267n
British and Foreign School Society, 102, 105
Bryce Commission (1895), 180

calculus 34, 62, 68, 77–8, 82, 98, 125, 129, 147, 154, 165–6, 177, 181, 218–20, 224, 226, 271n, 277n
calendar reform, 26, 241n
Cambridge University, 2, 3, 6, 8, 12, 13, 26, 29, 32ff, 34, 46, 60, 65, 75, 76ff, 79, 81, 112, 125ff, 135–6, 148, 176–7, 179, 268n
  colleges: Clare, 80; Emmanuel, 32, 34; Girton, 177; God's House (later Christ's), 5; Jesus, 5, 79, 150; King's 5, 263n; Magdalene, 32; Newnham, 177; Queens', 5; St Catharine's 5; St John's, 5, 24, 125–7; Sidney Sussex, 34; Trinity, 24, 34, 37, 76, 82, 142–3, 263n, 266n; Trinity Hall, 32
  fellowships, 263n
  influence of, 144, 277n
  Lectures Association, 177
  Lucasian Chair, 34, 47, 77–80, 248n, 251n
  Philosophical Society, 259n
  Press, 150, 154
  Sadleirian Chair, 127, 148, 266n
  Statutes, 12
  Teachers' Training Syndicate, 179, 274n
  Tripos (Senate House) Examination, 65, 77–8, 83, 98, 125–7, 131, 143–4, 177, 212–14, 252–3n, 266n
Centre for Curriculum Renewal and Educational Development Overseas (CREDO), 203

Church the, and education, 1–2, 239n
CIEM, *see* International Commission on Mathematical Instruction
City and Guilds Institute, 147
Civil Service entry, 253n
Civil War, 30, 32
Clarendon Commission Report (1864), 129ff, 263n
class system, 101, 114, 128, 279n
coaches, university, 125, 143
Cockcroft Report (1982), 210–11, 279n
Cockerton judgment, 151
College of education, *see* teacher-training college
College of Preceptors, 160, 179
Committee of the Privy Council on Education, 104, 106, 111, 118, 120
Commonwealth (1649–60), 45
compulsory education, 169
Congregational Board of Education, 106
conic sections, 50, 53, 62, 68, 78, 98, 123, 147, 175, 225–6
Copernican astronomy, 18, 242n
Council for Educational Technology (CET), 279n
Council for National Academic Awards (CNAA), 279n
County Major Scholarship, 274n
Crowther Report (1959), 279n
curriculum development, barriers to , 155, 165
Curriculum Study Group (Ministry of Education), 206
curve-stitching, 272n

Dainton Report (1968), 130, 279n
Dalton Plan, 182, 275n
day training colleges, 180, 274n
De Morgan Medal, 95, 259n
decimal arithmetic, 26
decimal coinage, 94
definitions, 23, 92
degree
  bachelor's, 2, 3, 34, 239n, of education, 201
  master's, 2, 3, 240n
  postgraduate, 188, 276n
  for women, 175ff
  see also Cambridge University Tripos Examination, examinations
Department of Science and Art, 135, 144, 168, 194
  regulations for schools, 146
development, stages of, 116
dialogue form (textbook), 19–20, 243n
dictionaries, mathematical, 67
dissenters, 46ff
dissenting academies, 47ff
  Attercliffe, 47, 248n
  Daventry, 54–5
  Kibworth, 48
  Market Harborough, 52
  Mile-End, 55
  Natland, 47
  Newington Green, 249n
  Newport Pagnall, 58
  Northampton, 52
  Rathmell, 47
  Warrington, 56

Ecole Polytechnique, 65, 87, 252n, 258n
Education Bill (1820), 85, (1917) 275n
  See also Acts
education, science of, 56ff
educational research, 116, 162, 187ff, 276n
Eldon judgment, 64, 124, 252n
Endowed Schools Commission, *see* Taunton Commission
engineers, mathematics for, 268n
Entebbe Project, 202–3
Euclid, *see* Name index
Examination Boards, 167
  Cambridge, 160, 225–6, 273n, 277n
  Joint Matriculation, 163
  Oxford, 160–1, 175, 273–4n
  Oxford and Cambridge, 135, 161–2, 232–3, 273n
examination papers
  arithmetic (LCC, 1923), 231–2
  Battersea Training School (1847), 217–20
  Cambridge Senior Local (1910), 225–6
  Cambridge Tripos (1802), 212–14
  County Minor (11+) (1920), 229–30
  London Matriculation (1838), 214–16
  Oxford and Cambridge Board School Certificate (1934), 232–3
examinations, 111, 133, 135, 152, 158, 160ff, 173, 175–6, 267n, 280n
  army entrance 128, 160
  Certificate of Secondary Education (CSE), 199
  London University Matriculation, 161, 176, 214–16, 273n
  Oxford and Cambridge scholarship, 162
  'scholarship' (County Minor, 11+), 171, 229–30, 273n
  School Certificate (and Higher), 162, 167, 187, 232–3
  university, 33, 65, 77–9, 83, 125, 239–40n
  see also Cambridge University Tripos
Examinations Council, 280n
examples, rôle of, 19, 21
exercises, 19, 243n

finger-reckoning, 1, 3, 15
fluxions, 62, 68, 77–8
four-colour problem, 139
French Revolution, 57, 80

Subject index

gauging, 27–8, 62
geometry, 1, 12, 15ff, 27, 30, 34, 40, 42, 48–51, 53–4, 62, 91, 98, 112, 116–17, 124–5, 130ff, 134ff, 141, 146–8, 150, 152ff, 158–60, 163, 166, 168, 175, 178, 181–2, 185, 198, 223, 225–6, 235–7, 272n
girls, education of, 4, 8, 64, 72, 172ff, 256n, 273n, 277n
Goldsmiths' College, London, 191
Governesses' Benevolent Institution, 172
graduates, data relating to, 145
Greenwich, Royal Naval College, 166, 265n
Gresham College, 33–4, 39, 40, 244–5n, 251n
guilds, trade, 5, 30, 32, 244n

Hadow Report (1926), 184–5, (1931) 185
Haeckel's Law, 258n
headmasters and the Church, 265n, 266n
Headmasters' Conference, 134, 136, 179–80
Hindu–Arabic notation, 1, 2, 14, 30
history of mathematics, the teaching of, 229, 278–9n
Home and Colonial Infant School Society, 272n
humanism, 9, 16, 30

Indian Civil Service, 128, 253n
in-service education, 63, 82, 155, 179, 195, 201, 211, 262n, 269n, 278–9n
Inspectorate, Her Majesty's, 104
Institute of Mathematics and its Applications (IMA), 279n
instruments, mathematical, 39, 42, 222, 266n
International Commission on Mathematical Instruction (ICMI or CIEM) 163–4, 166, 203, 271n
International Congress on Mathematical Education (ICME), 204
International Congress of Mathematicians (ICM), 163–4, 271n
International Mathematical Olympiad (IMO), 280n

James Report (1972), 278–9n
Jeffery Report (1944), 163, 193, 206, 234–7, 277n
Joint Mathematical Council (JMC), 279n

Labour Party, 184, 186
Lancashire mathematicians, degenerate, 260n
language, 22–3, 91
  see also vernacular
Literary and Philosophical societies, 73
Lockwood Report (1964), 279n
logarithms, 27, 33, 47
  four-figure tables of, 151, 154
London County Council, 184, 276n
London Day Training College, 180–1, 274n
London Infant School Society, 103
London Mathematical Society (LMS), 95ff, 133
London School Board, 170
London University, 33, 81, 82ff, 86, 94, 136, 161, 175–7, 180
  colleges: Bedford, 172, 176–7, 180–1; Birkbeck, 257n; Goldsmiths', 191; Imperial, 178, 266n, 268n; Kings, 94, 177, 179, 188, 190–1, 273n; Royal College of Science, 147; University, 82, 94ff, 126, 133–4, 176–7; Westfield, 177
  Institute of Education, 157, 180, 274n
  Matriculation Examination, 161, 176, 214–16, 273n
Lunar Society, Birmingham, 56, 250n, 258n

McNair Report (1944), 192
married women, 183, 269n, 275n
mastery, 20
mathematical ability, factors of, 276n
Mathematical Association (MA), 139, 147, 149, 155–7, 162, 164, 166, 168, 178, 188, 192, 194ff, 202, 204, 228–9, 266–8n, 274n, 277n
  Teaching Committee of, 149
mathematical laboratories, 153, 269n
mathematical practitioners, 12, 26, 33, 44, 76
mathematical schools, 35ff, 44–5, 47, 60, 64, 245–6n, 250n
mathematical societies, 92, 95, 258n
meaning in mathematics, 22–3, 156
mechanics, 38, 40, 42, 49, 51, 53, 58, 62, 68, 78, 110, 112, 117, 119–20, 130, 153, 159, 176–7, 217–18, 228
Mechanics' Institutes, 73, 99ff
  Alnwick, 97
  Birkbeck, 257n
  Glasgow, 257n
  Huddersfield, 100
  Liverpool, 100
  London, 93, 99, 257n
  Preston, 265n
  Rugby, 137
  York, 99–100
mensuration, 3, 27, 42, 62–3, 68, 110, 116, 160, 185, 197, 222, 234
Mercers' Company, 32, 244n
Merchant Taylors' Guild, 30
'Merton school', 240n
metaphor in mathematics, 80
metrication, 122
Midlands Mathematical Experiment (MME), 205–6
Military Academies, 64ff, 66, 70, 93, 99, 128–9, 251–3n
Mint
  Bristol, 7
  Irish, 7
mixed-ability teaching, 38, 130, 199, 278n
monasteries, dissolution of, 8
monitorial system, 102
motivation, 19

National Advisory Council for the Training and Supply of Teachers (NACTST), 193
National Society, 102, 105–6
navigation, 27, 30, 33, 35–6, 40, 62
negative numbers
  objections to, 80
  notation for, 159
Newcastle Commission Report (1861), 113, 120ff, 170
Newsom Report (1963), 277n, 279n
Norwood Report (1943), 193, 277n
notation, 1, 2, 14, 30, 120ff, 159, 243n
Nuffield Foundation Mathematics Project, 203, 205
numeration, 91, 120

Organisation for European Economic Cooperation (OEEC), 203, 206
overseas influence of English mathematics education, 67–8, 202–3
Oxford University, 2, 3, 6, 8, 12, 33–4, 44, 46, 75, 81, 87, 112, 127, 136, 177, 179, 206, 249n, 268n
  colleges: All Souls', 5, 6; Balliol, 160; Brasenose, 5; Cardinal (later Christ Church), 5; Corpus Christi, 5; Exeter, 47; Magdalen, 5, 272n; Merton, 240n; New, 4; Oriel, 75; Queen's, 135; Somerville, 72; University, 42
  Savilian Chairs, 33, 112, 245n, 247n, 249n

payment by results, 121–2, 170, 267n
Penny Cyclopaedia, 87
perils of a mathematical education, 51–2
periodicals, mathematical, 69ff, 95, 98
perspective, 3
philomaths, 12
Plowden Report (1967), 279n
Postgraduate Certificate in Education (PGCE), 201
practical mathematics, 90, 147, 153, 222–4, 242n, 269n, 272n
practical work, 142, 148, 153, 269n
principles for reform (Levett), 142
probability, 41–2, 94, 197, 207
progress, the idea of, 16
proof, 14, 21, 92
  see also rigour
psychology, 56–7, 116
  faculty, 113ff, 157, 174
  Gestalt, 189, 276n
  Herbartian, 158, 189
  Piagetian, 190, 195
  in teacher-training, 113, 181
pupil-teacher system, 171, 179, 261n, 273n

quadrivium, 1, 2, 43, 239n
Quarterly Journal of Education, 87
Queen's College, London, 172–3, 272–3n

reality and mathematics, 90, 156
Reformation, 6, 241–2n
regius professorships, 9
Regulations
  for DSA schools, 146
  for post-1902 secondary schools, 151
religion and mathematics, 49ff, 57
religious discrimination, 79, 81, 96, 127, 245n
Renaissance, 9
research, educational, 116, 162, 187ff, 276n
Restoration, 35, 45–6
Revised Code (1862), 121–2
rigour, 164–7
rote learning, 20, 89, 102, 196
Royal Astronomical Society, 85, 93, 95
Royal Commission on the Universities, 127
Royal Lancasterian Society, 102
Royal Society, 29, 36, 39ff, 45, 66, 76, 95, 118, 127, 143, 245–8n, 250n, 253–4n, 259n, 279n
Royaumont Seminar, 199
Rugby LEA, 184

Sandhurst, Royal Military College (formerly Marlow), 66, 99, 128, 253n
'sandwich' course, 37, 269n
School Curriculum Development Council, 280n
school industry cooperation, 280n
school-leaving age, 183, 191, 205, 272n
School Mathematics Project (SMP), 159, 203, 205, 278n
schools
  board, 169
  British, 102ff
  cathedral, 1, 239n
  central, 171, 173, 194, 208
  charity, 8, 53, 257n
  Church of England, 169; *see also* National
  comprehensive, 186, 193, 205, 276n
  continuation, 183–4, 275n
  dame, 101, 169
  Girls' Public Day School Trust (GPDST), 173
  grammar, 8–12, 63, 184–5, 194, 244n, 251–2n, 255–6n, 263n, 267n, 275n
  higher-elementary (higher grade), 146, 151, 267–8n
  infant, 170, 272n
  multilateral, 186, 193
  National, 102ff, 261n
  nursery, 183
  'petty', 9
  private, 55, 61, 72, 75, 272n
  proprietary, 87ff
  public (old), 129, 249n, 255n
  'science', 146
  secondary modern, 184–5, 194, 197, 275n, 277n

SPCK, 257n
Sunday, 102, 261n
technical, 184, 194, 208
town and village schools: Alnwick, 97; Barnsley, 263n; Batley, 54; Borough Rd, London, 105–6; Bradford, 263n; Bungay, 15; Charterhouse, 266n; Cheltenham College, 88; Cheltenham Ladies College, 173; Christ's Hospital, 35, 44–5, 47, 60, 245n, 250n, 254n; Christ's Hospital for Girls, Hertford, 182; City of London, 88, 264n, Clifton College, 128, 138ff, 265n; Collyer's School, Horsham, 245n; East Ham Girls, London, 172, 177; Eton, 124, 128–9, 137, 245n, 251n, 263n; Fulneck, 52; Greycoat Hospital, London, 181; Haberdashers' Aske's Hatcham Girls', London, 182, 273n; Harrow, 124, 129, 134, 263n, 269n; Heath, Halifax, 263n; Huntingdon, 30; Keighley, 263n; King Edward VI, Birmingham, 133, 138, 141, 255n, 266n; King William's, Isle of Man, 123, 262n; King's College School, London, 88; King's School, Pontefract, 263n; King's Somborne, 104, 261n; Kingswood, 52, 249n; Leeds, 252n, 263n; Liverpool Institute, 87; Manchester Grammar, 47, 251, 253n, 255n, 265n; Manchester High School for Girls, 174; Marlborough, 88, 128; Merchant Taylors', London, 31, 129; Neale's, London, 246n; Newcastle, 60, 64, 250n, 252n, North London Collegiate School for Girls, 173; Notting Hill High School for Girls, London, 136; Nottingham, 181; Plymouth, 244n; Queenswood, Stockbridge, 264–5n; Rossall, 88, 141; RNC Osborne, 154, 269n; Rugby, 126, 128ff, 137–8, 263n, 265n; St Paul's, London, 24, 29–32, 244n, 267n; St Peter's, York, 263n; Saunders', Rye, 246n; Sedbergh, 123–5, 127, 129, 262–3n; Sherborne, 274n; Sir Joseph Williamson's, Rochester, 246n; Skipton, 263n; Southampton, 249n; Tonbridge, 156, 269n; University College School, 87, 95, 264–5n; West Riding Proprietary School, Wakefield, 264n; Westminster, London, 249n; Winchester, 4, 124, 128, 137, 148, 153, 249n, 251n, 263n, 267n; York Minster, 239n
Schools Council, 206, 280n
Mathematics Curriculum Critical Review, 279n
Mathematics for the Majority Project, 279n
Continuation Project, 279n
Sixth Form Project, 279n
science teaching, 53, 56, 100, 122, 126, 129, 137–8, 261n, 265n
scrivenor, 4, 36, 63, 251n
Secondary Schools Examination Council (SSEC), 162, 276–7n
Senate House Examination, *see* Cambridge University Tripos
Senior Wranglers, 125
setting, 38, 130, 199, 278n
slide-rule, 39, 222, 232
Society for the Diffusion of Useful Knowledge (SDUK), 85ff
Society for Promoting Christian Knowledge (SPCK), 86, 257n
Spens Report (1938), 157, 186–7
Spitalfields Mathematical Society, 92, 95
squared paper, 147, 170, 222–3, 267–8n
stages of teaching (A, B, C), 153, 167, 200, 243n, 268n
state education, opposition to, 86, 101–2, 106, 260n
statistics, 162, 165, 174, 182, 197, 223, 228, 271n
Sunday School movement, 102
surveying, 47, 62, 68, 165, 228
syllabuses
arithmetic: payment by results, 1862, 121–2; Board of Education (1905), 224–5
Battersea Training School (1841), 107ff
Cambridge Tripos: (1772), 253n; (1810), 78; (1884), 220–1; (1911), 226–8
central schools (1910), 171
Fiske (1615), 27
Gregory (1700), 42–3
Hutton: Newcastle (1760), 61–2; Woolwich (1798), 68
Jeffery (1944), 234–7
Jennings (1718), 48–9
Jurin (1711), 60–1
Lancaster (1820, arithmetic), 103
Oxford University (1858), 160
Perry (1910), 222–4
Sedbergh School (1853), 123–4
Tate (1836), 98
training colleges (1856), 112
Watts (1741), 50–2.
symbolism, 1, 2, 18–19, 30, 77, 90–2, 120ff, 159, 243n

Taunton (Endowed Schools) Commission (1868), 131, 136, 176, 263n, 273n, 279n
teacher
associations, 73, 118, 133
see also AIGT, ATAM, ATM, MA
knowledge required of, 116–17
qualifications of, 36, 154, 276n;
qualities required by, 117–18
shortage of, 5, 102, 182, 199, 210, 276n, 278n
status of, 73, 101–2, 273n

teacher-training, 5, 31, 104ff, 111–13, 171, 179ff, 190ff, 199ff, 274–5n
colleges: Battersea, 106ff, 217–20, 261n, Cambridge, 179, Cheltenham, 261n, Chester, 106, 261n, Exeter, 106, Finsbury, 179, Glasgow, 106, Goldsmiths', 191, Homerton, 106, Kneller Hall, 118–19, 262n, Leicester, 191–2, Maria Grey, 179, Oxford, 106, St Mark's, Chelsea, 106, 111, 261n, Salisbury, 106, Whitelands, 106, 192, 199, York and Ripon, 261n; Zurich, 107
league tables of college 'results', 261n
teaching methods, 115ff, 128, 167, 171, 185, 210
Technician and Business Education Councils (TEC and BEC), 279n
tenure for university staff, 84
terminology, 22–3
tests, standardised, 231–2, 276n
timetables
Battersea Training School (1841), 107–9
Borough Rd School (1834), 105–6
Elizabethan grammar school, 10–11
Trades Union Congress, 186
trigonometry, 33, 42, 51, 53, 62–3, 68, 123, 125, 147, 158, 160, 175–6, 185, 223, 228, 236, 274n
Trinity House, 37–8
trivium, 1

unified mathematics, 159, 163, 193, 232–3
United Nations Educational, Scientific and Cultural Organisation (UNESCO), 203
universities
age of entrance, 4, 76
coaches, 125
education for women, 176ff
individual: Birmingham, 180, Cardiff, 148, 180, Durham, 37, 45, 47, 136, Edinburgh, 47, 66, 136, 179, 254n, 259n, Glasgow, 254n, Leeds, 177, Liverpool, 157, 177, 206, 266n, Manchester, 45, 177, 180, 274n, Nottingham, 180, Open, 209, 279n, Paris, 239n, Prague, 3, St Andrews 179, Southampton, 206, Vienna, 3
see also separate entries under Cambridge, London, Oxford

Vernacular, teaching in the 2, 16, 19, 25, 42ff, 53, 243n, 247n

West Point Military Academy, 252n
women
mathematics education of 175ff, 255–6n
married, 183, 269n, 275n
Woolwich, Royal Military Academy, 64ff, 70, 93, 128–9, 251–3n
writing master, 4, 36, 63, 124, 251n

York Courant, 98–9, 260n
York Medical School, 98